Advances in Sustainable Machining and Manufacturing Processes

Mathematical Engineering, Manufacturing, and Management Sciences
Series Editor: Mangey Ram, Professor, Assistant Dean (International Affairs), Department of Mathematics, Graphic Era University, Dehradun, India

The aim of this new book series is to publish the research studies and articles that bring up the latest development and research applied to mathematics and its applications in the manufacturing and management sciences areas. Mathematical tool and techniques are the strength of engineering sciences. They form the common foundation of all novel disciplines as engineering evolves and develops. The series will include a comprehensive range of applied mathematics and its application in engineering areas such as optimization techniques, mathematical modelling and simulation, stochastic processes and systems engineering, safety-critical system performance, system safety, system security, high assurance software architecture and design, mathematical modelling in environmental safety sciences, finite element methods, differential equations, reliability engineering, etc.

Differential Equations in Engineering
Research and Applications
Edited by Nupur Goyal, Piotr Kulczycki, and Mangey Ram

Sustainability in Industry 4.0
Challenges and Remedies
Edited by Shwetank Avikal, Amit Raj Singh, and Mangey Ram

Applied Mathematical Modeling and Analysis in Renewable Energy
Edited by Manoj Sahni and Ritu Sahni

Swarm Intelligence: Foundation, Principles, and Engineering Applications
Abhishek Sharma, Abhinav Sharma, Jitendra Kumar Pandey, and Mangey Ram

Advances in Sustainable Machining and Manufacturing Processes
Edited by Kishor Kumar Gajrani, Arbind Prasad, and Ashwani Kumar

Advanced Materials for Biomechanical Applications
Edited by Ashwani Kumar, Mangey Ram, and Yogesh Kumar Singla

Biodegradable Composites for Packaging Applications
Edited by Arbind Prasad, Ashwani Kumar, and Kishor Kumar Gajrani

Computing and Stimulation for Engineers
Edited by Ziya Uddin, Mukesh Kumar Awasthi, Rishi Asthana, and Mangey Ram

For more information about this series, please visit: https://www.routledge.com/Mathematical-Engineering-Manufacturing-and-Management-Sciences/book-series/CRCMEMMS

Advances in Sustainable Machining and Manufacturing Processes

Edited by
Kishor Kumar Gajrani
Arbind Prasad
Ashwani Kumar

CRC Press
Taylor & Francis Group
Boca Raton London New York

CRC Press is an imprint of the
Taylor & Francis Group, an **informa** business

Cover image: Getty

First edition published 2022
by CRC Press
6000 Broken Sound Parkway NW, Suite 300, Boca Raton, FL 33487-2742

and by CRC Press
4 Park Square, Milton Park, Abingdon, Oxon, OX14 4RN

CRC Press is an imprint of Taylor & Francis Group, LLC

ISBN: 978-1-032-08165-6 (hbk)
ISBN: 978-1-032-25687-0 (pbk)
ISBN: 978-1-003-28457-4 (ebk)

DOI: 10.1201/9781003284574

Typeset in Times
by SPi Technologies India Pvt Ltd (Straive)

This book is dedicated to all mechanical,
production, manufacturing,
and aerospace engineers

Contents

PART I Sustainable Machining

Preface

Humans impact the environment in numerous ways. Human civilization needs a healthy environment to ensure the survival of our planet. The balance between environment and technology is the need of today's world. Technology has made our life comfortable, but our environment is paying price. Nowadays, it has been realized that a healthy environment is the necessity of humans' and this planet's survival. Therefore, efforts are increasing toward adapting green and sustainable technology to reduce detrimental environmental impacts.

Sustainable machining and manufacturing processes are the need of today's world. Among various manufacturing processes, machining is one of the widely used. Hence, it has to be made sustainable. The chapters in this book are categorized under two broad sections: (1) sustainable machining and (2) manufacturing processes. Part I, includes the work of numerous researchers as a review of modeling and experimental work. Chapter 1 discusses challenges in machining advanced materials using conventional and nonconventional machining processes. Chapter 2 covers the challenges and probable opportunities during machining by advanced ceramic materials. Chapter 3 discusses various ways to characterize and evaluate eco-friendly cutting fluids. Chapter 4 covers dry machining using advanced textured cutting tools. Chapter 5 discusses advances in one of the important techniques of near-dry machining, namely, minimum quantity lubrication, its need, its significance, and economics and environmentally friendly ways for machining. Chapter 6 covers the application of nanofluids as cutting fluids during machining. Chapter 7 discusses using nanofluids for machining in the era of Industry 4.0 and its effect on environmental sustainability. Chapter 8 explores the use of ionic liquids as a potential sustainable green lubricant for machining in the era of Industry 4.0. The structure of various ionic liquids, their relative machining performance, and their overall environmentally sustainable aspects are covered. Chapter 9 discusses sustainable electrical-discharge machining and using sustainable dielectric while maintaining similar accuracy and precision during the process. Chapter 10 covers the effects of water jet pressure, flow rate, standoff distance, and abrasive grit size on depth of penetration, cutting rate, surface roughness, taper cut ratio, and top kerf width during sustainable abrasive jet machining. Chapter 11 explores the use of artificial neural networks to successfully predict various responses, such as surface roughness, cutting force, and tool wear, during machining. Chapter 12 discusses the machining and vibration behavior of Ti-TiB composites processed through powder metallurgy techniques. In Chapter 13, the numerical analysis of machining forces and shear angle during dry hard turning of AISI 4640 steel using Al_2O_3-coated tungsten-based cemented carbide cutting inserts is conducted to predict cutting force and shear angle. Chapter 14 explores the machining performance evaluation of titanium biomaterial alloys in computer numerical control turning using a cubic boron nitride tool insert.

Part II of this book discusses sustainability aspects in various manufacturing processes. Chapter 15 discusses the use of Industrial Internet of Things in manufacturing, various communication protocols, data management techniques, and software design models focusing on the Fourth Industrial Revolution. Chapter 16 explores the ways to improve forming characteristics in incremental sheet forming. The chapter aims to highlight and systematically review the recent strategies related to numerical techniques, such as finite element analyses, computer-aided design, tool path development, experimental setups, and hybrid techniques that have been proved to increase the quality of the formed ISF parts. Chapter 17 covers the similarities and differences in deformation mechanisms of polymers, metals, and their composites in dieless forming operations. Chapter 18 discusses the sustainable polishing of directed energy deposition–based cladding using micro-transferred arc.

This book will work as a reference book for researchers, practicing machine shop engineers, and managers. This book can also be used as a textbook for the postgraduate level and as an elective course book for the undergraduate level. *Advances in Sustainable Machining and Manufacturing Processes* provides a foundational link to more specialized research work in the domain of sustainable manufacturing.

Dr. Kishor Kumar Gajrani
Dr. Arbind Prasad
Dr. Ashwani Kumar

Acknowledgment

We would like to thank all the authors of various book chapters for their contributions. Our heartfelt gratitude to CRC Press (Taylor & Francis Group) and the editorial team for their support during the completion of this book. We are sincerely grateful to reviewers for their suggestions and illuminating views on each book chapter presented in book *Advances in Sustainable Machining and Manufacturing Processes.*

Editors

Dr. Kishor Kumar Gajrani is Assistant Professor in the Department of Mechanical Engineering at the Indian Institute of Information Technology, Design and Manufacturing, Kancheepuram, Chennai, India. He has an M.Tech and a PhD degree from the Department of Mechanical Engineering at the Indian Institute of Technology Guwahati. Thereafter, he worked as a postdoctoral researcher at the Indian Institute of Technology Bombay. He has been published in 27 international journals and book chapters of repute, as well as attended numerous conferences. His research interests are broadly related to the advancement of sustainable machining processes, advanced materials, tribology, coatings, green lubricants and coolants, and food packaging.

Dr. Arbind Prasad has completed his PhD in mechanical engineering from the Indian Institute of Technology Guwahati, Assam, India. He has filled four patents out of his research work. He has numerous international journal articles, book chapters, and reputed conference papers to his credit. Dr. Arbind Prasad has obtained various prestigious awards, such as the Sponsored Research Industrial Consultancy (SRIC) award from IIT Kanpur, the Best Oral Presentation from the American Chemical Society, and Best Paper awards from IIT Guwahati during the Research Conclave. He has been invited to deliver talks at various organizations of repute. He has coordinated various faculty development programs, short-term courses, symposiums, and national seminars and has completed research projects sponsored under various government schemes in India. He is currently working as Assistant Professor and Head (Mechanical Engineering) in the Department of Science and Technology, Government of Bihar, posted at Katihar Engineering College, Katihar, Bihar, India. His main areas of interest include manufacturing, machining, polymer composites, biomaterials, materials processing, and orthopedic biomedical applications.

Dr. Ashwani Kumar received a PhD (mechanical engineering) in the area of mechanical vibration and design. He is currently working as Senior Lecturer, Mechanical Engineering (Gazetted Officer Class II) at Technical Education Department, Uttar Pradesh (Government of Uttar Pradesh) India since December 2013. He has worked as Assistant Professor in the Department of Mechanical Engineering, Graphic Era University, Dehradun, India, from July 2010 to November 2013. He has more than 11 years of research and academic experience in mechanical and materials engineering. He is the series editor of book series *Advances in Manufacturing, Design and Computational Intelligence Techniques* published by CRC Press (Taylor & Francis). He is the associate editor for the *International Journal of Mathematical, Engineering and Management Sciences*, indexed in ESCI/Scopus and the Directory of Open Access Journals.

He is an editorial board member of four international journals and acts as a review board member of 20 prestigious (indexed in SCI/SCIE/Scopus) international journals with high impact factors, such as *Applied Acoustics*, *Measurement*, the *Journal of Engineering Science and Technology*, the *American Journal of Science &*

Engineering, *Journal of Mechanical Engineering*, and the *Latin American Journal of Solids and Structures*. In addition, he has published 85 research articles in journals, book chapters and conferences. He has authored/co-authored cum edited *SV-JME* of mechanical and materials engineering. He is associated with international conferences as an invited speaker/advisory board/review board member. He has delivered many invited talks in webinars, faculty development program, and workshops. He has been awarded as Best Teacher for excellence in academics and research. He has successfully guided 12 B.Tech, M.Tech, and PhD candidates through their theses.

In administration he is working as coordinator for the All India Council for Technical Education, extension of approval (E.O.A.), a nodal officer for PMKVY-TI Scheme (Government of India), and internal coordinator for CDTP scheme (Government of Uttar Pradesh). He is currently involved in the research area of machine learning, advanced materials, machining and manufacturing techniques, biodegradable composites, heavy vehicle dynamics, and Coriolis mass flow sensors.

Contributors

Ahmed Moustafa Abd-El Nabi
Production Engineering and Mechanical Design Department
Faculty of Engineering
Minia University
Minia, Egypt

Anupam Alok
National Institute of Technology Patna
Patna, India

Cem Alparslan
Faculty of Engineering and Architecture
Department of Mechanical Engineering
Recep Tayyip Erdoğan University
Rize, Turkey

Şenol Bayraktar
Faculty of Engineering and Architecture
Department of Mechanical Engineering
Recep Tayyip Erdoğan University
Rize, Turkey

A. S. Bolokang
Council for Scientific and Industrial Research
Manufacturing Cluster, Advanced Materials Engineering
Pretoria, South Africa

Phaneendra Kiran Chaganti
Department of Mechanical Engineering
Birla Institute of Technology and Science (BITS-Pilani)
Hyderabad, India

P. Chandramohan
Department of Mechanical Engineering
Sri Ramakrishna Engineering College
Coimbatore, India

Manas Das
Indian Institute of Technology
Guwahati, India

Milon Selvam Dennison
Kampala International University
Western Campus
Uganda

Kishor Kumar Gajrani
Department of Mechanical Engineering
Indian Institute of Information Technology, Design and Manufacturing
Kancheepuram, India

and

Centre for Smart Manufacturing
Indian Institute of Information Technology, Design and Manufacturing
Kancheepuram, India

R. Ganesh
Department of Production Engineering
Dr. Mahalingam College of Engineering and Technology
Pollachi, India

Sameer Ghanvat
Department of Mechanical Engineering
Dr. Babasaheb Ambedkar Technological University
Lonere, India

Nitesh Gothe
Department of Mechanical Engineering
Dr. Babasaheb Ambedkar Technological University
Lonere, India

Ahmed Mohamed Mahmoud Ibrahim
College of Mechanical and Vehicle Engineering
Hunan University
Changsha, China
and
Faculty of Engineering
Production Engineering and Mechanical Design Department
Minia University
Minia, Egypt

Neelesh Kumar Jain
Department of Mechanical Engineering
Indian Institute of Technology Indore
Indore, India

Kirubanidhi Jebabalan
Division of Materials Technology
Technical University of Liberec
Czech Republic

Hridayjit Kalita
Birla Institute of Technology
Ranchi, India

Manoranjan Kar
Department of Physics
Indian Institute of Technology Patna
Patna, India

Nikhil Khatekar
Department of Mechanical Engineering
PCT's A. P. Shah Institute of Technology
Thane, India

P. Kour
Department of Physics
Birla Institute of Technology Mesra
Patna, India

Amit Kumar
Department of Mechanical Engineering
National Institute of Technology Patna
Patna, India

Kaushik Kumar
Birla Institute of Technology
Ranchi, India

Manjesh Kumar
Indian Institute of Technology
Guwahati, India

Mechiri Sandeep Kumar
B V Raju Institute of Technology
Narsapur, India

Pravin Kumar
Department of Mechanical Engineering
Indian Institute of Technology Indore
Simrol, 453 552 (MP), India

S. Vinoth Kumar
Key Laboratory of High-Efficiency and Clean Mechanical Manufacture
National Demonstration Center for Experimental Mechanical Engineering Education
School of Mechanical Engineering
Shandong University
Jinan, China

Wei Li
College of Mechanical and Vehicle Engineering
Hunan University
Changsha, China

M. N. Mathabathe
Council for Scientific and Industrial Research
Manufacturing Cluster, Advanced Materials Engineering
Pretoria, South Africa

M. Abisha Meji
Kampala International University
Western Campus
Uganda

Satish Mullya
Annasaheb Dange College of Engineering and Technology,
Ashta, India

M. Neelavannan
Dr. Mahalingam College of Engineering and Technology
Coimbatore, India

Madan Mohan Reddy Nune
Department of Mechanical Engineering
Anurag University
Hyderabad, India

Upendra Maurya
Department of Mechanical Engineering
National Institute of Technology
Warangal, India

Raju Pawade
Department of Mechanical Engineering
Dr. Babasaheb Ambedkar Technological University
Lonere, India

Anand C. Petare
Central Workshop
Indian Institute of Technology Indore
Indore, India

I. A. Palani
Department of Mechanical Engineering
Indian Institute of Technology Indore
Indore, India

S. K. Pradhan
Department of Mechanical Engineering
Birla Institute of Technology Mesra
Patna, India

T. Ramkumar
Department of Mechanical Engineering
Dr. Mahalingam College of Engineering and Technology
Pollachi, India

Yinghui Ren
College of Mechanical and Vehicle Engineering
Hunan University
Changsha, China

Tribeni Roy
Birla Institute of Technology and Science Pilani (BITS-Pilani)
Pilani, India

and

School of Engineering
London South Bank University
London, United Kingdom

M. Selvakumar
Department of Automobile Engineering
Dr. Mahalingam College of Engineering and Technology
Pollachi, India

Parnika Shrivastava
Department of Mechanical Engineering
National Institute of Technology
Hamirpur, India

S. Kanmani Subbu
Department of Mechanical Engineering
Indian Institute of Technology Palakkad
Palakkad, India

Saurabh Thakur
Department of Mechanical Engineering
National Institute of Technology
Hamirpur, India

Abhay Tiwari
Department of Mechanical Engineering
Indian Institute of Technology Indore
Simrol, MP, India

V. Vasu
Department of Mechanical Engineering
National Institute of Technology
Warangal, India

Introduction

The ever-increasing trends of upcoming technology also lead to many concerns related to the environment, society, and economics in every field of engineering. Sustainability in any organization implies that something can be sustained indefinitely. Recently, governments and world leaders have been advocating for green and sustainable manufacturing initiatives. To support and promote such initiatives, researchers, scientists, engineers, and academic institutions have a responsibility to introduce educational programs related to sustainable manufacturing to prepare the future generation. Sustainable machining and manufacturing processes are a need of today's world. *Sustainable machining* is defined as the creation of products by cutting material that uses processes that are environmentally friendly, economically sound, and safe for employees, consumers, and communities, as well as can conserve energy.

To introduce these multidimensional machining and manufacturing processes into industries, as well as into the curriculum of future generations of researchers and engineers, is the ambition behind this book. This goal can be balanced by developing adequate economic, environmental, and social criteria, with analysis of their independencies and application of that analysis for guiding technological innovation in respective economic, environmental, and societal frameworks.

This book provides a lucid way for readers to understand the advances in sustainable techniques for machining and manufacturing applications. The book consists of 18 chapters dedicated to the advances in sustainable machining and manufacturing processes. The chapters discuss the challenges faced when machining advanced materials, the use of eco-friendly cutting fluids, and how they affect the machined component characteristics and our environment. This book also covers topics such as dry and near-dry machining, machining with advanced textured cutting tools, minimum quantity lubrication, nanofluids and ionic fluids in the era of Industry 4.0, sustainable electrical-discharge machining, abrasive jet machining, and artificial neural network–based machining. This book includes machining and vibration behavior of composites and finite element analysis during machining with advanced ceramic coated carbide tools. Furthermore, the use of the Industrial Internet of Things in manufacturing, sheet forming, and deformation mechanism, as well as sustainable polishing using micro-plasma transferred arc, is discussed.

This book addresses the challenges and solutions for sustainable machining and manufacturing processes. It discusses prevailing trends and suggests research findings for industries to move toward sustainable development by improving economic and social perspectives, as well as reducing the detrimental effects to the environment. Overall, the aim of this book is to catalogue the latest achievements in the modern machining and manufacturing industry that can be helpful for future generations.

Dr. Kishor Kumar Gajrani
Dr. Arbind Prasad
Dr. Ashwani Kumar

Part I

Sustainable Machining

1 Challenges in Machining of Advanced Materials

M. N. Mathabathe and A. S. Bolokang

Council for Scientific and Industrial Research,
Manufacturing Cluster, Advanced Materials Engineering,
Pretoria, South Africa

CONTENTS

1.1 INTRODUCTION

The fuel efficiency and performance improvement in the aerospace application has led to the development of advanced materials, namely, from steel to nickel to titanium alloys; materials with lower density, for example, aluminum to carbon fiber composites; and ceramic and metal matrix composites (CMCs and MMCs, respectively), which are restituting part of elevated temperature alloys subject to engine application en route to the end of the 20th century [1].

Advanced structural materials can be defined as complex shapes or materials combinations to attain properties linking to functionality, for example, smart materials. These are materials designed for good mechanical, electrical, or thermal properties; high-efficiency energy conversion; materials with embedded sending systems for reliability and safety; and smart materials vehicles and large space structures are subject to a high-strength-per-mass ratio [2]. Researchers studied the design and fabrication of specific structures to enhance materials property three-dimensional (3D) printed structures, such as curved [3], honeycomb [4], cell shapes [5] or hexachirales [6].

DOI: 10.1201/9781003284574-2

The thermal, mechanical and electrical properties, which also enhance the performance of polymer matrix composites, make carbon nanotubes (CNTs) attractive for industry use. However, identifying prospective applications for space use is a challenge for these materials [7]. On one hand, the chief challenges formed by today's establishments involve comprehending and assimilating the rapidly advancing technological machining methods fabricated by industry. The integration and application of these advances prepare the industry for the next set of leading technical improvements. The present work focuses on a vast area of machining techniques and their challenges faced by advanced materials.

1.2 MACHINING PROCESS AND MATERIALS

1.2.1 Cutting Tool

Advanced materials, such as superalloys, are a challenge for machining due to their inherent high hardness, low thermal conductivity, and great resistance to shearing. However, they require elevated cutting temperatures and high cutting forces to avoid severe tool wear [8].

The major tools employed in the fabrication of Ni-based superalloys are ceramic, carbide tools, and polycrystalline cubic boron nitride (PCBN). The latter is commonly used due to its superior machining capabilities for "difficult to cut" materials [9]. The PCBN is designed without a chip breaker, which is a flat rake, resulting in more resistance to breaking the chip during the cutting process. However, its shortcomings are that the surface of the workpiece is effortlessly damaged by the chip winding workpiece [9]. It was suggested that applying high-pressure cooling may overcome these drawbacks [10–11], while the importance of tool failure is attributed to analyzing the damage of tool material [12].

1.2.2 Material Selection

Advanced ceramic materials, for example, are clustered into two groups, that is, (1) conductive—such as the typical zirconium diboride (ZrB_2), metal nitrides (TiN/ZrN), boron carbide (B_4C), titanium diboride (TiB_2), and other similar materials—and (2) nonconductive silicon nitride (Si_3N_4), alumina (Al_2O_3), zirconia (ZrO_2), and silicon carbide (SiC). However, adding advanced ceramic materials or conductive particles to the latter, such as B_4C, TiC, Si, CaO, TiB_2, Si_3N_4 + TiC, Si_3N_4 + TiN, ZrO_2 + Cao, Al_2O_3 + TiC, ZrO_2 + Y_2O_3 + TiN, and Al_2O_3 + TiN, makes them conductive advanced materials [13–14]. However, the conventional machining shortcomings of advanced ceramic materials are attributable to their elevated hardness and brittleness [13]. Additionally, their processing/manufacturing performance is not cost-effective, particularly the expense incurred during the polishing stage [13]. Therefore, the surface of advanced materials may suffer damage during machining, resulting in stress concentration and cracks, thus affecting the mechanical strength of components [14].

Ceramic phases, that is, Al_2O_3/Al_2O_3 or SiC/SiC, in the same structure of CMCs mitigate the aggregate residual stresses emanating from the manufacturing process. Further, the coefficient of thermal expansion (CTE) mismatch and processing temperature affects the residual stresses amid matrix and fibers, characterized by X-ray diffraction (XRD), Raman spectroscopy, or micro-hardness tests [15].

1.2.3 Types of Machining Techniques

Many machining techniques effectuated unrelenting tool wear associated with longer cutting time, the high cutting force resulting in higher machining cost, thus bearing machining challenges of advanced materials [12]. Furthermore, when processing advanced ceramics, for example, by abrasive machining and grinding, the product components experience plastic deformation and considerable residual stress, friable layers, and surface cracks [16, 17]. Table 1.1 outlines some machining techniques, both conventional and nonconventional.

The susceptible factors in the performance of electrical-discharge machining (EDM) are a low material removal rate (MRR) and high tool wear, which may be conquered by other advantages when compared with other nonconventional techniques [18].

On one hand, dry machining is conducted without the facilitation of cutting fluids. This process has gained widespread over the past years and is increasingly used by manufacturers that fabricate products with metals [19]. However, the technique exhibits drawbacks, such as (1) equipped for materials workpiece and machining approaches, which do not yield favorable results; (2) elevated temperatures at the tool–workpiece interface on the cutting edge; (3) the tool heating up due to the amplified temperature and losing its hardness [19]; (4) the surface integrity and dimensional accuracy of the workpiece are altered [20]; and (5) only materials that pose good machinability can be employed for dry machining [21].

1.3 TOOL WEAR/LIFE SPAN AND COMMERCIAL METAL CUTTING

Some advanced materials, such as the $40CrNi_2Si_2MoVA$ (300 M) steel, employed in aerospace because of its good strength, fracture toughness, elevated transverse plasticity, good corrosion resistance, and fatigue performance have their inherent drawbacks. The 300 M steel suffers poor thermal conductivity, resulting in a large cutting force and an elevated temperature, difficult chip control, and easy tool wear during the cutting process [30]. However, Zhang et al. [31] used cryogenic minimum quantity of lubricant (CMQL) technology to analyze the varying tool wear lubricants and cutting force amid high-speed turning of 300 M steel. The results indicated that this method reduced tool wear and prolonged tool life by the reduction of cutting force, cutting temperature at the knifepoint and friction amid the tool and workpiece. Furthermore, the authors studied the genetic algorithm under CMQL conditions which optimized cutting parameters and presented theoretical attribution for tool wear control at high-speed cutting 300 M steel. The ultimate tool wear prediction is demonstrated in Equation 1.1:

$$VB = 0.0362v^{0.1054} f^{0.0567} a_p^{0.0119}, \quad (1.1)$$

TABLE 1.1
Feature/Operation of Ceramic Metal Matrix Composites on the Machining Techniques

Machining Techniques	Information Machined	Challenges	Concluding Remarks	References
Conventional				
Orthogonal cutting	• Defined cutting edges • Fiber-reinforced orientations, i.e. parallel, across, and transverse	Ductile and brittle behavior transpire due to ceramic machining at small uncut chip thickness leading to grain fracture and slip-plane mechanisms	Roughness/coarseness of machined surface is susceptible to microcracking mechanisms of the particles, leading to residual stress as the scratch load increases	[22]
Milling	• Surface finish – roughness and morphology	Observed ductile to brittle transition in the matrix during machining. Fiber removal due to pullout mechanism.	Concluded that surface roughness increased at a penetration depth of >4 μm, resulting in larger grooves due to brittle fracture	[23]
Drilling	The tool rotates along its axis	Entry and exit delaminations are induced machining damage due to high thrust forces employed-conventional drilling (CD)	Rotary ultrasonic machining (RUM) had an average reduced thrust force (~10%–15%), resulting in less significant exit delamination than CD	[24]
Grinding	Favored finishing operation for hard/brittle materials to achieve dimensional accuracies	Three grinding methods showed different results performed on C/Si materials CG provided a surface roughness (Ra) ~2–4 times lower than IG, while the UAG produced much higher values of Ra, as a result of the induced impact on the abrasive grains -caused cracks to propagate	Grinding holes >1-mm diameter are useful for successful machining of slots and surface finish employing the typical conventional mechanical techniques using cubic boron nitride and diamond tools	[25]
Nonconventional				
Abrasive waterjet (AWJ)	Cut and shape hard metals	Reduced surface quality at the jet exit because of kerf taper angle effect and brittle fracture removal due to loss of energy of the jet	The technique can be effective in machining slots, holes, and through cuts in Al_2O_3/Al_2O_3 and SiC/SiC CMCs, provided the operating parameters are optimized	[26]
Pulsed laser ablating (PLA)	Hole making in ultra-hard materials	The main drawback is heat-affected zone (HAZ)	Minimization of thermal damage, the PLA is recommended provided the pulse duration is enhanced, i.e., milli-, nano-, pico-, or femtosecond laser ablation.	[27, 28]
Electrical-discharge machining (EDM)	Cuts and holes in electrically conductive materials difficult to cut	Temperature built up due to process sparks, affecting the machined surface resulting in residual stress levels amid the fibers and matrix.	Debris removal is important to prevent damage to the workpiece. Additionally, deep flushing and tool vibration enhanced surface quality.	[29]

where *VB* is the amount of flank wear and v, f, and a_p are the cutting speed, feed, and depth of cutting, respectively. However, if the influence of a factor is not significant on the dependent variable, then the coefficient of the factor should be zero, that is, $bi = 0$, utilizing a t test as shown in Equation 1.2:

$$ti = \frac{bi / \sqrt{cii}}{\sqrt{Q/(n-m-1)}}, i = 1,2,\ldots\ldots,m. \tag{1.2}$$

The tool wear coefficient level as determined by Zhang et al. [31] was 0.05(α = 0.05), and $t(\alpha/2)$ $(n - p - 1)$, and the ultimate results showed that $t0.025(n - p - 1) =$ 2.22814. This indicated the significant effect of the cutting speed on the forecast value, with subsequent feed, and the depth of cutting.

To achieve economic tool life with subsequent surface conditions, cutting parameters must be carefully chosen. For example, on powder metal Ni alloys [32], these are generally

- Strain <0.01 mm,
- Surface roughness <0.8 μm,
- Nonparent material required,
- No redeposited material or layer, and
- No light contrast amorphous or recast layer.

For the cutting process completion, cutting strategies, namely, cutting parameters and tool geometries, must be properly analyzed. Wear evaluation is subject to an assessment conducted on cutting tools that are worn displaying wear features. In particular, flank wear and chipping of cutting edges cannot be utilized to measure tool wear according to ISO 3685 standard [32]. However, Abele et al. [33] employed statistical experiments/five-axis milling that involved a merger between the flank or tip and the cutting tool radius. Key process variables to consider using this method are axial depth of cut, cutting speed or feed rate, and to predict tool wear behavior.

A lot of research in the cutting process covers tool wear of tungsten carbide and carbide tools, but little has been covered on polycrystalline cubic boron nitride (PCBN) tool wear under high-pressure cooling. This was described by [34], that the workpiece deploys increased pressure on the tool flank, while its contact area is small, resulting in the flank being worn out. However, the flank face cutting area poses challenges because of insufficient coolant while the wear is escalating affecting the machined surface integrity. A study by Wu et al. [34] demonstrated the wear morphology and profile of the PCBN tool. Conversely, the wear model from the study provided a source to cutting superalloys subjected to cooling at high pressures to minimize machining and tool wear. The model was also subsequently reciprocal to cutting parameters, cutting time, and cooling pressure.

1.4 MACHINABILITY

Machinability of materials is clearly expressed as the effortless practice of materials removal (chips) by cutting tool edge, employing conventional machining operations to yield a suitable cost-effective surface quality [35].

The machinability of NiTi shape memory alloy (SMA) is a challenge due to the existence of intermetallics, resulting in strain hardening effects and poor surface texture [36]. Two varying approaches to machine NiTi SMAs have been adopted by Kong et al. [37], the results indicated that abrasive water jet machining was superior to plain water jet machining regarding a more controlled depth and surface texture. Likewise, Frotscher et al. [38] outlined that abrasive water jet machining was an appropriate machining process compared to the micromachining process, which was able to decrease the thermal effect and cutting time of the machining process. Some other unconventional machining processes such as the wire-electrical-discharge machining (WEDM), laser beam and EDM could be utilized for machining the NiTi alloys but have drawbacks, such as heat-affected zone, microvoids, and recast layer. To overcome these drawbacks, Manjaiah et al. [39] consolidated the parameters of the water jet machining to improve surface roughness and the kerf angle of the composites. It was concluded that the reinforcement of the composite in wt.% improved the integrity of the surface machined by clearing particles from the surface matrix [39]. Table 1.2 shows the conducted machinability studies of some of the advanced materials.

1.5 MACHINING PROCESS SELECTION

Machining process selection and technologies have become demanding for advanced materials. The prerequisites to consider are precision machining and reduced surface roughness (quality), large material removal rate (productivity), decreased tool wear (tool cost). The study reported by Feucht et al. [45] indicated the ultrasonic technology influence and the integrated machining of hard-to-machine advanced materials. Table 1.3 lists the summary of some of the simulated model approaches used for cutting mechanisms.

1.5.1 CHALLENGES RELATED TO MACHINING

Prerequisites in the CMC structures affected by the machining process are elevated tensile or compressive residual stresses, processing temperature, and CTE mismatch [50]. On one hand, the machining of CMCs is challenging because of their (1) brittle behavior, (2) high hardness, (3) heterogeneous structure composed (matrix, fibers, porosities), and (4) orthotropic mechanical and thermal behavior [50].

1.5.2 PRACTICAL ASPECTS AND DEVELOPMENTS

Contemporarily, manufacturing industries are harnessing the utilization of limited quantity lubrication and dry machining, ascribed to their ecological and economic conveniences. The minimum quantity lubrication (MQL) assisted machining is used

TABLE 1.2
Machinability Studies on Some of the Advanced Materials

Approach	Material	Application	Type of Electrode/ Cutting Fluids	Remarks	Reference
Electric discharge machining (EDM)	Ti-6Al-4V	Biomedical, automotive, and aerospace	Graphite, aluminum, copper, and brass	Discharge current elevated values cause coarser surface integrity. Unlike other electrodes, graphite resulted in high surface roughness, particularly at elevated discharge current values	[40]
High-pressure cooling (HPC)	Inconel 718	Nuclear reactors, pumps, spacecraft, gas turbines, and rocket motors	Vegetable oil–based	The utility of cutting oils viz. their environmental, economic, and societal pillars combined with the surface texture of tools of the HPC method can enhance the high-speed machinability and productivity of superalloys.	[41]
Wire electrical discharge (WED)	AA2024/Al_2O_3/BN hybrid composite	Automobile, structural and aerospace industries	Molybdenum	From the optimization method, it was concluded that the pulse on-time makes a significant impact on the desired performance measures during machinability of the composite	[42]
Wire electrical discharge (WED)	Hybrid metal matrix composite (HMMC): Al LM6 as matrix, silicon carbide, and dunite added as reinforcements	Military components, automotive, and aerospace	Brass	The machinability analysis on performance has shown that pulse ON duration is the prevalent variable for achieving the performance measures desired.	[43]
Wire electrical discharge (WED)	Al/AlCoCrFeNi$Mo_{0.5}$ MMC	Engineering materials for automotive industry	0.25 mm diameter copper wire	The surface roughness, KW, and MRR were minimal. The lower effectiveness range because of reinforcement over the MMCs machinability is a positive indicator for the contemporary utilization of the novel material.	[44]

KW = kerf width, MRR = material removal rate.

TABLE 1.3
Simulated Model Approaches Used for Cutting Mechanisms

Machining Approach	Theoretical Model	Type of Material	Process Parameters	Remarks	Author
PCBN turning	Adaptive genetic algorithm	GH4169 superalloy	Initial population size, i.e., m, P_c, P_m	Surface roughness during the turning process can be estimated, and the maximum error amid measured and predicted value is 0.107	[46]
Femtosecond laser processing	COMSOL software	Diamond microgrooves	Inlet velocity and size cross-section shape with a depth of 1000 microns	It was concluded that rectangular microgrooves have good heat dissipation compared to triangular and trapezoidal structures	[47]
PCBN tooling	Chip breaking model	GH4169 superalloy	$f = 0.15$ mm/r, $ap = 0.4$ mm, $v_c = 160$ m/min	Bending moment increases, and the crimp radius decreases, leading to a reduction of feed rate and depth of cutting during the high-pressure cooling process. Ultimately, the breaking performance of the chip is improved	[8]
CNC machining	Stereoscopic/spherical geometry method (SGM), and Projection method (PM)	For advanced materials	Position angle θ and orientation angle ϕ of drilling and milling modular fixture	The procedure and principle of solving spatial angle in the modular fixture by SGM, and PM was well executed	[48]
Intelligent machining combined with sensor-based control systems	Interactive search method (ISM), multi-objective genetic algorithm (MOGA), and genetic algorithms (GA)	AISI 1064 steel	Cutting speed, depth of cut, and feed rate	The results showed that ISM exhibited optimal outcomes in the field of manipulation of machining processes parameters	[49]
Abrasive water jet cutting	SPH algorithm and Lagrange model numerical simulation	Q235, X60, X80 and 304 stainless steels	Pressure 40 Mpa, target distance 5 mm, and 80 mesh garnet abrasive	The erosion effect was experienced during cutting due to stress and friction, impact deformation, resulting in the depth of cutting increasing with an increase in cutting pressure.	[18]

where f is the cutting force, ap is the cutting depth, and v_c is the deflection.

in auto parts, for example. It has been used in the crankshaft (drilling oil holes) and the block, which is typically a challenging approach as a result of diameter and large length [19].

A new novel concept of energy efficiency grade evaluation (EEGE) has been formulated by Ma et al. [51]. This approach is developed in various stages: (1) the enlisting of the inherent energy efficiency (IEE), (2) the advancement of quantitative approach in machining systems, and (3) creation of IEE evaluating indicator system from both the inherent specific energy (ISE) and inherent energy utilization. The EEGE method is a new tool for analyzing the energy efficiency of the machining system. However, future works objectives should be on establishing fundamental databases and interval threshold standards and discovering the application of the EEGE method to configure high-energy-efficiency machining systems [51].

Nanomachining technology using electron beam processing was studied by [52]. The structure of the self-organized surface of the nanomaterials was etched by engaging varying etching speeds, forces, depths, and probe cyclic times; however, the proportionality amid the etching depth and force, while the self-organized nanomaterials are progressively raised. The author indicated that selection of suitable parameters is a possibility to produce a linear structure with a width of approximately 60 nm and a depth of approximately 8 nm on the self-organized surface of the nanomaterial and a dotlike structure with a point spacing and height of 70 nm and 4 nm, respectively.

One of the complex systems is intelligent manufacturing systems in whereby enhancement approaches are combined with sensor-based control systems. Researchers have ascertained the viability of sensors transmitting networks with each other during the cutting operation. However, the objective was to devise a procedure that qualifies a CNC lathe spindle and smart feed drive to react to variations in signals fed to them from a series of external sensors, able to detect disturbances throughout the cutting process. Regrettably, no prototype was developed for this proposed system. Instead, this was achieved via a simulation of lathes created to substantiate logic and coding of the sensory communication network [53].

On the other hand, advanced engineering structures with embedded sensors form the basis of progressing attempts in structural health monitoring (SHM) systems [54]. However, there are certain dependencies not required for SHM designers which are treated as noises in the signals created by sensors. Some of these dependencies that affect sensing properties are (1) mechanical and thermal loading, (2) signal processing method, (3) integration configuration between material and surface-bonded, (4) fabrication process utilized for integration, and (5) base metal properties [54]. Albeit, piezoelectric materials offer high durability, providing good sensing mechanisms. Optical sensors, although costly, are efficient in quantifying strain and temperature simultaneously, resulting in overall superior performance. Generally, sensor fabrication is a countless in situ quantification of SHM variables but attributed to laboratory applications, for example, characterization and identification of complex structures in terms of deformation and failure mechanisms. Additionally, the emergence of multimaterial three-dimensional (3D) printing methods eases the integration of sensor materials into 3D printed composites. Consecutively, contributing greater understanding of the deformation interfacial mechanisms of composite materials used in the architecture industry [54].

1.6 CONCLUSION

Advanced materials for current and future aerospace applications, for example, are invented to withstand conditions such as environmental damage (oxidation and corrosion), creep strain, dwell crack growth, elevated temperature yield stress, and microstructure stability, exclusive of rises in cost and density. However, advanced materials, such as the advanced Ni alloys, solicit enhanced process of machining to acquire modified materials mechanical properties, including cost improvement outcome.

Tool failure mechanism research study as part of the challenges faced during machining is prevalent in the comprehension of tool structure and applications. Recently, tool failure exploration included both the macro problems of tool failure and microcrack propagation within the material before the appearance of tool surface crack defects. However, the intense cutting process outlining the damage mechanism of the carbide tool was explored, namely, (1) crack initiation, (2) crack propagation, (3) damage accumulation, and (4) tool breakage.

REFERENCES

1. Klocke, F., Soo, S. L., Karpuschewski, B., Webster, J. A., Novovic, D., Elfizy, A., Axinte, D. A., & Tönissen, S. (2015). Abrasive machining of advanced aerospace alloys and composites. *CIRP Annals – Manufacturing Technology*, *64*(2), 581–604. https://doi.org/10.1016/j.cirp.2015.05.004.
2. Gates, T. S. H. (2003). Computational materials: Modeling and simulation of nanostructured materials and systems. In *44th American Institute of Aeronautics and Astronautics, Structures, Structural Dynamics and Materials Conference*, Langley Research Center, Hampton, Virginia, US.(p. 1534).
3. Galantucci, L. M., Lavecchia, F., & Percoco, G. (2008). Study of compression properties of topologically optimized FDM made structured parts. *CIRP Annals*, *57*, 243–246.
4. Abramovitch, H., Burgard, M., Edery-Azulay, L., Evans, K. E., Hoffmeister, M., Miller, W., Scarpa, F., Smith, C. W., & Tee, K. F. (2010). Smart tetrachiral and hexachiral honeycomb: Sensing and impact detection. *Composites Science and Technology*, *70*, 1072–1079.
5. Miller, W., Smith, C. W., Scarpa, F., & Evans, K. E. (2010). Flatwise buckling optimization of hexachiral and tetrachiral honeycombs. *Composites Science and Technology*, *70*, 1049–1056.
6. Prall, D., & Lakes, R. S. (1997). Properties of a chiral honeycomb with a poisson's ratio of—1. *International Journal of Mechanical Sciences*, *39*, 305–314.
7. Vartak, D. A., Satyanarayana, B., Munjal, B. S., Vyas, K. B., Bhatt, P., & Lal, A. K. (2020). Potential applications of advanced nano-composite materials for space payload. *Australian Journal of Mechanical Engineering*, 1–9. https://doi.org/10.1080/14484846.2020.1733176.
8. Wu, M., Li, L., Liu, X., Cheng, Y., & Yu, Y. (2018). Analysis and experimental study of the cutting mechanism in machined nickel-base superalloys GH4169 with PCBN tools under high pressure cooling. *Integrated Ferroelectrics*, *189*(1), 105–120. https://doi.org/10.1080/10584587.2018.1456175
9. Mamalis, A. G., Kundrák, J., & Horváth, M. (2002). Wear and tool life of CBN cutting tools. *The International Journal of Advanced Manufacturing Technology*, *20*(7), 475–479.

10. Kramar, D., Sekuli, M., Jurkovi, Z., & Kopa, J. (2013). The machinability of nickel-based alloys in high pressure jet assisted (HPJA) turning. *Metalurgija*, *52*(4), 512–514.
11. Schwartz, M. (1992). *Handbook of structural ceramics*. New York: McGraw Hill.
12. Tuersley, I. P., Jawaid, A., & Pashby, I. R. (1994). Review: Various methods of machining advanced ceramic materials. *Journal of Materials Processing Technology*, *42*(4), 377–390.
13. Ming, W., Jia, H., Zhang, H., Zhang, Z., Liu, K., Du, J., Shen, F., & Zhang, G. (2020). A comprehensive review of electric discharge machining of advanced ceramics. *Ceramics International*, *46*(14), 21813–21838. https://doi.org/10.1016/j.ceramint.2020.05.207.
14. Samant, A. N., & Dahotre, N. B. (2009). Laser machining of structural ceramics a review. *Journal of the European Ceramic Society*, *29*(6), 969–993.
15. Ayrikyan, A., Prach, O., Khansur, N. H., & Webber, K. G. (2018). Investigation of residual stress in lead-free BNT-based ceramic/ceramic composites. *Acta Materialia*, *148*, 432–441.
16. Zhang, B., Zheng, X. L., Tokura, H., & Yoshikawa, M. (2003). Grinding induced damage in ceramics. *Journal of Materials Processing Technology*, *132*(1), 353–364.
17. Kirchner, H. P. (2010). Damage penetration at elongated machining grooves in hotc-pressed Si_3N_4. *Journal of the American Ceramic Society*, *67*(2), 127–132.
18. Hua, W., Zhang, W., Li, J., Cai, K., & Jiang, Y. (2020). Feasibility of abandoned ammunition materials cold cutting experimental study and numerical prediction. *Integrated Ferroelectrics*, *207*(1), 1–11. https://doi.org/10.1080/10584587.2020.1728660.
19. Goindi, G. S., & Sarkar, P. (2017). Dry machining: A step towards sustainable machining – Challenges and future directions. *Journal of Cleaner Production*, *165*, 1557–1571. https://doi.org/10.1016/j.jclepro.2017.07.235.
20. Klocke, F., Gierlings, S., Brockmann, M., & Veselovac, D. (2011). Influence of temperature on surface integrity for typical machining processes in aero engine manufacture. *Procedia Engineering*, *19*, 203–208.
21. Thakur, A., & Gangopadhyay, S. (2016). Dry machining of nickel-based super alloy as a sustainable alternative using TiN/TiAlN coated tool. *Journal of Cleaner Production*, *129*, 256–268.
22. Spriet, P. (2014). CMC applications to gas turbines. In *Ceramic matrix composites: Materials, modeling and technology*. Bansal, N.P., & Lamon, J., Hoboken, NJ: John Wiley & Sons, Inc. (pp. 591–608).
23. Yuan, S., Li, Z., Zhang, C., & Guskov, A. (2017). Research into the transition of material removal mechanism for C/SiC in rotary ultrasonic face machining. *The International Journal of Advanced Manufacturing Technology*, *95*(5–8), 1751–1761.
24. Ding, K., Fu, Y., Su, H., Chen, Y., Yu, X., & Ding, G. (2014). Experimental studies on drilling tool load and machining quality of C/SiC composites in rotary ultrasonic machining. *Journal of Materials Processing Technology*, *214*(12), 2900–2907.
25. Diaz, O. G., Axinte, D., Butler-Smith, P., & Novovic, D. (2018). On understanding the microstructure of SiC/SiC ceramic matrix composites (CMCs) after a material removal process. *Materials Science and Engineering*, *713*(26), 1–11.
26. Hashish, M., Kotchon, A., & Ramulu, M. (2015). Status of AWJ machining of CMCS and hard materials. In *International Technical Conference on Diamond, Cubic Boron Nitride and Their Applications*. Indianapolis, IN (pp. 1–14).
27. Tuersley, I., & Hoult, T. (1998). The processing of SiC – SiC ceramic matrix composites using a pulsed Nd-YAG laser: Part II the effect of process variables. *Journal of Materials Science*, *3*, 963–967.
28. Li, W., Zhang, R., Liu, Y., Wang, C., Wang, J., Yang, X., & Cheng, L. (2016). Effect of different parameters on machining of SiC/SiC composites via pico-second laser. *Applied Surface Science*, *364*, 378–387.

29. Wei, C., Zhao, L., Hu, D., & Ni, J. (2013). Electrical discharge machining of ceramic matrix composites with ceramic fiber reinforcements. *The International Journal of Advanced Manufacturing Technology*, *64*(1–4), 187–194.
30. Huiping, Z., Hongxia, Z., & Yinan, L. (2014). Surface roughness and residual stresses of high speed turning 300 M ultrahigh strength steel. *Advances in Mechanical Engineering*, *2014*, 1–7. https://doi.org/10.1155/2014/859207.
31. Zhang, H. P., Zhang, Z., Zheng, Z. Y., & Liu, E. (2020). Tool wear in high-speed turning ultra-high strength steel under dry and CMQL conditions. *Integrated Ferroelectrics*, *206*(1), 122–131. https://doi.org/10.1080/10584587.2020.1728633.
32. Kappmeyer, G., Hubig, C., Hardy, M., Witty, M., & Busch, M. (2012). Modern machining of advanced aerospace alloys-Enabler for quality and performance. *Procedia CIRP*, *1*(1), 28–43. https://doi.org/10.1016/j.procir.2012.04.005.
33. Abele, E., Lieder, M., Hölscher, R., & Hubig, C. (2012). Qualitative description of tool wear characteristic. *Wt-Online*.
34. Wu, M., Wu, S., Chu, W., Liu, K., & Cheng, Y. (2020). Experimental study on tool wear in cutting superalloy under high-pressure cooling. *Integrated Ferroelectrics*, *207*(1), 208–219. https://doi.org/10.1080/10584587.2020.1728681.
35. Alabdullah, M., Polishetty, A., Nomani, J., & Littlefair, G. (2019). An investigation on machinability assessment of Al-6XN and AISI 316 alloys: An assessment study of machining. *Machining Science and Technology*, *23*(2), 171–217. https://doi.org/10.1080/10910344.2018.1486415.
36. Lin, H. C., Lin, K. M., & Chen, Y. C. (2000). A study on the machining characteristics of TiNi shape memory alloys. *Journal of Materials Processing Technology*, *105*, 327–332.
37. Kong, M. C., Axinte, D., & Voice, W. (2011). Challenges in using waterjet machining of NiTi shape memory alloys: An analysis of controlled depth milling. *Journal of Materials Processing Technology*, *211*(6), 959–971.
38. Frotscher, M., Kahleyss, F., Simon, T., Biermann, D., & Eggeler, G. (2011). Achieving small structures in thin NiTi sheets for medical applications with water jet and micro machining: A comparison. *Journal of Materials Engineering and Performance*, *15*, 776–782.
39. Manjaiah, M., Narendranath, S., Basavarajappa, S., & Gaitonde, V. N. (2014). Wire electric discharge machining characteristics of titanium nickel shape memory alloy. *Transactions of Nonferrous Metals Society of China (English Edition)*, *24*(10), 3201–3209.
40. Ahmed, N., Ishfaq, K., Moiduddin, K., Ali, R., & Al-Shammary, N. (2019). Machinability of titanium alloy through electric discharge machining. *Materials and Manufacturing Processes*, *34*(1), 93–102. https://doi.org/10.1080/10426914.2018.1532092.
41. Mahesh, K., Philip, J. T., Joshi, S. N., & Kuriachen, B. (2021). Machinability of Inconel 718: A critical review on the impact of cutting temperatures. *Materials and Manufacturing Processes*, *36*(7), 753–791. https://doi.org/10.1080/10426914.2020.1843671.
42. Naidu, B. V. V., Varaprasad, K. C., & Prahlada Rao, K. (2021). Machinability analysis on wire electrical discharge machining of stir casted AA2024/Al_2O_3/BN hybrid composite for aerospace applications. *Materials and Manufacturing Processes*, *36*(6), 730–743. https://doi.org/10.1080/10426914.2020.1854466.
43. Manikandan, N., Balasubramanian, K., Palanisamy, D., Gopal, P. M., Arulkirubakaran, D., & Binoj, J. S. (2019). Machinability analysis and ANFIS modelling on advanced machining of hybrid metal matrix composites for aerospace applications. *Materials and Manufacturing Processes*, *34*(16), 1866–1881. https://doi.org/10.1080/10426914.2019.1689264.

44. Karthik, S., Prakash, K. S., Gopal, P. M., & Jothi, S. (2019). Influence of materials and machining parameters on WEDM of Al/AlCoCrFeNiMo 0.5 MMC. *Materials and Manufacturing Processes*, *34*(7), 759–768. https://doi.org/10.1080/10426914.2019.1594250.
45. Feucht, F., Ketelaer, J., Wolff, A., Mori, M., & Fujishima, M. (2014). Latest machining technologies of hard-to-cut materials by ultrasonic machine tool. *Procedia CIRP*, *14*, 148–152. https://doi.org/10.1016/j.procir.2014.03.040.
46. Li, L., Wu, M., Liu, X., Cheng, Y., & Yu, Y. (2017). The prediction of surface roughness of PCBN turning GH4169 based on adaptive genetic algorithm. *Integrated Ferroelectrics*, *180*(1), 118–132. https://doi.org/10.1080/10584587.2017.1338881.
47. Dou, J., Cui, J., Fang, X., Dong, X., Ullah, N., & Xu, M. (2020). Theoretical and experimental study on machining rectangular microgroove of diamond by femtosecond laser. *Integrated Ferroelectrics*, *208*(1), 104–116. https://doi.org/10.1080/10584587.2020.1728722.
48. Xu, L., Chen, K., & Du, W. W. (2020). The application and the calculation of deflection angle of the adjustable modular fixture which has space double angle. *Integrated Ferroelectrics*, *210*(1), 175–188. https://doi.org/10.1080/10584587.2020.1728858.
49. Radovanović, M. (2019). Multi-objective optimization of multi-pass turning AISI 1064 steel. *The International Journal of Advanced Manufacturing Technology*, *100*(1–4), 87–100.
50. Diaz, O. G., Luna, G. G., Liao, Z., & Axinte, D. (2019). The new challenges of machining ceramic matrix composites (CMCs): Review of surface integrity. *International Journal of Machine Tools and Manufacture*, *139*, 24–36. https://doi.org/10.1016/j.ijmachtools.2019.01.003.
51. Ma, F., Zhang, H., Gong, Q., & Hon, K. K. B. (2021). A novel energy efficiency grade evaluation approach for machining systems based on inherent energy efficiency. *International Journal of Production Research*, *59*(19), 6022–6033. https://doi.org/10.1080/00207543.2020.1799104.
52. Li, D. (2020). Surface structure etching law of self-organizing nanomaterials based on AFM and SEM. *Integrated Ferroelectrics*, *207*(1), 49–61. https://doi.org/10.1080/10584587.2020.1728664.
53. Imad, M., Hopkins, C., Hosseini, A., Yussefian, N. Z., & Kishawy, H. A. (2021). Intelligent machining: A review of trends, achievements and current progress. *International Journal of Computer Integrated Manufacturing*, 1–29. https://doi.org/10.1080/0951192X.2021.1891573.
54. Montazerian, H., Rashidi, A., Milani, A. S., & Hoorfar, M. (2020). Integrated sensors in advanced composites: A critical review. *Critical Reviews in Solid State and Materials Sciences*, *45*(3), 187–238. https://doi.org/10.1080/10408436.2019.1588705.

2 Machining by Advanced Ceramics Tools

Challenges and Opportunities

P. Kour and S. K. Pradhan
Birla Institute of Technology, Mesra, Patna Campus, India

Amit Kumar
National Institute of Technology Patna, Patna, India

Manoranjan Kar
Indian Institute of Technology Patna, Bihta, Patna, India

CONTENTS

DOI: 10.1201/9781003284574-3

2.1 INTRODUCTION

In recent competitive scenarios, the most modern and proficient technologies are highly demanding by manufacturing companies for making the equipment and devices [1]. High precision and appropriate surface roughness of the products are produced by machining processes [2]. The machining procedure of cutting tools is experiencing tough production circumstances for multifaceted geometries of products [3]. The hardness of these materials is developed and improved over the period. The hardness of the cutting tools for turning, drilling, and milling type of old-style or conformist methods of machining processes is higher than the processed material hardness. Machinery will have a short life if the hardness of the material is low and, hence, increases the cost of machining [4]. High-speed machining for a long duration should have enhanced mechanical properties, which depends on the advanced cutting tool.

Machining of materials is the process used to obtain a higher surface finish, close tolerance, and complex geometrical shapes. Cutting tools are one of the most important machining tools that can remove the materials from the workpiece to get the required size, configuration, and finish. Machining tools are the mother of machines as without them no component can be finished. John Wilkinson's boring machine was the first machine tool invented in AD1775, which paved the way for James Watt's steam engine. One of the important parts of machining is material cutting technology. In this process, by controlling the removal of material from the workpiece, it takes a preferred final shape and size of cutting tools. Here the cutting tools are used to form a chip that is removed from the workpiece.

Different sharp cutting tools such as turning, drilling, milling, and so on are used to remove the materials. Grinding is used in abrasive processes to remove the materials. Besides sharp-cutting tools, many nontraditional processes such as water jet cut, laser, and electrical-discharge machining (EDM) are used to confiscate material. Single-point tools and multiple-cutting-edge tools are the two basic types of metal cutting tools that are classified depending on one or multiple cutting edges, respectively. The rake face is the surface of the tool. The angle between normal to the machining direction and the rake face is called the rake angle. For highly brittle tool materials, rake angles are zero and negative to give extra strength to the tool's tip.

A freshly produced work surface can be protected from abrasion, which can destroy the finish through the flank of the tool. It offers a clearance between a freshly produced work surface and the tool. The relief angle is the angle at which the flank surface is oriented. The clearance angle is the angle that the base of the tools makes with the machined surface. The strength of the tool tip is decreased with an increase in the clearance angle. So the angle is of the order of 5–6°. The factors that affect the material cutting are the work material, the geometry of the cutting tool, the material of the cutting tool, the feed rate, the speed of cutting, the cutting fluid, and the depth of the cut used.

Cutting tools must be made of a material harder than the material to be cut. A hard tool is used for plastic deformation of metal material in material-removing technology to obtain the required physical shape and properties of the tools. The tool must be able to withstand the heat generated in the metal material-cutting process or non-metal-material-cutting process. In recent scenarios, high carbon steel, ceramics, diamonds, and the like are used for different cutting tools in metal-processing industries.

The life of the tool is specified by the actual cutting time to failure, the length of work cut to failure, the volume of metal removed to failure, the number of components produced, and the cutting speed for a given time to failure. Based on Taylor concept of optimum cutting speed for best productivity [5],

$$VT^n = C, \tag{2.1}$$

where T is the tool life in minutes and V is the cutting speed, with C and n as experimentally identified constants. These constants depend on the material and work of the tool. The broad extension of this formula to be applied in a wide range is [6]

$$V T^n f^x d^y = C, \tag{2.2}$$

Where f and d are the rate of feed and depth of cut, respectively, and C, n, x, and y are the experimentally identified constants. The tool gets worn out due to long-term usage called tool wear. The factor responsible for this tool wear is very high stress, very high temperature, virgin metal, very high stress gradient, and very high-temperature gradient. Cutting is categorized according to dimension as macro- and microcutting.

In macrocutting, the cutting tool would not be sharp. The edge radius of the cutting tool is found to be in the range of 0.02–0.05 mm. Uncut chip thicknesses of 0.1 mm and greater are used [7]. In mircocutting, features are usually in the range of 25–500 μm, and they are controlled by the tool size. The radius of the cutting edge of the micro tools is in the range of 1–10 μm. Cutting tools are associated with many challenges. These tools should have good mechanical properties, such as a good cutting edge, to avoid plastic deformation [8]. To be used at a high temperature, its hardness should be high at an elevated working temperature. To oppose the maximum mechanical load, its fatigue resistance should be high. It should show good fracture toughness, and to preserve precision, these cutting tools show high stiffness. Besides this, it also shows good thermal properties. To transport the temperature away from the cutting edge, its thermal conductivity should be high with its resistance. The chemical composition of these tools should be stable. To avoid buildup at the cutting edge, it should be sufficiently lubricated. So, in the machining process, the characteristics of cutting tools are very significant. The quality and the cost of the product depend on materials, geometry, design, and life. Advanced cutting tool materials are needed for the new alloys. Cutting tools based on ceramic materials to be used for superalloys, high-speed machining process shows high hardness and good recital in wear resistance [9]. Ceramics not only are hard but also, without any chemical decomposition, can sustain at a high temperature (more than 1600 °C) [10]. As compared to carbide, which can take about 870 °C, ceramics can acquire extra heat used to soften it up to 2200 °C [11]. A high-speed dry-machining condition needs cutting tools based on ceramics. So its involvement in metal processing has been increasing day by day. As compared to cutting tools based on steel for high-speed machining and hard alloys, these ceramics-based tools have considerable wear and heat resistance. The cutting speed of 1.5 to 8 times has been achieved on ceramic-based tools as compared to carbide tools. The tools used for cutting have to face different challenges and need to be improved. The vital requirements for machining with ceramics are a high cutting speed and square or round inserts with a large nose radius. An adequate and uninterrupted power supply should be provided. Tools also have a negative rake angle, large nose radius, large side cutting edge angle on the ceramic, deeper cut with light deed rate. Tools have also avoided the use of coolant. Tools holders are kept at the minimum overhang.

2.2 CUTTING TOOL BASED ON CERAMIC MATERIALS

Superalloy materials and cast irons are very hard to cut. Cutting tools based on ceramics are not only used for cutting but also for machining these materials [11]. As compared to carbide-based cutting tools, these ceramic tools have wear resistance, corrosion, and high hardness [12]. Also, they have cutting speed high, high strength for cutting hard-to-cut material, and very high resistance to abrasive wear and cratering [13]. On the other hand, it is associated with the limitation of small transverse rupture strength and brittleness [13]. In the beginning, due to low thermal conductivity, ceramic cutting tools suffer from low resistance to thermal and mechanical shock and low toughness [14]. With high-speed machining, those defects then turn out to be somewhat more distinct. Therefore, the limited standard of depth of cut is used to reduce the cycle time [15]. Intermittent cutting tools based on ceramic also count troubles. The color of the ceramics is gray, black, or white, depending on its manufacturing by hot pressing and cooled-pressed technique. Compared to hot-pressed ceramics, the hardness of cold-based ceramics is higher [13]. Mainly cutting tools based on ceramic belong to four families like the family of Si_3N_4 and Al_2O_3 ceramics, cermet materials, and SiAlON-based ceramics [14]. Silicon nitride (Si_3N_4)– and aluminum oxide (Al_2O_3)–based ceramics are the most used cutting tools [11]. These major tools are associated with other materials, such as aluminum tools that have magnesium, titanium, chromium, or zirconium oxides [10]. The main composition of cermets is 70% ceramic and 30% titanium carbide. They are transformed into the shape of bullets by being hot-pressed at a high temperature and pressure. After sintering, the ceramics are processed to the shape to be used in the cutting tools.

2.2.1 Ceramic Tools Based on Aluminum Oxide

Alumina or aluminum oxide has shown a strong ionic bond that gives rise to desirable material characteristics. Machining tools for high-strength steels are mainly used, with aluminum oxide as its base materials [16]. The high hardness makes them a desirable candidate for cutting tools, but the small damage tolerance and brittleness hinder its application [17]. But at a high temperature, alumina-based ceramic shows high hardness and high wear resistance, with comparatively small chemical reactivity with steels and other materials. To increase the hardness, at high temperatures, ceramic oxide disperses these hard particles. These dispersed particles produce crack deflection, crack impediment, and crack branching that give rise to an increase in bending strength and fracture toughness. So the resistance to adhesive and abrasive wear increases due to an increase in hardness and toughness. Resistance to the thermal shock and the ability of thermal shock cycling increase due to a small thermal expansion and higher thermal conductivity. For high-temperature cutting, that is, above 800 °C, the depth of cutting, the cutting speed, the feed rate, and the cutting condition should be taken into account as there is titanium nitride and titanium carbide particle loss in the reinforcing properties due to high-temperature oxidation [18]. Aluminum oxide (Al_2O_3) and zirconium oxide–dispersed aluminum oxide Al_2O_3/ZrO_2 are the main forms of ceramic aluminum oxide [19]. These Al_2O_3/ZrO_2-based tools are yellow to gray/white. These tools show resistance to thermal deformation and wear with large chemical internees. During phase transformation, these tools are hardened. The feeding rate of these tools during continuous cutting semi-finishing, and finishing is very low. Nanoparticles and SiC whiskers are used to prepare a new type of Al_2O_3-based ceramic tool material that is synergistically toughened [20]. Its flexural strength increases with an increase in temperature and then a decrease with the temperature. A decrease in elasticity modulus and oxidation of SiC results in a decrease in the mechanical properties of these composites. Healing the defects and micro-cracks at high temperatures results in an enhancement of its mechanical properties. Fracture toughness and flexural strength are enhanced by the incorporation of micro- and nanosized TiC. Al_2O_3 ceramic incorporating in graphene platelets (GPLs) shows 30% increase in flexural strength and 27% in fracture toughness compared to monolithic Al_2O_3 [21].

2.2.2 Ceramics Tools Based on Silicon Nitride

To overcome the drawback of Al_2O_3-based cutting tools, a new type of ceramics-based cutting tools, that is, silicon nitride Si_3N_4–based ceramics, were proposed during the 1980s. The flexural strength of ceramic tools is an important mechanical property. A high flexural strength of 700 to 1100 MPa has been reported for silicon nitride ceramics. They also have higher fracture toughness and exceptional thermal shock resistance. Cracks do not develop easily compared to other steady cutting tools. [22]. Various kinds of silicon nitride–based cutting tools are hot-pressed silicon nitride (Si_3N_4 HIP), reaction-bonded silicon nitride (Si_3N_4 RB), and SiAlON, among others [19]. Due to their good wear resistance, hardness, toughness, resistance to thermal shock and elevated temperature strength, silicon nitride–based cutting tools were explored in previous decades [23]. Because it is thermally decomposed at high temperatures due to a low diffusion rate, it needs an additive for sintering. So various additives are investigated to be used to increase the density of Si_3N_4 during the sintering of the liquid-phase mechanism [24]. Ceramic tools based on alloys of SiAlON-containing aluminum, silicon, nitrogen, and oxygen are simpler to manufacture compared to monolithic silicon nitride. It also shows good mechanical properties such as high red hardness, better notch and wear resistance, and good thermal shock resistance. Flexural strength is also increased by incorporating CeO_2, Nd_2O_3, Sm_2O_3, and Dy_2O_3 in silicon nitride due to the small elongation of the grain with a denser microstructure. Also, the depressed grain growth due to the incorporation of RuO_2 in silicon nitride resulted in enhanced thermal conductivity [24].

2.2.3 Ceramic Tools Based on Composite Materials

Base-material-machining properties are enhanced by reinforcing them to produce composite material. Composite of Al_2O_3/TiC matrix with GPLs shows enhanced microstructure and fracture toughness, but its hardness is decreased [25]. Inserting additional components results in improved performance of the cutting tool material [26]. The cryogenic technique is also used to enhance the physical and mechanical properties of the cutting tools. In this technique, the materials are cooled below room temperature, kept at that temperature for a long period, and then are brought back to normal temperature by heating. Here, the cooling temperature and the number of hours kept at that temperature affect the material's properties [27]. Fractural strength was increased in GPL-incorporated silicon nitride composite due to an increase in resistance to crack deflection, cracking of graphene platelet, overlapping cracks, and branching [28]. The transverse rupture strength of cemented carbonitride was increased by replacing cobalt with nickel and results in a small decrease in Vickers hardness [29].

2.2.4 Enhance the Cutting Properties with Coatings

The performance and properties of the cutting tools can be enhanced by coating on their surface [30]. Chemically inert, antifriction, hard, and thermally isolating layer over the composite structure shows a reduction in the interactions among the tool and chips. Also, friction is reduced; as a result, one can get better wear resistance in a broad temperature range of cutting [31]. Mechanical properties, such as tool life, machining process with high cutting speed are obtained by single- or multilayer coating. Ceramic tools without a defective element with high hardness and heat resistance are used for the metalworking industry. Several layers of coating techniques are being used to enhance the mechanical properties and the life of cutting tools [32]. The coating not only improves the surface properties, but the contact area's thermomechanical properties are also reduced. Dry-modeling technologies reduce environmental impact, health hazards with major economic efficiency [33]. Multilayer nanoscale composite coatings have direct control on the contact characteristics and enhance performance with a long tool life [32, 34].

2.2.5 Textured-Surface Ceramic Cutting Tools

One of the ways to enhance the performance of cutting tools is the differently shaped textured patterns. These textured surfaces are mainly used to lubricate the surface. A lubricant reservoir is created with the use of the surface-texturing process. Microholes, dimples, grooves, and channels with textures of different forms full of a lubricant are formed on the face of the tools. These lubricant pools reduce not only friction but also the wear at the interface of the chip tool [35]. The microstructures of specific shapes and sizes are created in the cutting edge of the tools used for lubrication. So appropriate microstructure morphology can reduce the cutting temperature, cutting force, and wear of rake face during the dry-cutting process [36]. Friction wear of cutting force is decreased in the nano-textured tool [37]. In a ceramic plate microstructure, graphite grain is formed and run out directly from the cutter, resulting in a decrease of the friction force and temperature by covering the rake face during tool wear-out. Cutting temperature, cutting forces, tool wear, and coefficient of friction are decreased due to lubrication by molybdenum disulfide (MoS_2) in the micro- and nanoscale structure of Al_2O_3/TiC [38]. As compared to linear structure, square textured tools show enhanced performance in terms of surface finish and cutting force. The reduced contact area of the tool chip results in reduced friction [39]. Adding a lubricant powder coating to the metal surface to lubricate the surface results in an enhanced microstructure, mechanical properties, and wear resistance and better antifriction properties [40]. Tool geometry and its performance are affected by continuous or intermittent cutting [41]. It is concluded that the design of the cutting tools can be tailored such as the construction of differently shape textured patterns and self-lubricating surfaces.

2.3 EFFECTS OF METHODS OF MANUFACTURING ON THE PROPERTIES OF CERAMIC CUTTING TOOLS

The manufacture of ceramic and composite-based tools has undergone various steps. To prepare the ceramics of desired properties, first, it is synthesized by using a suitable method; then it undergoes drying, and the binder is then burned out by heating. The obtained powder is then sintered into the desired product. The optimum properties of the tools are shown in the good-sintered ceramic as having a high density and uniform, fine microstructure. The ceramics have high density due to the atomic transport mechanism and growth of the particle [42]. The ceramics are sintered either by the contact method (fuel heating and electric resistance) or the noncontact method (microwave, radio frequency, induction) [43]. The sintering of the ceramics has been performed by hot pressing, spark plasma sintering (SPS), and other methods. The mechanical properties of the ceramics are hindered by the high temperature and large holding time [44]. Sintering temperature can be controlled through the time of holding, setting the ramp rate, the voltage, the pulse current, and the duration of the pulse. It affects phase stability, densification kinetics, grain growth, and stoichiometry [45]. The grain growth can be minimized, resulting in a high density and hardness by increasing the velocity of the sintering process. Nonconventional microwave sintering now attracts the interest of the scientific community due to the small consumption of energy and cost of production, shorter processing time, and high rate of heating. Moreover, it results in enhanced physical properties [46].

2.3.1 Contact Manufacturing

2.3.1.1 Hot Pressing

Mechanical properties, such as durability, wear resistance, and others, and the speed of the cutting tool should be enhanced. So powder metallurgy process produces ceramic-based cutting tools [47]. In the hot-pressing method, high temperatures and high pressures are used to produce sintering and creep. Cutting tools based on nanosized aluminum oxide and tungsten monocarbide and prepared by hot-pressing methods show better durability [48]. Optimizing the temperature, pressure, time, and the concentration of TiS_2 in $Al_2O_3/TiB_2/TiSi_2$–based cutting tools prepared by the vacuum hot-pressing method shows better microstructure and mechanical properties [49]. A composite prepared by the SPS process shows the formation of a thinner microstructure and a relatively higher density, resulting in increased fracture toughness, flexural strength, and hardness [50]. The composite of aluminum-based carbide is prepared by the hot-pressing method. It shows enhanced fracture toughness due to a decrease in the residual stress in the matrix. Also, crack deflection, cracks bridged by particles, and intergranular grain mechanisms consume energy during the interaction [51].

2.3.1.2 Spark Plasma Sintering

Another method having potential for material processing is SPS. A pulsed direct current is the source, and the Joule effect of heating is used for sintering to get a denser nanostructure. The power used, processing cycle time, and thermal conductivity are the factors that control the preparation of the materials. They depend on the temperature difference between the edges of the workpiece and the core and the distribution of heat [52]. Sputtering at high temperatures occurs when spark plasma with the pressure of spark oxide films and adsorptive gases eradicated surface powder impurities. The electromagnetic field produces the first migration of the ions by high-speed diffusion [53]. The high-temperature state at the confined region is performing as evaporation, joining, recrystallization, and the phenomenon of solidification. Another sintering method for preparing super-hard polycrystalline diamond, ceramic composites, cubic boron nitride, refractory substances, and nanopowder is the high-pressure SPS (HP SPS) method. Four steps are used for materials preparation in SPS sintering. Materials are first to undergo vacuum treatment, and then they are warmed up and get pressed. At the point of interaction of the powder, a spark discharge occurs with the rise in temperature of the local zone to several thousand. Temperature measured in SPS is usually lower than the actual sintering temperature due to the method of measurement, that is, either by a thermocouple attached to the wall of the graphite matrix or by pyrometer attached to the surface but not directly in the powder [54]

2.3.2 Noncontact Manufacturing Methods

2.3.2.1 Microwave Sintering

In the noncontact method of high-temperature sintering, one of the most used cost-effective processes is microwave sintering, which takes less energy and has a shorter processing time. Ceramics need high processing temperatures. Microwave sintering makes it a cost-efficient process. It also produces improved characteristics due to diffusion kinetics and reaction [43, 46]. Microwave sintering is a small energy-consumption process with a shorter duration of sintering. It also has a high rate of sintering, a high degree of densification with improved mechanical properties, and a flexible processing technique for net shape material. Several mechanisms, such as resistive heating, dielectric heating, electromagnetic heating, and bipolar rotation, are responsible for the interaction of microwave radiation with the absorbing material.

Enhanced densification, energy efficiency, environmentally friendly, lower sintering temperature, uniform volumetric heating, fine grain due to fast heating with minor and equiaxed pores in the sintered green compacts in microwave heating, resulting in better mechanical properties and performance of the product [55]. Pure alumina is sintered by microwave heating, which is difficult to sinter through other processes to complete densification due to its extremely refractory nature.

2.3.2.2 Self-Propagation High-Temperature Synthesis

Combustion synthesis (CS), or self-propagating high-temperature synthesis (SHS), belongs to the self-sustained combustion process, producing nearly precious material and compounds. SHS includes reactions of compounds, self-sustained reactions, elemental powders synthesis, porous combustion, and gaseous oxidizer in solid reactive media, and termite-type reactions are applied for the SHS of advanced materials [56]. Organic or inorganic compounds undergo a combustion reaction process in SHS [57]. Compressed powders are ignited either in an inert atmosphere or in an air-producing chemical reaction so that enough heat energy is released for a self-sustaining reaction [58]. It is associated with some major shortcomings, such as alkali cations, corrosion by water vapor, low electrical conductivity, and mechanical properties that are uneven in oxidation. SHS process is relatively simple and requires small energy. A high-quality product can be produced by this process. Among other noncontact methods, the SHS process is more favorable [59].

2.4 EFFECT OF DIFFERENT PROCESSING CONDITIONS ON CERAMIC CUTTING TOOLS

Corrosion resistance, fatigue strength, reduction in wearing, and friction can be enhanced by controlling workpiece hardness, coating materials, and conditions of cutting [60]. Cost-effective production can be achieved by the introduction of modern technologies [1]. The temperature during cutting or machining affects directly to the working condition. Product quality depends on the surface quality or surface roughness. Roughness is affected by the grip of chips on the boundaries of the cutting tool, the sensitivity, and the vibration of the machine tool [61]. The rate of feeding is an important consequence of the surface quality. The cutting force influences the cutting depth. The hardness of machined component material and cutting condition will affect the tool life, cost, and quality of the product. Improved surface roughness, longer tool life with less cutting force obtained by ceramic end milling tools compared to the commercial cemented carbides [62]. Fatigue behavior predicts the failure of ceramic tools. In machining operations, cutting fluids are introduced to enhance lubricity. This will produce cooling to diminish variability of the process and life of the tool [63]. But it is classified by many foundations and countries as dangerous waste and enforces management if they have certain alloys and oil. Coolant is needed for hard-to-cut materials machining. For a cutting tool's hardness to be increased, self-lubricating materials are needed for the design, and cutting tools with a textured pattern are used to overcome the drawback. Cutting conditions are affected by dry or wet machining [64]. Machine operation determines the machining parameter [1]. So appropriate equations should be established to show the enhanced condition of machining depending on the material, the geometry of the cutting tool, and the roughness of the surface of the machined component. Proper cutting tools selection depends on performance and durability. Cutting performance depends on the cutting forces and ultimately the cost of the product. So perfect tool geometry is essential as these sharp cutting tools experience high stress during the cutting process [65].

2.5 CONCLUSION

Cutting tools and their quality are the main problems in machining. Cutting tools should have good mechanical properties, high hardness, and reasonable life with suitable performance. Ceramic tools based on carbide show enhanced surface quality and longer tool life. Modern carbide tools could be the next generation of ceramic cutting tools. The mechanical properties of ceramic tools, such as bending strength, crack resistance, and hardness, have been achieved by cutting-edge technologies for applications. A cost-effective cutting tool with improved performance has huge importance. Several parameters affect the cutting tool materials in a precise cutting process. These parameters are dependent on each other. Ceramic cutting tools are very promising candidates for future technology. But the mechanical properties still need to be enhanced. So their mechanical and tribological properties need to be developed in different cutting conditions.

2.6 FUTURE SCOPE

Ceramic-based cutting tools, however, show good mechanical properties with excellent performance compared to traditional tools but still are associated with some shortcomings that need to be addressed. The major drawback that these ceramics are associated with is high brittleness and low toughness. On the other hand, with the constant improvement in the technology of cutting tools based on ceramic in current years, one needs to step into a new era of fully developed cutting tools based on ceramics that will play a vital role in the field of machining in the future. Here are some future steps by which new ceramic technology can overcome its drawbacks:

1. Ceramic tools coated with TiN/TiAN by a chemical vapor deposition, hot pressing, sol-gel, or physical vapor deposition process are appropriate for making a dry-cutting tool for high-speed use. They also can survive relatively large cutting to obtain good-quality surfaces. It can be used for dry and high-speed cutting, abrasion resistance, and longer life. So in the future, different coatings may be used to enhance the toughness and brittleness of the ceramic tools.
2. Whisker reinforcement can enhance the strength and toughness of ceramics. A few defect-incorporated whiskers of a suitable amount can be used in

whisker-reinforced ceramic. These whisker-reinforced ceramic tool materials will show enhanced fracture toughness accomplished by phase transformation. It holds a good influence on thermal shock resistance and toughness resistance. So different whisker compositions with different volume fractions on the toughness and brittleness of the ceramics need to be explored.

3. During high-speed cutting resistance to damage can be enhanced through functionally gradient ceramic tools prepared by surface nitriding and vacuum sintering. Changes in different sintering parameters and surface treatment may improve the mechanical properties of cutting tools to be used in a robust environment.
4. Rare earth elements have the unique properties of diminishing pores, filtering the interface, and increasing the bonding strength and toughness of the interface. So ceramic cutting tools modified by rare earth elements may need to be explored.

REFERENCES

1. Hrelja, M., Klancnik, S., Irgolic, T., Paulic, M., Balic, J., Brezocnik, M. (2014). Turning parameters optimization using particle swarm optimization. *J. Proc. Eng.*, 69, 670–677.
2. El-Hofy, H.A. (2005). *Advanced machining processes*. McGrawHill, New York, USA.
3. Baptista, R., Simoes, J.F.A. (2000). Three and five axes milling of sculptured surfaces. *J. Mat. Process. Technol.*, 103, 398–403.
4. Hamdy, K., Mohamed, M.K., Abouelmagd, G. (2013). New electrode profile for machining of internal cylindrical surfaces by electrochemical drilling. *Int. J. Control. Autom. Syst.*, 2, 2165–8277.
5. Taylor, F.W. (1907). On the art of cutting metals. *Trans. ASME*, 28, 31–350.
6. Kronenberg, M. (1966). *Machining science and applications*, Pergamon Press, Oxford, UK.
7. Oxley, P.L.B. (1963). Mechanics of metal cutting for a material of variable flow stress. *ASME. J. Eng. Ind.*, 85, 339–345.
8. Gutiérrez-González, C.F., Suarez, M., Pozhidaev, S., Rivera, S., Peretyagin, P., Solís, W., Díaz, L.A., Fernandez, A., Torrecillas, R. (2016). Effect of TiC addition on the mechanical behaviour of Al_2O_3-SiC whiskers composites obtained by SPS. *J. Eur. Ceram. Soc.*, 36, 2149–2152.
9. Zhao, B., Liu, H., Huang, C., Wang, J., Cheng, M. (2017). Fabrication and mechanical properties of Al_2O_3-SiCw TiCnp ceramic tool material. *Ceram. Int.*, 43, 10224–10230.
10. Mohammadpour, M., Abachi, P., Pourazarang, K. (2012). Effect of cobalt replacement by nickel on functionally graded cemented carbonitrides. *Int. J. Refract. Met. Hard Mater*, 30, 42–47.
11. Grzesik, W. (2017). Cutting tool materials. *Adv. Mach. Processes Metallic Mater.*, 2 35–63.
12. Kitagawa, T., Kubo, A., Maekawa, K. (1997). Temperature and wear of cutting tools in high-speed machining of Inconel 718 and Ti-6A1-6V-2Sn. *Wear*, 202, 142–148.
13. Li, L., He, N., Wang, M., Wang, Z.G. (2002). High-speed cutting of Inconel 718 with coated carbide and ceramic inserts. *J. Mater. Process. Technol.*, 129, 127–130.
14. Schneider Jr., G. (2009). Cutting Tools Application. http://www.Toolingandproduction.com/web/home.php

15. Zhao, J. (2014). 25-the use of ceramic matrix composites for metal cutting applications. *Adv. Ceram. Matrix Compos*, 2, 623–654.
16. Jack, D.H. (1986). Ceramic cutting tool materials. *Mater. Des.*, 7, 267–273.
17. Gonzalez-Julian, J., Schneider, J., Miranzo, P., Osendi, M.I., Belmonte, M. (2011). Enhanced tribological performance of silicon nitride-based materials by adding carbon nanotubes. *J. Am. Ceram. Soc.*, 94, 2542–2548.
18. Cheng, M., Lia, H., Zhao, B., Huang, C., Yao, P., Wang, B. (2017). Mechanical properties of two types of Al_2O_3/TiC ceramic cutting tool material at room and elevated temperatures. *Ceram. Int.*, 43, 13869–13874.
19. Whitney, E.D. (1995). *Ceramic cutting tools: Materials, development, and performance*, William Andrew Publishing, Norwich, USA.
20. Stephenson, D.A., Agapiou, J.S. (2016). *Metal cutting theory and practice*. 3rd edn., CRC Press Taylor & Francis Group, New York, USA.
21. Grigoriev, S.N., Fedorov, S.V., Hamdy, K. (2019). Materials, properties, manufacturing methods and cutting performance of innovative ceramic cutting tools – A review. *Manufacturing Rev.* 6, 19, 2330.
22. Wang, D., Xue, C., Cao, Y., Zhao, J. (2017). Fabrication and cutting performance of an Al_2O_3/TiC/TiN ceramic cutting tool in turning of an ultra-high-strength steel. *Int. J. Adv. Manuf. Technol.*, 91, 1967–1976.
23. Smirnov, B.I., Nikolaev, V.I., Orlova, T.S., Shpeizman, V.V., Arellano-Lopez, A.R., Goretta, K.C., Singh, D., Routbort, J.L. (1998). Mechanical properties and microstructure of an Al_2O_3-SiC-TiC composite. *Mat. Sci. Eng. A*, 242, 292–295.
24. Liu, W., Chu, Q., He, R., Huang, M., Wu, H., Jiang, Q., Chen, J., Deng, X., Wu, S. (2018). Preparation and properties of TiAlN coatings on silicon nitride ceramic cutting tools. *Ceram. Int.*, 44, 2209–2215.
25. Švec, P., Brusilová, A., Kozánková, J. (2008). Effect of microstructure and mechanical properties on wear resistance of silicon nitride ceramics. *Mater. Eng.*, 16, 34–40.
26. Cheng, Y., Zhang, Y., Wan, T., Yin, Z., Wang, J. (2017). Mechanical properties and toughening mechanisms of graphene platelets reinforced Al_2O_3/TiC composite ceramic tool materials by microwave sintering. *Mat. Sci. Eng. A*,680, 190–196.
27. Gandotra, S., Singh, J., Gill, S.S. (2011). Investigation of wear behaviour on coated and non-coated carbide inserts subjected to low temperature treatment. *J. Metall. Eng.*, 1, 1–16.
28. Kumar, S., Khedkar, N.K., Jagtap, B., Singh, T.P. (2017). The effects of cryogenic treatment on cutting tools. *IOP Conf. Series: Mater. Sci. Eng.*, 225, 1–9.
29. Dusza, J., Morgiel, J., Duszová, A., Kvetková, L., Nosko, M., Kun, P., Balázsi, C. (2012). Microstructure and fracture toughness of Si_3N^{4+} graphene platelet composites. *J. Eur. Ceram. Soc.*, 32, 3389–3397.
30. Chung-Cheng, T., Hong, H. (2002). Comparison of the tool life of tungsten carbide coated by multi-layer TiCN and TiALCN for end mills using Taguchi method. *J. Mater. Process. Technol.*, 123, 1–4.
31. Bouzakis, K.-D., Michailidis, N., Skordaris, G., Bouzakis, E., Biermann, D., Saoubi, R.M. (2012). cutting with coated tools: Coating technologies, characterization methods and performance optimization. *CIRP Ann. Manuf. Technol.*, 61, 703–723.
32. Vereschaka, A.S., Grigoriev, S.N., Tabakov, V.P., Sotova, E.S., Vereschaka, A.A., Kulikov, M.Y. (2014). Improving the efficiency of the cutting tool made of ceramic when machining hardened steel by applying nano-dispersed multi-layered coatings. *Key Eng. Mater.*, 581, 68–73.

33. Gurdial Blugan, G., Strehler, C., Vetterli, M., Ehrle, B., Duttlinger, R., Blösch, P., Kuebler, J. (2017). Performance of lightweight coated oxide ceramic composites for industrial high speed wood cutting tools: A step closer to market. *Ceram. Int.*, 43, 8735–8742.
34. Long, Y., Zeng, J., Wu, S. (2014). Cutting performance and wear mechanism of Ti-Al-N/Al-Cr-O coated silicon nitride ceramic cutting inserts. *Ceram. Int.*, 40, 9615–9620.
35. Gajrani, K.K., Sankar, M.R. (2017). State of the art on micro to nano-textured cutting tools. *Mater. Today Proc.*, 4, 3776–3785.
36. Feng, Y., Zhang, J., Wang, L., Zhang, W., Tian, Y., Kong, X. (2017). Fabrication techniques and cutting performance of micro textured self-lubricating ceramic cutting tools by in-situ forming of Al_2O_3-TiC. *Int. J. Refract. Met. Hard Mater.*, 68, 121–129.
37. Sugihara, T., Enomoto, T. (2012). Improving anti-adhesion in aluminum alloy cutting by micro strip texture. *Prec. Eng.*, 36, 229–237.
38. Lian, Y., Deng, J., Yan, G., Cheng, H., Zhao, J. (2013). Preparation of tungsten disulfide (WS2) soft-coated nano-textured self-lubricating tool and its cutting performance. *Int. J. Adv. Manuf. Technol.*, 68, 2033–2042.
39. Xing, Y., Deng, J., Zhao, J., Zhang, G., Zhang, K. (2014). Cutting performance and wear mechanism of nanoscale and microscale textured Al_2O_3/TiC ceramic tools in dry cutting of hardened steel. *Int. J. Refract. Met. Hard Mater.*, 43, 46–58.
40. Rathod, P., Aravindan, S., Venkateswara, R.P. (2016). Performance evaluation of novel micro-textured tools in improving the machinability of aluminum alloy (Al 6063). *Proc. Technol.*, 23, 296–303.
41. Wu, G., Xu, C., Xiao, G,, Yi, M., Chen, Z., Xu, L. (2016). Self-lubricating ceramic cutting tool material with the addition of nickel coated CaF_2 solid lubricant powders. *Int. J. Refract. Met. Hard Mater.*, 56, 51–58.
42. Cui, X., Guo, J., Zheng, J. (2016). Optimization of geometry parameters for ceramic cutting tools in intermittent turning of hardened steel. *Mater. Des.*, 92, 424–437.
43. Cinert, J. (2018). Study of mechanisms of the spark plasma sintering technique, PHD Thesis, Czech Technical University, Prague, Czech Republic.
44. Matizamhuka, W.R. (2016). Spark plasma sintering (SPS) an advanced sintering technique for structural nanocomposite materials. *J. S. Afr. Inst. Min. Metall.*, 116, 1171–1180.
45. Saheb, N., Iqbal, Z., Khalil, A., Hakeem, A.S., Al-Aqeeli, N., Laoui, T., Al-Qutub, A., Kirchner, R. (2012). Spark plasma sintering of metals and metal matrix nanocomposites: A review. *J. Nanomater.* DOI:10.1155/2012/983470.
46. Munir, Z.A., Anselmi-Tamburini, U., Ohyanagi, M. (2006). The effect of electric field and pressure on the synthesis and consolidation of materials: A review of the spark plasma sintering method. *J. Mater. Sci.*, 41, 763–777.
47. Borrell, A., Salvador, M.D. (2018). Chapter 1: Advanced ceramic materials sintered by microwave technology. In *Sintering technology method and application*, Intech Open., 3–24, DOI: 10.5772/intechopen.78831
48. Gołombek, K., Domrzalski, L.A. (2007). Hard and wear resistance coatings for cutting tools. *J. Achiev. Mater. Manuf. Eng.*, 24, 107–110.
49. Gevorkyan, E., Lavrynenko, S., Rucki, M., Siemiatkowski, Z., Kislitsa, M. (2017). Ceramic cutting tools out of nanostructured refractory compounds. *Int. J. Refract. Met. Hard Mater.*, 68, 142–144.
50. Li, M., Huang, C., Zhao, B., Liu, H., Wang, J., Liu, Z. (2017). Mechanical properties and microstructure of Al_2O_3-TiB_2- $TiSi_2$ ceramic tool material. *Ceram. Int.*, 43, 14192–14199.

51. Zou, B., Huang, C., Song, J., Liu, Z., Liu, L., Zhao, Y. (2012). Effects of sintering processes on mechanical properties and microstructure of TiB_2-TiC + 8 wt% nano-Ni composite ceramic cutting tool material. *Mater. Sci. Eng. A* 540, 235– 244.
52. Yin, Z., Huang, C., Zou, B., Liu, H., Zhu, H., Wang, J. (2014). Effects of particulate metallic phase on microstructure and mechanical properties of carbide reinforced alumina ceramic tool materials. *Ceram. Int.*, 40, 2809–2817.
53. Tiwari, D., Basu, B., Biswas, K. (2009), Simulation of thermal and electric field evolution during spark plasma sintering. *Ceram. Int.*, 35, 699–708.
54. Ramirez, C., Miranzo, P., Belmonte, M., Osendi, M.I., Poza, P., Vega-Diaz, S.M., Terrones, M. (2014). Extraordinary toughening enhancement and flexural strength in Si_3N_4 composites using graphene sheets. *J. Eur. Ceram. Soc.*, 34, 161–169.
55. Mishra, S.K., Pathak, L.C. (2009). Self-propagating high-temperature synthesis (SHS) of advanced high-temperature ceramics. *Key Eng. Mater.* 395, 15–38.
56. Kim, H., Kawahara, M., Tokita, M. (2000). Specimen temperature and sinterability of Ni powder by spark plasma sintering. *J. Jpn. Soc. Powder Powder Metal.*, 47, 887–891.
57. Levashov, E.A., Mukasyan, A.S., Rogachev, A.S., Shtansky, D.V. (2017). Self-propagating high-temperature synthesis of advanced materials and coatings. *Int. Mat. Rev.*, 62, 203–239.
58. Borovinskaya, I., Gromov, A., Levashov, E., Maksimov, Y., Mukasyan, A., Rogachev, A. (2017). *Concise encyclopedia of self-propagating high-temperature synthesis*, 1st edn, Elsevier Science, Amsterdam, Netherlands.
59. Yi, H.C., Moore, J.J. (1990). Self-propagating high-temperature (combusting) synthesis (SHS) of powder-compacted materials. *J. Mater. Sci.*, 25, 1159–1168.
60. Matli, P.R., Abdul-Shakoor, R., Mohamed, A.M.A., Gupta, M. (2016). Microwave rapid sintering of Al-metal matrix composites: A review on the effect of reinforcements, *Microstructure. Mech. Prop. Metals*, 6, 143–162.
61. Keblouti, O., Boulanouar, L., Azizi, M.W., Yallese, M.A. (2017). Effects of coating material and cutting parameters on the surface roughness and cutting forces in dry turning of AISI 52100 steel. *Struct. Eng. Mech.*, 61, 519–526.
62. Ucun, I., Aslantasx, K., Gokcxe, B. Bedir, F. (2014). Effect of tool coating materials on surface roughness in micromachining of Inconel 718 super alloy. *Proc. IMechE. Part B: J. Eng. Manuf.*, 228, 1–13.
63. Wang, B., Liu, Z. (2016). Cutting performance of solid ceramic end milling tools in machining hardened AISI H13 steel. *Int. J. Refract. Met. Hard Mater.*, 55, 24–32.
64. Jianxin, D., Wenlong, S., Hui, Z., Pei, Y., Aihua, L. (2011). Friction and wear behaviours of the carbide tools embedded with solid lubricants in sliding wear tests and in dry cutting processes. *Wear*, 270, 666–674.
65. Sugihara, T., Tanaka, H., Enomoto, T. (2017). Development of novel CBN cutting tool for high speed machining of Inconel 718 focusing on coolant behaviours. *Proc. Manuf.*, 10, 436–442.

3 Characterization and Evaluation of Eco-Friendly Cutting Fluids

Phaneendra Kiran Chaganti
BITS-Pilani, Hyderabad, India

Madan Mohan Reddy Nune
Anurag University, Hyderabad, India

CONTENTS

DOI: 10.1201/9781003284574-4

3.1 INTRODUCTION

Cutting fluid is one of the most significant components utilized in the machining industry, and it has been in use for more than a century. Since then, sophisticated cutting-fluid development approaches have been pushed. Several types of cutting fluids, such as mineral, vegetable, animal, and emulsion oils, were gradually introduced over time. In general, cutting fluids were designed with increased machining performance, anticorrosion and antibacterial properties, wettability, and fluid stability in mind. Most machining industries were worried about the volatility of petroleum products and the disposal of oil-based cutting fluids at the end of life [1]. This push led to the introduction of synthetic or semisynthetic water-based metalworking fluids, which account for about 80–90% of all machining operations currently. Due to new regulations, the detrimental effects of oil-based semisynthetic cutting fluids have become more obvious. The U.S. National Institute for Occupational Safety and Health estimates that 1.2 million workers are affected by oil-based cutting fluids each year and cites health-related difficulties, such as dermatitis and lung disease, as a result of their exposure [2]. Scientists have been working on alternatives as a result of these issues. Synthetic cutting fluids that are not based on oil are effective in lowering temperatures and improving machining performance [3–5]. On the other hand, due to the presence of harmful chemical additions, they are quite expensive to treat after usage [6, 7]. In the current machining industry, non-oil-based metalworking fluids have a considerable impact, and their use during machining enhances surface smoothness, tool life, and productivity [7–13]. The increased demand on manufacturers and researchers to remove or minimize the usage of oil-based cutting fluids is influenced by growing awareness of green manufacturing and industry focus on environmentally friendly products. Choosing a suitable cutting fluid while keeping green manufacturing in mind can lower manufacturing costs and increase productivity [14–16]. The pressurized nano-cutting fluid has been shown to generate a wear-protective barrier between the tool and work interface, reducing tool wear [17–19]. Using a nano-cutting fluid jet has been shown to lower the amount of heat generated in the machining zone [20, 21]. Silicon dioxide (SiO_2), polytetrafluoroethylene (PTFE), molybdenum disulfide (MoS_2), aluminum oxide (Al_2O_3), calcium fluoride (CaF_2), tungsten disulfide (WS_2), copper oxide (CuO), and titanium dioxide (TiO_2) are some of the most commonly employed nanoparticles in cutting fluids [22–24]. However, the concentration of nanoparticles in the base fluid affects the coefficient of friction, and high nanoparticle concentrations may not be cost-effective or environmentally friendly [25]. A mixture of soft lamellar structure amorphous solid nanoparticles such as SiO_2- and MoS_2-based hybrid nano-cutting fluids demonstrate great load-carrying capability. Between these two, even a tiny concentration of SiO_2 nanoparticles in cutting fluid results in a 21.8% drop in coefficient of friction and an 8.6% reduction in wear volume [26]. Other difficulties, such as deposition and diffusion at high temperatures, a high surface area–to–volume ratio, and an agglomeration of nanosized particles on machined parts, must be addressed in addition to performance [27].

Escherichia coli, *Staphylococcus aureus*, *Salmonella typhi*, and *Salmonella typhimurium* are frequent microorganisms discovered in oil-based cutting fluids. Irritation of the skin, lungs, nose, and throat has been linked to exposure to these bacteria residing in oil-based cutting fluids [28]. These constraints discussed earlier highlight the significance of characterizing and evaluating of an eco-friendly cutting fluid for worker safety as well as a higher quality machining process prior to introducing into commercial market.

3.2 CHARACTERIZATION OF NOVEL CUTTING FLUIDS

A newly developed or modified cutting fluid needs to be characterized to understand the usability of a cutting fluid to a given machining process. This chapter describes a generalized procedure for evaluating the developed or modified cutting fluid. The minimum number of tests that need to be conducted are shown in Figure 3.1.

The characterization of a new or modified cutting fluid is classified into two broad categories, namely, basic and advanced. This classification is done based on the equipment used and the expertise required for the test. The details of these tests for evaluating a cutting fluid are given in the following sections.

3.3 BASIC CHARACTERIZATION STUDIES

3.3.1 DENSITY

The density of a material is found by dividing the mass of the material by the volume occupied. The density of a cutting fluid plays a significant role in circulating cutting fluid in the cutting fluid circulatory system. The density of a cutting fluid determines the consistency and quality of developed or modified cutting fluid. The density of a solid/liquid is generally determined based on gravimetric buoyancy or gravimetric displacement principles; both work on Archimedes' principle. The general operating procedure to determine the density of a cutting fluid is shown in Figure 3.2.

In the first step, the known volume sinker is placed in a beaker, and the weight is measured in the presence of air. In the second, step the cutting fluid is poured into the beaker and the known volume sinker is kept in the liquid and the total weight is measured. The density of the liquid is determined using Equation 3.1:

$$\rho_{CF} = a.\frac{(X - Y)}{V_{sinker}}.\rho_0, \tag{3.1}$$

where ρ_{CF} is the density of cutting fluid, a is the correction factor, X is the weight of the sink in air, Y is the weight of the sink in cutting fluid, V_{sinker} is the volume of the sinker, and ρ_0 is the density of air.

The eco-friendly cutting fluids have the same viscosity as the oil-based emulsion. When it comes to producing a greasing layer on the cutting surface, viscosity is crucial.

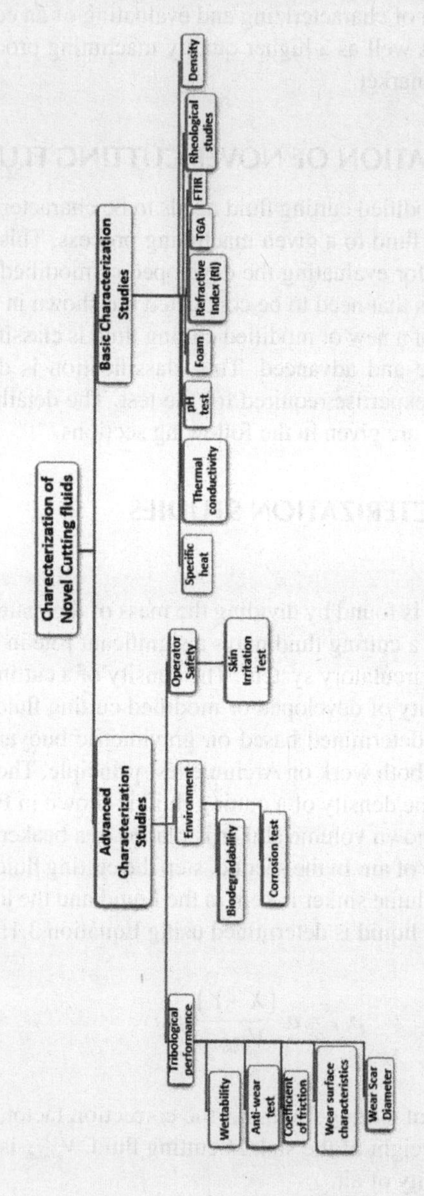

FIGURE 3.1 Hierarchy of various studies to be conducted for novel eco-friendly cutting fluids.

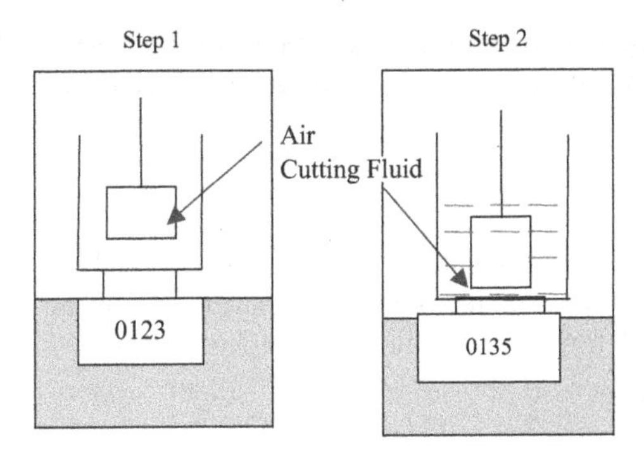

FIGURE 3.2 Density measurement procedure using a digital density meter.

3.3.2 Viscosity and Rheological Studies

The purpose of the rheological test was to determine the fluid flow properties. The viscosity is the ratio of stress applied to the shear rate. The rheometer is the instrument used to determine the viscosity and viscoelasticity of a fluid. The working principle of a standard rheometer is shown in Figure 3.3. The rheometer works on either controlled strain rate or controlled stress. The strain rate is generally controlled with angular displacement of the motor, and the applied stress is determined based on the torque of the motor.

The stress calculation for stress output rheometer is done using Equation 3.2:

$$\sigma = \tau . K_{\sigma}, \tag{3.2}$$

where σ is the stress (Pa), τ is the torque (N-m), and $K\sigma$ is the stress constant.

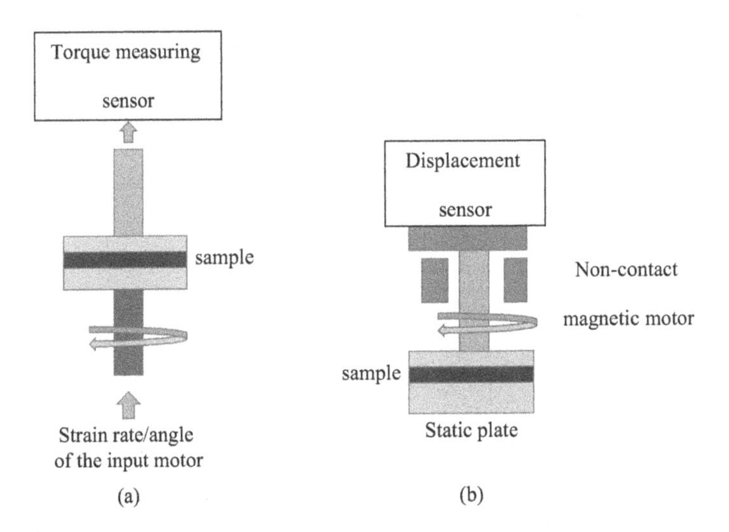

FIGURE 3.3 Working principle of a rheometer (a) stress output (b) strain output.

The stress constant depends on the geometry of the plates. In the strain output-based measurements, the percentage strain can be calculated using Equation 3.3:

$$\%\gamma = \theta . K_{\gamma} . 100, \tag{3.3}$$

where γ is the strain, θ is the angular displacement (rad), and $K\gamma$ is the strain constant.

Using motor spindle angular velocity, the shear rate of the cutting fluid can be calculated using the formula given in Equation 3.4:

$$\dot{\gamma}(CF) = K_{\gamma} \cdot \Omega, \tag{3.4}$$

where $\dot{\gamma}(CF)$ is the shear rate on the cutting fluid, $K\gamma$ is the shear constant, and Ω is the angular velocity of motor spindle in rad/s.

The viscosity of the cutting fluid can be calculated using Equations 3.2 and 3.4. The viscosity is the ratio of stress (σ) to shear rate ($\dot{\gamma}(CF)$). The viscosity measured on a rheometer ranges from 100000–0.00001 Pa-s. To fit them into a graph a logarithm scale is generally used. In general, the oils have a viscosity in the range of 0.01–3.82 Pa-s. The American Society for Testing and Materials (ASTM) standard for determining the dynamic viscosity of a cutting fluid is D2983. The high temperature at the cutting zone in turn increases the temperatures of the cutting fluids. Hence, it is required to determine the kinematic viscosity of the cutting fluid at different temperatures using ASTM D341 standard. The general trends for an eco-friendly nano-cutting fluids are shown in Figure 3.4.

A linearity in the shear stress curve is generally observed for commercially used water emulsions. In contrast, a nonlinearity is generally observed for nano-cutting fluids. The nano-cutting fluids can undergo higher shear stress even at high shear rates. It is desirable as higher shear rates are present at some locations of the cutting zone. These tests are conducted at different temperatures to understand the change in viscosity with variation in temperature.

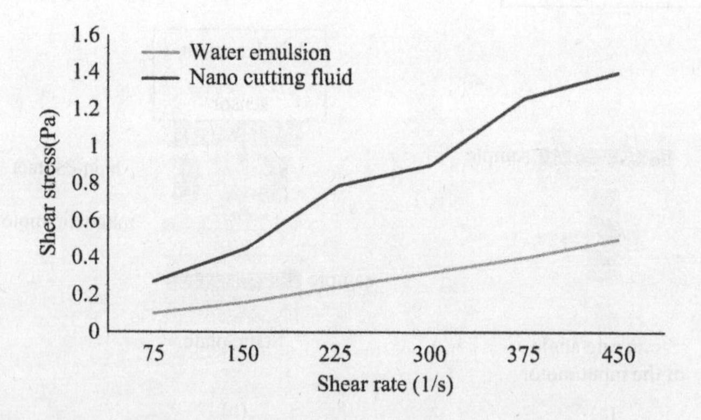

FIGURE 3.4 Shear stress versus shear rate for an oil emulsion and nano-cutting fluid at 20 °C.

3.3.3 Specific Heat

Different cutting fluids absorb different amounts of energy from the cutting zone when heated up, and they give different amounts of energy when they are cooled down. This property of a material is called specific heat capacity. This property is generally measured using instruments such as a calorimeter or differential scanning calorimeter (DSC). The basic equation used for calculating the specific heat capacity using these instruments is given in Equation 3.5:

$$c = \frac{E}{m.\Delta T}, \tag{3.5}$$

where c is the specific heat capacity (J/kg^{-1}C^{-1}), m is the mass of the cutting fluid (kg), E is the energy provided to heat the specimen (Joules), and ΔT is the temperature difference between hot and cold specimens.

The procedure used for the DSC test is that the cutting fluid is taken in a small pan, it is heated in temperature steps, and a graph is obtained for different temperature ranges. A sample output graph from a DSC is demonstrated in Figure 3.5 [29]. In this graph, the endothermic and exothermic regions are observed to understand the nature of the cutting fluid. If it is exothermic, it means that it will absorb the heat and be better for a cutting fluid.

3.3.4 Thermal Conductivity

The measure of warmth shipped during the machining interaction is controlled by the warm conductivity of metalworking fluid (MWF). There are devices available in the market to measure the warm conductivity of the eco-friendly cutting liquids. For warm conductivity and the effusivity of materials, these instruments utilize the modified transient plane source (MTPS) approach [30].

A 10-ml test of the MWF was set in an empty chamber and fixed a defensive loop. A predetermined current was transmitted through a winding warming component to

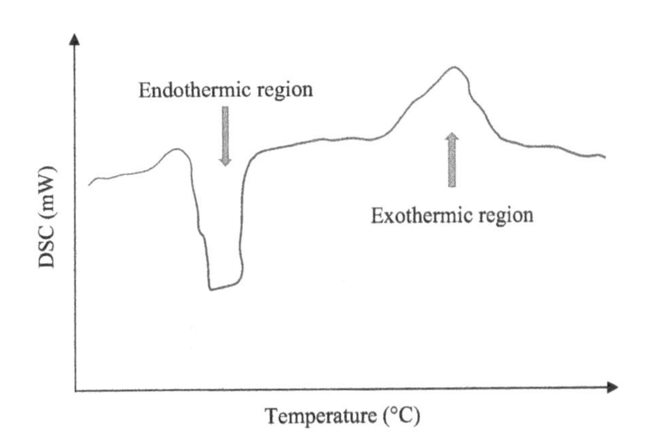

FIGURE 3.5 DSC example result output.

produce a limited quantity of warmth. The pace of expansion in the sensor voltage was utilized to compute the example's warm conductivity. In general, these tests are conducted at different temperatures and volume fraction of cutting fluids. The thermal conductivity of nano-eco-friendly cutting fluids is in the range of 0.6–0.7 W/Mk.

3.3.5 Stability and Biodegradability Test

The chemical reaction of a cutting fluid with other compounds decides the stability of the cutting fluid. The oil-based cutting fluids keeps up with soundness by framing electrostatic or steric unpleasant boundaries with the assistance of a surfactant or emulsifier. The stability of eco-friendly cutting fluids can be tested in the biological oxygen demand (BOD). Inside the incubator, a regulated atmosphere with a temperature of 27 °C was maintained for the duration of the test, which lasted 24 hours. In addition, using the ultraviolet (UV)–visible spectroscopy equipment is used to test the stability of SiO_2 nanoparticles in the nano-cutting fluids.

Biodegradation is the process of transforming any material through chemical breakdown by bacteria or enzymes in microorganisms. *Biodegradability* refers to a solution's ability to decompose naturally. Chemical oxygen demand (COD) and BOD are commonly used to assess the biodegradability of cutting fluids. Oil-based cutting fluids keep up their strength by framing electrostatic or steric boundaries with the assistance of a surfactant as per ISO 15705 norm. The standard used for BOD test is OECD 301F. The sample of cutting fluid needs to be taken in a BOD bottle, and it is hatched in an incubator for 5 days. The BOD of the bottles is measured at frequent intervals to understand the biodegradability. It is preferred to use a prefixed oxygen measurement sensor for the bottle cap; otherwise, the bottle needs to be taken out of the incubator to measure the oxygen. Sample data are presented in Figure 3.6 to understand interpreting the results obtained in BOD/COD test.

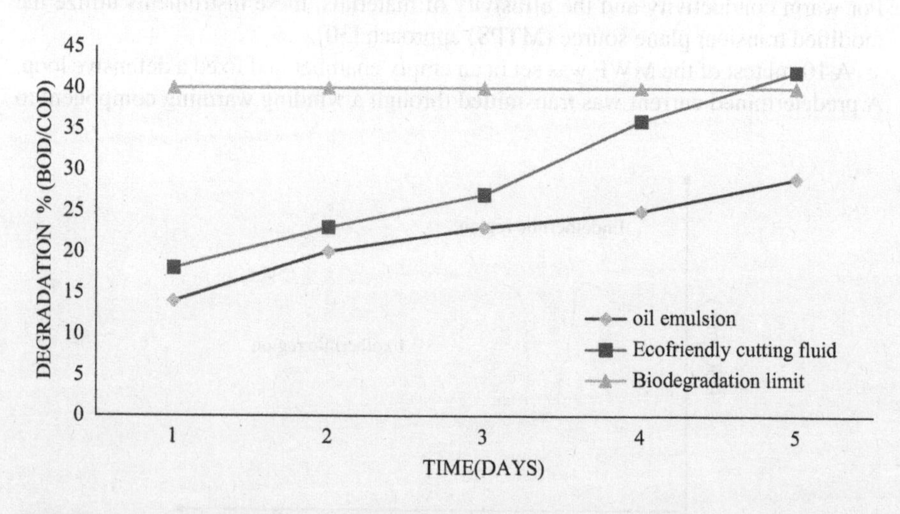

FIGURE 3.6 Biodegradation percentage of eco-friendly and oil emulsion–based cutting fluid.

Figure 3.6 shows the COD/BOD numbers for each of the five days. In the MWF samples, the amount of oxygen demand was measured. An oxygen-consuming amphibian corruption method was utilized to assess the biodegradability of cutting fluid. "BOD" is used to determine the amount of dissolved oxygen in organic matter under aerobic condition, while the "COD" can determine oxygen requirement for organic matter in water. The degree of biodegradability of a cutting fluid is determined by "the ratio of BOD to COD." According to the literature [31–33] the biodegradation percentage for synthetics, semi-engineered, and manufactured liquids should be 40% larger. Chemicals that pass the 5-day biodegradability test are regarded as entirely biodegradable.

As indicated in Figure 3.6, the degradation percentages (%) of eco-friendly cutting fluids and oil-based cutting fluid were compared. The eco-friendly cutting fluids degraded 41.2% in the 5-day test, whereas the oil-based MWF degraded 29.3%. The standard limit for biodegradability of a cutting fluid is 40%. If the cutting fluid percentage crosses the set limit, it is easily biodegradable.

3.3.6 pH Test

The alkaline or acidic nature of the eco-friendly cutting fluids is determined by the pH of the cutting fluids. A pH of 7 or higher is required for an MWF; otherwise, it may react with metals in applications such as storage tanks, uncut or chopped metal chips, and so on, causing corrosion. The pH of the eco-friendly cutting fluids was determined in this study with a pH meter. The electro-kinetic characteristics of "alumina and copper powder" suspensions in liquid solution directly connected to their stability. To generate large repulsive forces, a well-dispersed suspension with a high surface-charge density can be obtained. As a result, determining the zeta potential and absorbency has become a crucial step in determining the dispersion behavior of nanoparticles in a liquid media. Figure 3.7 illustrates the consequence of pH on the receptiveness and potential of zeta of sodium dodecylbenzenesulfonate (SDBS) dispersant nano-suspensions. Figure 3.7 demonstrates a strong correlation between "zeta potential and absorbency for both nanofluids"; the greater the absolute value of zeta potential, the greater the porosity. At pH 2, both nanofluids' "absolute zeta potential and particle absorbency are at their lowest values," indicating that the electrostatic repulsion between particles is insufficient to overcome the attraction force between particles, resulting in poor dispersion stability. Because both the absolute magnitude and the particle surface of zeta potential increase as pH rises, the electrostatic shock power between particles becomes strong enough to thwart Brownian motion-induced fascination and crash.

Increased electrostatic power can likewise result in all the free particles by expanding the space between particles beyond the range of hydrogen, thereby decreasing the likelihood of molecule thickening and subsiding. The potential of "zeta" upsurges as the pH rises, increasing the stability of the nanoparticle dispersion. Based on the zeta potential and absorbency values, pH "alumina 8.0" and pH "copper 9.5" possibly chosen as the functioning pH for two nanofluids. This procedure helps to determine the perfect pH value for a nano cutting fluid.

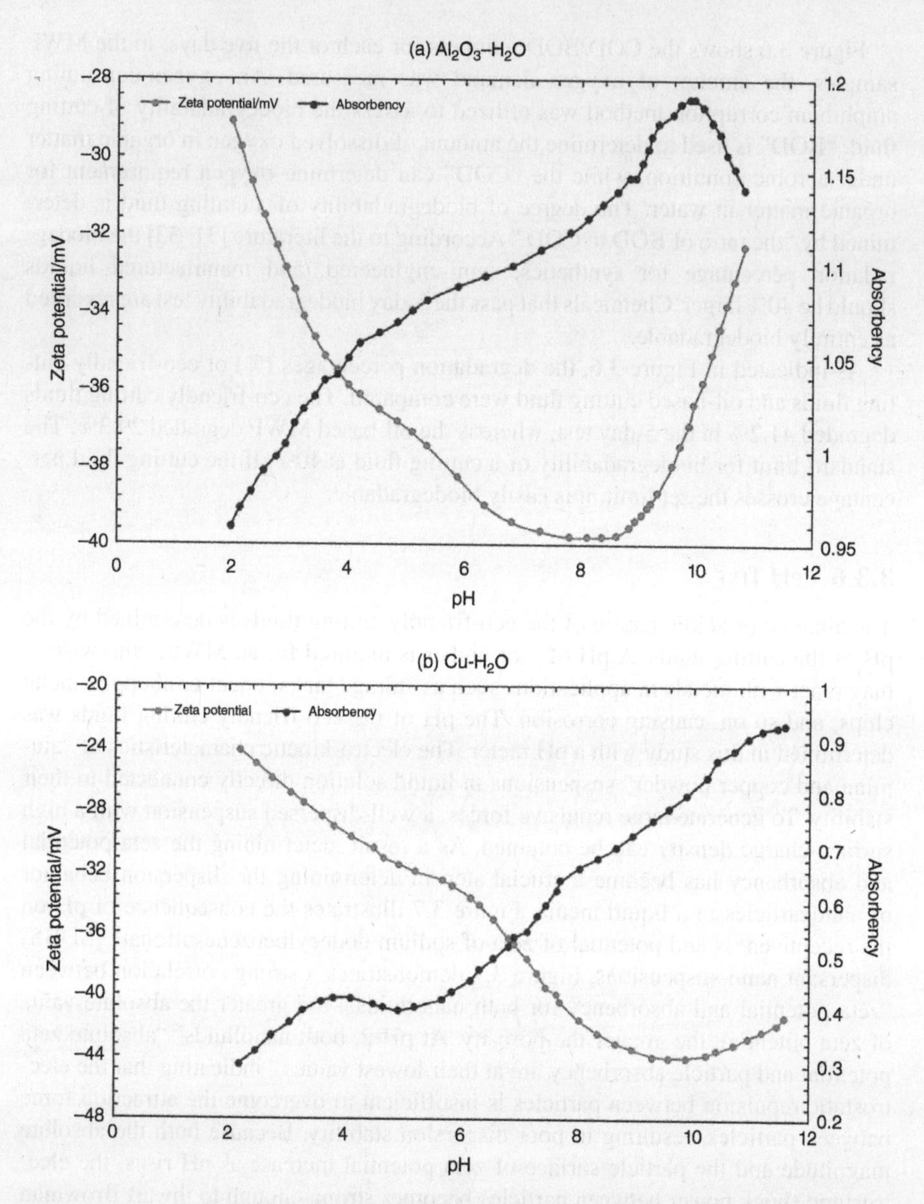

FIGURE 3.7 Behavior of pH on "zeta potential": (a) Al_2O_3-H_2O and (b) Cu-H_20 [34].

3.3.7 Foam Test

For a cutting fluid, the development of foam is undesired as it causes fluids loss and cavitation effect in the pump system. Factors that influence foam production include levels of cutting fluid in the reservoir, cutting fluid concentration in the reservoir, the flow rate of cutting fluid through the pump housing intake pipe and high fluid velocities. As a rule, two sorts of tests are utilized to confirm froth solidness: handshake and rapid testing. The sample of cutting fluid is made at a 1:20 proportion and barely shaken in a 100-ml graduated chamber with an absolute limit of 50 ml for the handshake test. The beginning and last volumes are tested for a general reduction in froth over the long run after a foreordained stretch. The handshake test was utilized to decide the froth dependability in this examination [35]. The standard for conducting the foam test is ASTMD3601. In general, this test is not conducted on a freshly prepared sample; rather, aging the cutting fluid is strongly recommended. The most recommended time for aging is 24 hours, but in some instances, agitation or aeration technique are used to accelerate the reaction.

3.3.8 Refractive Index

Utilizing overabundance constituents in the fluid builds the odds of greater harmfulness and brings down the warmth move rate, the development of froth, and so on; the abundance of constituents can be deciphered utilizing the refractive record. Refractive index (RI) is a dimensionless number that depicts how light spreads through the medium. The RI of the eco-friendly cutting liquids and semisynthetic greases were practically in a same in reach: 1.3–1.4.

Figure 3.8 depicts the outcome of UV–visual spectroscopy scanning of a sample nano-cutting fluid. Spectrophotometry measures the colloidal stability of nanoparticles in terms of spectrum absorption. These specific nanoparticles absorb light between 300 and 400 nm in wavelength [36, 37]. The peak absorbance of nanoparticles in the produced cutting fluid prepared at 0 hour, after 12 hours, and after 24 hours was 370-nm, 369.5-nm, and 369-nm wavelengths, respectively, as shown in Figure 3.8. The graph shows that the nanoparticles in generated cutting fluid are stable since they have approximately the same absorbance value corresponding to their storage period at the same peak wavelength.

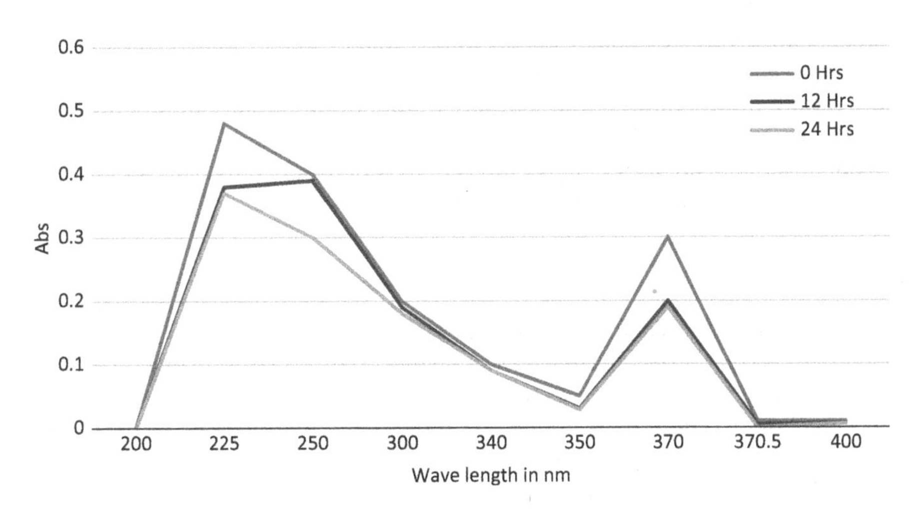

FIGURE 3.8 Ultraviolet–visual spectrum of sample eco-friendly cutting fluid for a wavelength from 200–400-nm wavelength.

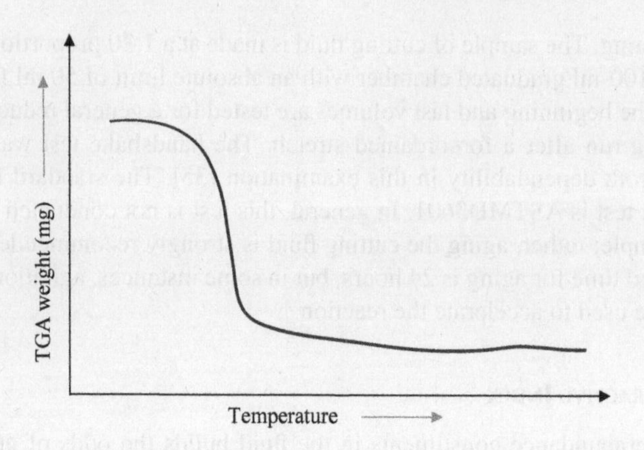

FIGURE 3.9 The detailed thermogravimetric curve of eco-friendly cutting fluids [29].

3.3.9 Thermogravimetric Analysis

Thermogravimetric analysis (TGA) is used to measure the thermal stability of a cutting fluid. In this method, a sample of cutting fluid is taken in a pan and is heated at a constant rate while maintaining the controlled atmosphere. The mass change of the sample will be continuously observed. The micro-furnace installed in the TGA will ensure that the temperatures increase up to 1000–1500 °C.

A sample TGA curve for a cutting fluid looks similar to Figure 3.9. In this curve, the *x*-axis represents the temperature, and *y*-axis represents the weight of the sample during the variation in temperature. Wherever there is a sudden drop in the mass, the corresponding temperature is an important parameter as it represents the main constituent. The other percentage of mass changes may be some volatile matters or the minor constituents in the cutting fluid. In general, if nanoparticles are present in the cutting fluid, the mass loss temperature will be very high, giving more stability to the cutting fluid at high temperatures of the cutting zone.

Due to their superior molecular mass and thermal stability, nanoparticles in a fluid have previously been shown to be capable of maintaining temperatures beyond 600 °C [38]. Despite increasing the temperature to 1000 °C for eco-friendly cutting fluids, the fraction build damage never reached 100% as a result of this. Cutting fluids that are more stable may lessen the amount of waste produced.

3.3.10 Fourier-Transform Infrared Spectroscopy Analysis

Fourier-transform infrared (FTIR) spectroscopy is used to determine the organic, polymeric, and inorganic constituents in a liquid. The modern eco-friendly cutting fluids are made using different organic and or inorganic material combinations. This FTIR analysis is useful to determine the stability of constituent compounds after usage of cutting fluid for a certain duration.

The FTIR instrument sends infrared rays (IR) through the cutting fluid sample, and the range of IR is 10000–100 cm^{-1}. Some portion of this radiation is absorbed by

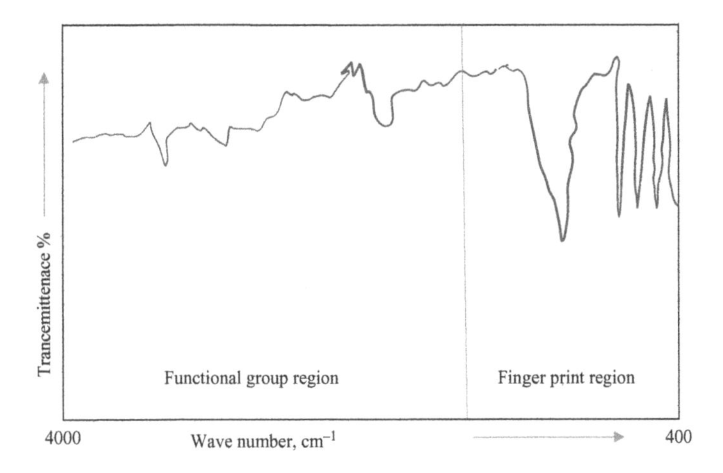

FIGURE 3.10 Studying the steadiness of eco-friendly cutting fluids solution [29].

the atoms and creates thermal instability; in turn, the received signal ranges between 4000–400 cm^{-1} on the spectrum. The signal received from a chemical compound is unique and is called a spectral fingerprint.

The transmittance (percentage) versus wavenumber for the eco-friendly cutting fluids can be acquired using FTIR and a sample representative graph is shown in Figure 3.10. This curve is designed to separate the functional group region and the fingerprint region. Most people would consider the fingerprint region to be a non-functional zone, which contains argon and hydrogen molecules that contain protons and ions. These molecules connect because of internal molecular vibrations, resulting in ionic compounds.

FTIR analysis can be used to determine the physical stability (existence of specific chemical compound) in eco-friendly cutting fluids. A sample pellet with a diameter of 10 mm having a mixture of eco-friendly cutting fluid and KBr (potassium bromide) is made. Alkali halides, such as KBr, are commonly used to create pellets that are exposed to compression.

A full conduction treatment was given to the fingerprint region. Waves per centimeter (w/c) is the number of waves in a given length of centimeter, and the units are cm–^{1}. Frequency and energy are inversely proportional to the wavelength, and the wave number (WN) is relational to frequency and energy. Hydrogen bonding is indicated by the presence of hydrogen bonding in the eco-friendly cutting fluids in the WNs 3675 and 3425. In a similar method, other compounds' presence in the used cutting fluid can be confirmed. Functional elements have an associated stretching zone identified by the numbers 2925, 2854, and 1618. When it comes to stretching, there are two basic types: symmetric and asymmetric. When atoms vibrate toward each other in symmetric stretching, they pull together; when they vibrate away from each other, they push apart.

3.4 ADVANCED CHARACTERIZATION STUDIES

3.4.1 Tribological Performance Studies of Developed Cutting Fluid

The tribological features of eco-friendly cutting fluids are crucial in defining their lubricating capability.

3.4.1.1 Anti-Wear Test

The eco-friendly cutting fluid needs to be tested for anti-wear properties as they are important to avoid tool wear in the machining process. The standard used for assessing the anti-wear properties is "ASTM D4172." As per this standard, for a standard nanoparticle-based cutting fluid, the standard setting in a four-ball test instrument are a normal force of 147 N at a temperature of 75 °C and a rotational speed of 1200 rpm. Each test needs to be conducted for an average of 60 minutes. The weights corresponding to the force are determined based on preliminary studies, and for a nanoparticle, a similar study on graphite atoms in literature [34] used 147 N. The toolmakers microscope is generally used to measure the wear on the three stationary balls. By averaging the wear scar diameter (WSD) on three balls, the average WSD can be obtained. The smaller the WSD, the better the cutting fluid, and the larger the WSD, the poorer the lubricant property of the cutting fluid.

3.4.1.2 Coefficient of Friction

The coefficient of friction (CoF) has a significant impact on transmission efficiency; the lower the value, the higher the transmission efficiency. The standard used for this experiment is ASTM D5183. By estimating the shaft force, the frictional powers were observed progressively. In this experiment, the CoF was estimated using Equation 3.6 [39, 40]:

$$\mu = \frac{T\sqrt{6}}{3W_r}, \tag{3.6}$$

where μ is the CoF, T is the applied torque in N-m, and Wr is the applied force in N.

3.4.1.3 Wear Surface Characteristics

The four-ball tester is used to measure the effect of load on the wear of a steel ball in the presentence of the lubricant. The wear properties of eco-friendly cutting fluids can be investigated using a four-ball tester under boundary pressure settings. The four-ball wear tester consists of three balls being maintained stationary in a ball pot and a fourth ball being held in a revolving spindle. The balls utilized in this experiment were AISI 420 stainless steel balls with a diameter of 12.7 mm and a hardness of 64 to 66 HRC. Before each trial, the balls were carefully cleaned. The amount of MWF used in each test was 10 ml.

The upper ball was in contact with the lower three balls while it was lubricated with a thin film of lubricant. The steel balls undergo plastic deformation; this can be measured and then observed under the toolmakers microscope.

The plastic deformation or WSD of the steel balls after the experiment needs to be estimated using a creator's magnifying lens in the toolmakers microscope. The CoF of eco-friendly cutting fluids is observed to be much less compared to conventional oil-emulsion cutting fluids. The aftereffects of the eco-friendly cutting liquids' frictional coefficient variety have been explored in the literature [41].

3.4.2 Wettability Study

Cutting fluid wettability testing is a crucial step in determining its lubricating and cooling qualities.

The contact angle measurement serves to find the interfacial energy (surface tension) of a liquid's substrate. In Figure 3.11, there are three surface tension components: solid–vapor, liquid–vapor, and solid–liquid interfaces (known in some circles as solid–liquid, liquid–liquid, and solid–solid) observed while testing a cutting fluid for the contact angle. The expression to find the contact angle is given in Equation (3.7) [42]:

$$\cos\theta = \frac{\sigma_{sv} - \sigma_{sl}}{\sigma_{lv}}, \tag{3.7}$$

where θ is the contact angle, σ_{sv} is the interfacial energy between the solid and the vapor region, σ_{lv} is the interfacial energy between the liquid and the vapor region, and σ_{sl} is the interfacial energy between the solid and the liquid region.

With a lower contact angle, the wetting area per unit volume rises, which improves greasing up and heat convection at the solid–fluid interface. It has been proved that nano-cutting fluids have shown lower cutting angles compared to conventional cutting fluids [43–45]. In general, if the contact angle is 0°, then the cutting fluid is well spread on the solid surface. If the contact angle is between 0–90°, then the cutting fluid is hydrophilic. If the contact angle is above 90°, then the cutting fluid is hydrophobic, and this condition is completely undesirable.

3.4.3 Corrosion Study

The cutting fluids in machining generally is an emulsion in which water is the major portion. Hence, it is very critical to test the cutting fluid for corrosive atmosphere, especially its reaction with metals. Industries often use a corrosion test technique such as the Herbert method, which follows the IP 125 standard. A flat ground cast-iron

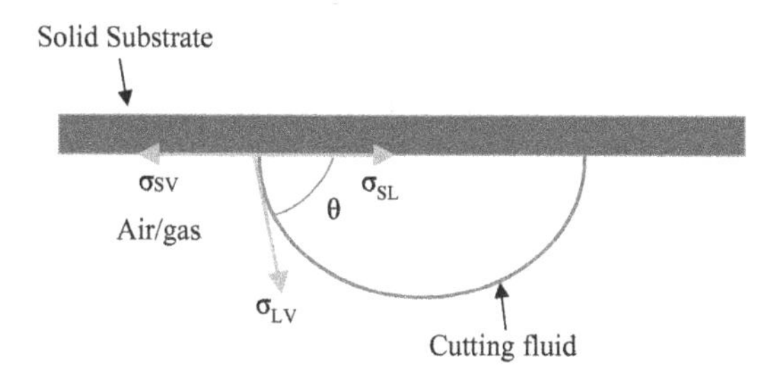

FIGURE 3.11 Contact angle measurement.

plate is coated with steel chips, and a dilute cutting fluid is applied until a given concentration is measured in the Herbert corrosion test. According to the corrosion that sets in after a 24-hour lapse time at room temperature in a testing chamber, the chips and, especially, the plate are analyzed. The test results are analyzed by identifying the black spots and rust on the metal parts. The standard Herbert corrosion filter paper test was utilized to evaluate corrosion properties. Yellow metals were tested for corrosion by immersing test strips in an aqueous fluid at 50 °C. After testing for 3 hours, the surface of the strips was inspected visually. A change in color (i.e., staining, darkening, color changes) indicates corrosion [46]. As the corrosion property is crucial in the structure of machine tool and cutting tool, this test needs to be conducted before releasing the cutting fluid into the market.

3.4.4 Acute Skin Irritation Test

The acute skin irritation test is used to evaluate the developed cutting fluid for the operator's health, if the operator is exposed to cutting fluid for longer durations. This test is generally conducted based on OECD 404 standard. New Zealand albino rabbits are generally used for conducting this test. The cutting fluid sample of 0.5 ml is applied to the predetermined sample of rabbits using a bandage. They are kept in a controlled environment chamber maintained at room temperature with a relative humidity between 30–70% RH. The bandage is removed after 4 hours, and the skin on these sample rabbits is observed after 24, 48, and 72 hours for erythema and edema. In general, the parameters for erythema and edema are ranked on a scale, and 0 is the recommended scale for the safety of an operator.

3.5 SUMMARY

In this chapter, the characterization tests required for evaluating an ecofriendly cutting fluid were explained. The tests were classified under two broad categories, namely, basic and advanced characterization techniques. The basic characterization techniques determine the basic physical properties of a cutting fluid, starting with density and pH value. The advanced characterization techniques included tribological properties, environmental friendliness, and operatory safety. This chapter gave a step-by-step procedure and standards for conducting tests and analyzing the obtained results. It will be useful for evaluating the newly developed or modified cutting fluid for characterization and eco-friendliness.

REFERENCES

1. Ningsih Y R (2017), Utilization of sulphurized palm oil as cutting fluid base oil for broaching process, *IOP Conference Series: Earth and Environment Science*, 60, 012008.
2. National Institute for Occupational Safety (1998), *Criteria for a recommended standard: Occupational exposure to metalworking fluids*, 98–102.
3. King N, Keranen L, Gunter L, Sutherland J (2001), Wet Versus Dry Turning: A Comparison of Machining Costs, Product Quality, and Aerosol Formation, SAE Technical Paper 2001-01-0343, ISSN:0148-7191.

4 Feng S C, Hattori M (2000), Cost and Process Information Modeling for Dry Machining, *Proceedings of the International Workshop on Environment and Manufacturing*, https://tsapps.nist.gov/publication/get_pdf.cfm?pub_id=821121 (Accessed August 18, 2021.

5 Fratila D, Evaluation of near-dry machining effects on gear milling process efficiency, *Journal of Cleaner Production*, 17, 839–845.

6 Nandgaonkar S, Gupta T V K, Joshi S (2016), Effect of water oil mist spray (WOMS) cooling on drilling of Ti6Al4V alloy using ester oil based cutting fluid, *Procedia Manufacturing*, 6, 71–79.

7 US Dept of labor, Safety and Health Topics (2018), *Metalworking fluids – safety and health best practices manual*, Occupational Safety and Health Administration.

8 Yu Y, Guo Y, Wang L, Tang E (2010), Development of environmentally friendly water-based synthetic metal-cutting fluid, *Modern Applied Science*, 1, 53–58.

9 Brown M, Fotheringham J D, Hoyes T J, Mortier R M, Orszulik S T, Randles S J, Stroud P M (1997), Synthetic base fluids, in R M Mortier, M F Fox, and S T Orszulik, eds., *Chemistry and technology of lubricants*, Dordrecht: Springer Netherlands, 35–74.

10 Tripathi A, Vinu R (2015), Characterization of thermal stability of synthetic and semi-synthetic engine oils, *Lubricants*, 3, 54–79.

11 Debnath S, Reddy M M, Yi Q S (2014), Environmental friendly cutting fluids and cooling techniques in machining: A review, *Journal of Cleaner Production*, 83, 33–47.

12 Ramanaa M, Srinivasulub K, Raoc G (2011), Performance evaluation and selection of optimal parameters in turning of Ti-6Al-4V alloy under different cooling conditions, *International Journal of Innovative Technology and Creative Engineering*, 1, 10–21.

13 Jayal A D, Balaji A K, Sesek R, Gaul A, Lillquist D R (2007), Machining performance and health effects of cutting fluid application in drilling of A390.0 cast aluminum alloy, *Journal of Manufacturing Process*, 9, 137–146.

14 Byers J P (2017), *Metalworking fluids*, 3rd ed. CRC Press.

15 Pusavec F, Krajnik P, Kopac J (2010), Transitioning to sustainable production – Part I: Application on machining technologies, *Journal of Cleaner Production*, 18, 174–184.

16 Rubio E M, Agustina B, Marín M, Bericua A (2015), Cooling systems based on cold compressed air: A review of the applications in machining processes, *Procedia Engineering*, 132, 413–418.

17 Jiang C-C, Cao Y-K, Xiao G-Y, Zhu R-F, Lu Y-P (2017), A review on the application of inorganic nanoparticles in chemical surface coatings on metallic substrates, *RSC Advances*, 7, 7531–7539.

18 Xie H, Jiang B, He J, Xia X, Pan F (2016), Lubrication performance of MoS_2 and SiO_2 nanoparticles as lubricant additives in magnesium alloy-steel contacts, *Tribology International*, 93, 63–70.

19 Naves V T G, Da Silva M B, Da Silva F J, Evaluation of the effect of application of cutting fluid at high pressure on tool wear during turning operation of AISI 316 austenitic stainless steel, *Wear*, 302, 1201–1208.

20 Amrita M, Srikant R R, Sitaramaraju A V (2014), Evaluation of cutting fluid with nanoinclusions, *Journal of Nanotechnology Engineering Medicine*, 4, 031007.

21 Singh R K, Sharma A K, Dixit A R, Mandal A, Tiwari A K (2017), Experimental investigation of thermal conductivity and specific heat of nanoparticles mixed cutting fluids, *Materials Today Proceedings*, 4, 8587–8596.

22 Donnet C., Erdemir A (2004), Solid lubricant coatings: Recent developments and future trends, *Tribology Letters*, 17, 389–397.

23 Sharma A K, Tiwari A K, Dixit A R (2016), Characterization of TiO_2, Al_2O_3 and SiO_2 nanoparticle based cutting fluids, *Materials Today Proceedings*, 3, 1890–1898.

24. Najiha M S, Rahman M M, and Kadirgama K, Performance of water-based TiO_2 nanofluid during the minimum quantity lubrication machining of aluminium alloy, AA6061-T6, *Journal of Cleaner Production*, 135, 1623–1636.
25. Tosum N, Rostam S, Rasul S (2016), Use of Nano Cutting Fluid in Machining, *Fourth International Conference on Advanced Mechanical Automation Engineering – MAE 2016*, Rome, Italy, 17–21.
26. Xie H, Jiang B, Hu X, Peng C, Guo H, Pan F (2017), Synergistic effect of MoS_2 and SiO_2 nanoparticles as lubricant additives for magnesium alloy-steel contacts, *Nanomaterials* 7, 1–16.
27. Aureli F, D'Amato M, De Berardis B, Raggi A, Turco A C, Cubadda F (2012), Investigating agglomeration and dissolution of silica nanoparticles in aqueous suspensions by dynamic reaction cell inductively coupled plasma-mass spectrometry in time resolved mode, *Journal of Analytical Atomic Spectrometry*, 27, 1540.
28. Gilbert Y, Veillette M, Duchaine C (2010), Metalworking fluids biodiversity characterization, *Journal of Applied Microbiology*, 108, 437–449.
29. Nune M M R, Kiran C P (2019), Development, characterization, and evaluation of novel eco-friendly metal working fluid, *Measurement*, 137, 401–416.
30. Singh R K, Sharma A K, Dixit A R, Mandal A, Tiwari A K (2017), Experimental investigation of thermal conductivity and specific heat of nanoparticles mixed cutting fluids, *Materials Today: Proceedings*, 4, 8587–8596.
31. Nune M M R, Kiran C P (2020), Performance evaluation of novel developed biodegradable metal working fluid during turning of AISI 420 material, *Journal of Brazilian Society of Mechanical Science and Engineering*, 42, 1–16.
32. Adams M C, Gannon J E, Bennett E O (1979), BOD and COD studies of synthetic and semisynthetic cutting fluids, *Water Air Soil Pollution*, 11, 105–113.
33. Gajrani K K, Ram D, Ravi Sankar M (2017), Biodegradation and hard machining performance comparison of eco-friendly cutting fluid and mineral oil using flood cooling and minimum quantity cutting fluid techniques, *Journal of Cleaner Production*, 165, 1420–1435.
34. Huang J, Wang X, Long Q, Wen X, Zhou Y, Li L (2009), Influence of pH on the stability characteristics of nanofluids, *2009 Symposium on Photonics and Optoelectronics*, Wuhan, China, 1–4.
35. Lin J, Wang L, Chen G (2011), Modification of graphene platelets and their tribological properties as a lubricant additive, *Tribology Letters*, 41, 209–215.
36. Husnawan M, Saifullah M G, Masjuki H H (2007), Development of friction force model for mineral oil base stock containing palmolein and anti-wear additive, *Tribology International*, 40, 74–81.
37. Ing T C, Mohammed Rafiq A K, Azli Y, Syahrullail S (2012), The effect of temperature on the tribological behavior of RBD palm stearin, *Tribology Transactions*, 55, 539–548.
38 Yunus R, Fakhru'l-Razi A, Ooi T L, Iyuke S E, Perez J M (2004), Lubrication properties of trimethylolpropane esters based on palm oil and palm kernel oils, *European Journal of Lipid Science Technology*, 106, 52–60.
39. Sharif M Z, Azmi W H, Redhwan A A M, Zawawi N N M, Mamat R (2017), Improvement of nanofluid stability using 4-Step UV-vis spectral absorbency analysis, *Journal of Mechanical Engineering*, 4, 233–247.
40. Nandanwar R, Singh P, Haque F Z, Lee S H (2015), Synthesis and characterization of SiO_2 nanoparticles by sol-gel process and its degradation of methylene blue, *Chemical Science Journal*, 5, 1–10.

41. Angayarkanni S A, Philip J (2013), Role of adsorbing moieties on thermal conductivity and associated properties of nanofluids, *Journal of Physical Chemistry C*, 117, 9009–9019.
42. Vidal F A C, Ávila A F (2014), Tribological investigation of nanographite platelets as additive in anti-wear lubricant: A top-down approach, *Journal of Tribology*, 136, 031603.
43. Kaufman Y, Chen, S Y, Mishra H, Schrader A M, Lee D W, Das S, Donaldson S H, Israelachvili J N (2017), Simple-to-apply wetting model to predict thermodynamically stable and metastable contact angles on textured/rough/patterned surfaces, *Journal of Physical Chemistry C*, 121, 5642–5656.
44. Eshraghi E, Kazemzadeh Y, Qahramanpour M, Kazemi A (2017), Investigating effect of SiO_2 nanoparticle and sodium-dodecyl-sulfate surfactant on surface properties: Wettability alteration and IFT reduction, *Journal of Petroleum Environment Biotechnology*, 8, 6–10.
45. Weston J S, Jentoft R E, Grady B P, Resasco D E, Harwell J H (2015), Silica nanoparticle wettability: Characterization and effects on the emulsion properties, *Industrial Engineering Chemistry Research*, 54, 4274–4284.
46. Pedišić L, Šarić M, Bielen S (2003), Application possibilities of new AW/EP additive types in watermiscible metalworking fluids, *Industrial Lubrication Tribology*, 55, 23–31.

4 Advances in Textured Cutting Tools for Machining

Anand C. Petare, Neelesh Kumar Jain, and I. A. Palani
Indian Institute of Technology Indore, Simrol, India

S. Kanmani Subbu
Indian Institute of Technology Palakkad, Palakkad, India

CONTENTS

4.1 INTRODUCTION

The rapid development of automobile technologies and consumer products in the middle of the 20th century leads to the use of machining processing for metal removal to produce desired geometry. The growth in civil and military production enhanced the use of high-speed steel (HSS) and cemented carbide cutting tools for turning, facing, threading, drilling, boring, sawing, and milling operations [1]. Increasing demand for high-speed machining, large production rates, close tolerance with high-quality surface finish with less roughing operation led to the development of a machine tool, cutting tool geometry and materials, cooling and lubrication systems, and controlling

DOI: 10.1201/9781003284574-5

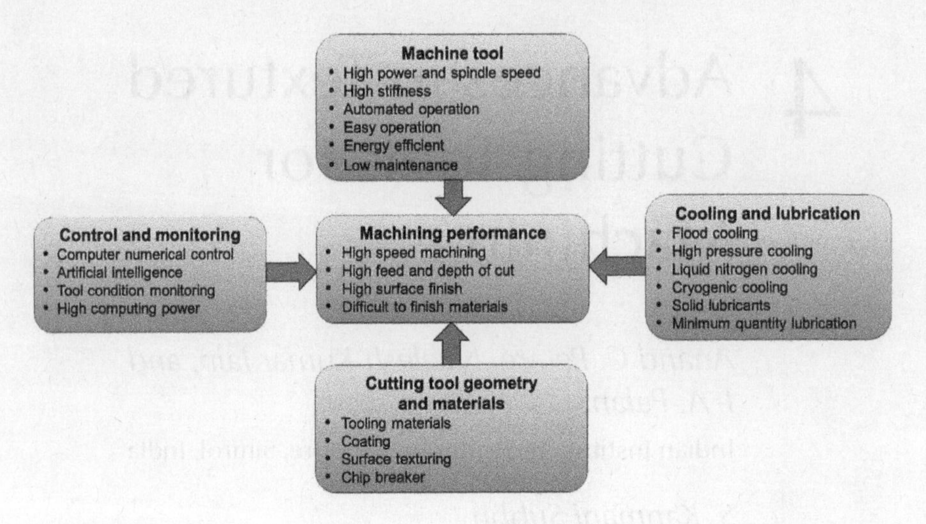

FIGURE 4.1 Different aspects of machine tool technology to enhance machining performance [2, 3].

and monitoring of machining operations [2, 3]. Figure 4.1 depicts different aspects of machine tool technology for enhancing machining performance. Machining contributed to 25% of overall manufacturing costs, and 35% to 50% of the power supplied to the machine is wasted in overcoming the friction at the tool tip and workpiece interaction [4, 5]. Furthermore, advancement in technology required precise machining of difficult to machine materials such as tool steel, case-hardened steel, titanium, nickel, aluminum alloys, and metal-matrix composites for various applications in automotive, aerospace, nuclear, biomedical, pneumatics, dies and mold, micro-, and nanodevices [6]. The machining of these materials at high speed, feed, and depth of cut using a conventional cutting tool causes excessive heat generation due to friction at the chip–tool interface. This results in severe wear at the creator, nose, and flank areas of the tool and produces a poor-quality surface on the workpiece. To overcome frictional heat generation and improve lubrication during machining, cutting oil is introduced at the tool tip–w orkpiece interface. It reduces tool wear and surface roughness by improving frictional characteristics and lubrication properties, but its disposal and storage cause environmental hazards. Furthermore, when machining difficult-to-machine materials, the contact time of the cutting oil with the tool tip–workpiece interface is very short; hence, a very small amount of heat is removed, resulting in high surface roughness and tool wear. In this situation, increasing the flow of cutting oil inside the cutting zone and storing cutting oil over the cutting edge of the tool can help improve surface finish and tool life [7]. Also, improvement in the tool geometry, chip flow pattern, and tool–workpiece interaction mechanism can enhance the surface finish and tool life by reducing cutting forces and friction [8].

Nowadays, surface texturing technology gaining widespread attention to improve wear characteristics, lubrication retention, friction, fatigue behavior, load-carrying capacity, and resistance to corrosion of engineering components, such as the piston ring, piston, cylinder liner, camshaft, bearing, mechanical seals, gears, and biomedical implants [9–12]. With the cutting tool having a specified geometry and interaction patterns with the workpiece surface, by modifying the flank, face, heel, and nose radius; the cutting edges; the tool angles; the tool surface; the contact length and clamping method, the tribological performance, heat transfer mechanism, and surface finish can be improved. In this direction, many studies reported that the application of texturing on cutting tools improve their performance by (a) creating micro pools to store cutting fluid for absorbing heat produced at tool–workpiece interaction and simultaneously reducing the temperature of the cutting tool and improving the surface quality of workpiece and tool life, (b) reducing the interaction between the chip and tool surface thereby reduces tool wear, (c) creating a tiny reservoir for entrapping worn debris generated during machining reduces tool wear and damage to the finished surface, and (d) increasing heat dissipation area of the tool, which improves the flow of cutting oil and natural air, helps reduce tool temperature, cutting forces, friction, and tool wear [11]. Applications of various machining processes to create different types of textures on turning inserts, milling cutters, and drills for machining of mild steel, alloy steel, stainless steel, aluminum alloy, titanium alloys, and nickel-based alloys reported worldwide and constant research in cutting-tool texturing are in underway. This chapter summarized recent advances in different types of texturing processes, their capabilities, advantages, and limitations. The influence of texturing on the performance of cutting tool for machining advanced materials, such as titanium alloys, stainless steel, glass, and aerospace alloys, for reducing friction, tool wear, cutting temperature and cutting forces, surface quality, and morphology of chip reviewed for quick reference to the readers.

4.2 TEXTURING PROCESSES

To create textures on engineering components and cutting tools various machining processes such as micro-plasma transferred arc (μPTA), laser surface texturing (LST), micro grinding (uG), micro-electrical-discharge machining (μEDM), focused ion beam machining (FIB), ultrasonic machining (USM), micro indentation (μI), and chemical etching (ChE) were used by researchers [13]. The details of each process and its working principles are discussed in the following paragraphs; also, their capabilities, advantages, and limitations are compared in Table 4.1.

4.2.1 Micro-Plasma Transferred Arc

In the PTA process, a pilot arc is initiated between two negatively charged tungsten electrodes and a positively charged copper concentric nozzle. The pilot arc causes the ionization of plasma gas, which formed a high-powered plasma between the plasma gun and substrate. The developed high-power plasma is directed to pass between the gun and the substrate. The temperature of plasma can reach up to 25,000 °C, which can melt the filler and deposit it on the substrate. Figure 4.2 illustrates its working principle schematically. It offers a high rate of deposition with low power consumption and low cost [11]. The advancement in digital power supply systems and process control offers the operation of PTA at the micro level by maintaining a very low current of the order of 0.1 mA with fine control. This enables using PTA at the

TABLE 4.1
Comparison of Different Texturing Processes

Process	Capabilities	Advantages	Limitations
Micro-plasma transferred arc	• Can produce high-instance temperature up to 25,000 °C • Flexible operation in continuous or pulsed mode	• High deposition rate • Controlled and focused arc that gives good material depositions	• High equipment cost • Skilled operator required • Noise generation • High maintenance cost
Laser surface texturing	• Complex-shaped textures can be easily fabricated • Faster than other texturing processes • Hardened materials can be easily textured	• Highly accurate and precise • High texturing speed • Less material distortion • Not affected by the hardness of materials	• Heat-affected zone causes thermal damage to the work surface • Formation of burrs and bulges on the work surface damages surface integrity
Micro grinding	• Capable to mirror finish difficult-to-machine materials • $R_a = 0.01–0.02$ µm	• Simple and easy operation • Flexibility to prepare form grinding wheel according to finishing requirements	• Cost of the grinding wheel is very high • Excessive breaking of grinding wheel • Slow process
Micro-electrical-discharge machining	• All electrically conductive materials can be textured • $R_a = 0.015–0.80$ µm • Can develop textures on thin sections, brittle, and fragile materials easily	• Less wear and breakage of the tool • Low tooling and fixture cost. • Machined surface is free from burrs, and no post finishing is required	• Specialized control system and power supply required • Only conductive materials can be machined • Required to fabricate complementary-shaped tool
Focused ion beam machining	• Can create a round or square hole and complex patterns easily • Deposition size less than 50 nm can be maintained easily	• Flexible process for local deposition and not required complex mask structure • Easy to control size, height, and position structural patterns	• Irregular sputtering causes damage to workpiece • High beam temperature results in thermal damage • Ion implantation results in variable machined surface
Ultrasonic machining	• Can machine harder above 40 Rockwell hardness at C-scale (HRC). • $R_a = 0.25–0.75$ µm • Minimum holes up to 76 µm can be drilled	• Fragile, brittle, and hard material can be machine easily • No physical change in the structure of work material • No burrs and distortion	• High tool wear • Low material removal rate • Unable to drill deep holes

(*Continued*)

TABLE 4.1 (Continued)

Process	Capabilities	Advantages	Limitations
Micro indentations	• Indentation force can change	• Simple and quick technique • No delamination of the coating	• Indenter probe may break if the surface is uneven • Only square pyramid-shaped textures can be made
Chemical etching	• Can machine brittle and hard materials also • A complex pattern can be engraved easily	• Low cost of tooling • Multiple parts can be machined simultaneously • Delicate components can be machined • No burrs formation on finished components	• Unable to cut sharp corners • Expensive process • Long machining time • Less dimensional accuracy

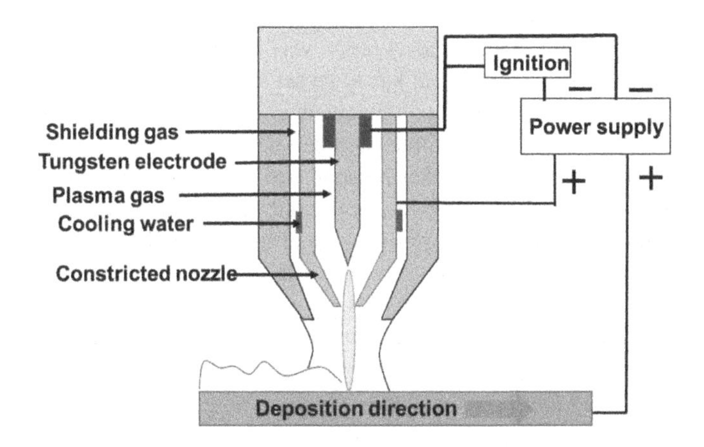

FIGURE 4.2 Schematic of working principle of plasma transferred arc (PTA) process.

micro level, and the process referred to as μPTA. The working principle of plasma generation in the μPTA process is the same as that in the PTA process. The steady and precise controlled arc offers deep material penetration with less distortion and thermal damage. This process is best suited for creating micro features by depositing metal that is required in small-parts manufacturing and repairing defective/damaged dies and gears. [12].

4.2.2 Micro Grinding

μG is an extensively used process for fine finishing of small-sized grooves, pins, miniature gears, bushes, gun drills, hob cutter, and micro components and to resharpen the cutting edges of tools. It uses a highly precise grinder with fine control of depth

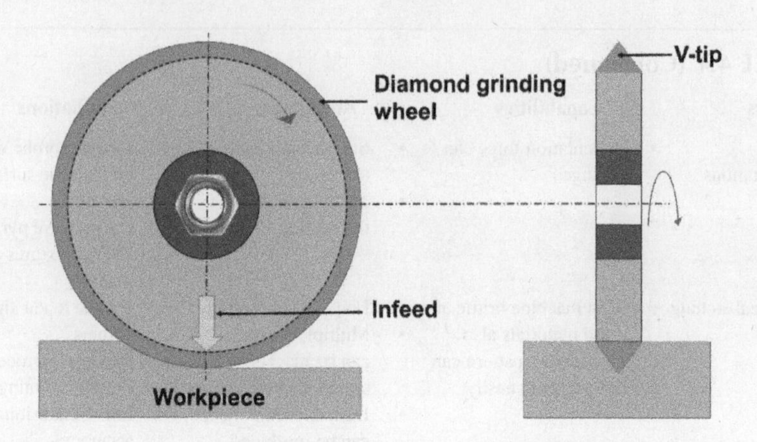

FIGURE 4.3 Schematic of working principle of the micro-grinding process.

of cut and accurately dressed grinding wheel used for finishing micro components. A formed grinding wheel impregnated with micro-abrasive particles, resin, and other binding materials with the desired contour was developed using the electrolytic dressing technique. The depth of cut has to be set to less than 100 nm to achieve the surface finish of order 10 nm. Figure 4.3 depicts a schematic of its working principle. Micro grooves with a very small dimension can be easily produced by this process. The use of a small depth of cut makes it capable of mirror finishing of brittle materials. Complicated and intricate grooves, deep holes, and cavities are unable to finish by micro grinding due to the narrow-sized tip of the wheel and high grinding force [13, 14]. The facility of maintaining tip size of grinding enhanced the use of μG for creating micro grooves and microcavities on carbide cutting tools, ferrite, silicon wafers BK7 glass, ceramics, germanium immersion grating (GIG) element for spectrograph, and hard coatings [15–18].

4.2.3 Micro-Electrical-Discharge Machining

EDM is a nontraditional machining method in which electrical discharge is developed between an electrically conductive workpiece and tool. Figure 4.4 depicts a schematic of the working principle of EDM. Generally, the tool is referred to as cathode (negative pole), and the workpiece is referred to as anode (positive pole). These both anode and cathode are connected to the pulse power supply unit and the dielectric medium is passed between the interelectrode gap (IEG). A high-order direct-current voltage of 40 to 400 V is applied in the IEG it causes the removal of electrons from the cathode and ions from the anode. When the voltage in the IEG reaches more than the breakdown voltage of the dielectric medium, it causes the ionization of the dielectric molecules and starts releasing secondary electrons, and a high-discharge zone is formed in the IEG. The continuous strike of electrons and ions causes its conversion into kinetic energy and generates a short duration spark, which is enough for melting and vaporizing the workpiece material.

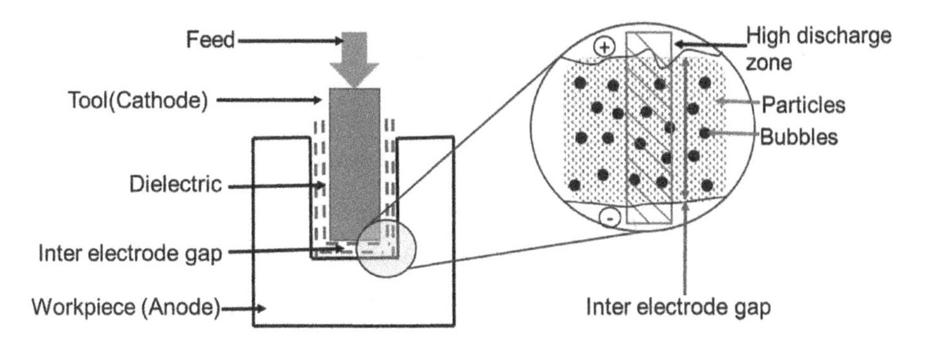

FIGURE 4.4 Schematic of working principle of electrical discharge machining.

μEDM is a small version of EDM for machining micro components and products for various mechanical applications. The working principle of μEDM is the same as that of EDM only the electrode size, power supply system, and resolution of axes are different [19]. The servo system of μEDM has a very high sensitivity, which makes it enable to minimize IEG up to 1 μm. This helps μEDM for machining miniature gears, micro holes, microinjection mold, micro shaft, micro inserts, biomedical implants, printer ink injector, fuel injector nozzles, and defense applications, among others [20]. The μ-EDM process is suitable for machining all electrically conductive materials and is difficult to machine, such as titanium alloys, high-speed steel (HSS), cemented carbide, tungsten, nickel-based superalloy, ceramics, and metal matrix composites [21].

4.2.4 Focused Ion Beam Machining

The FIB machining process is frequently used in manufacturing semiconductor devices, microelectronics, high-resolution texturing/embossing, chip processing, repairing lithographic masks, and additive manufacturing. It facilitates high-precision milling and deposition, which makes it capable of fabricating micro- and nano-features on objects with a high resolution. [22]. It is a sputtering-based molecular manufacturing process in which the material removal takes place in the form of molecules or atoms from the workpiece surface. Figure 4.5 depicts the schematics of the working principle of the two-lens FIB process.

A liquid metal ion (LMI) is used as a source for generating a focused ion beam, and it can maintain a minimum ion beam size of ø5 nm. This LMI comprises a tungsten needle connected to the storage of gallium metal; it generates a very fine ion beam of gallium. The fine beam advances further through a mass separator, which permits the stable ion beam to pass farther. A long drift tube is connected next to the mass separator to make the ion beam vertical by removing inclined and undirected ions. At the bottom of the drift tube, objective lenses are located that decrease the beam spot size and enhance the focus of the beam. Then the beam is passed through the beam deflector, which hits ions over the substrate for sputtering or material

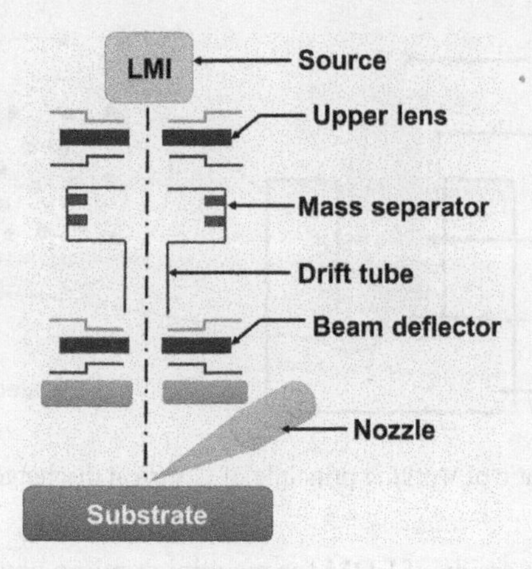

FIGURE 4.5 Schematics of the working principle of two-lens focused ion beam machining.

removal [23]. The high primary current accelerates ions very rapidly, which causes easy material removal from the substrate surface this allows precision milling of the substrate up to 1-μm accuracy. Easy control of position, size, and sharpness of the beam makes it a very powerful tool for machining micro- and nanostructures [24–25].

4.2.5 Ultrasonic Machining

USM is a nonconventional machining method developed by L. Balamuth for machining brittle and hard materials. It uses an ultrasonic transducer to convert high-frequency electrical power into linear mechanical vibrations [26]. A concentrator or horn coupled with a transducer transfers mechanical vibrations to abrasive particles and water-containing slurry through the tool. The continuous impact of abrasive particles over the workpiece surface causes material removal [27]. Figure 4.6 depicts schematics of the working principle of ultrasonic machining.

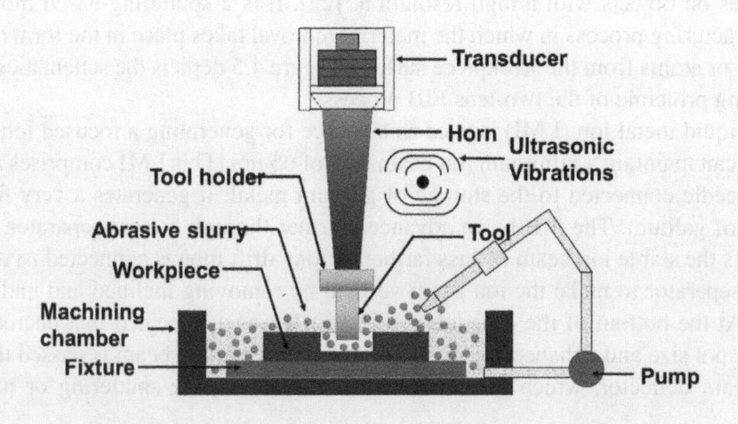

FIGURE 4.6 Schematics of the working principle of ultrasonic machining.

4.2.6 Micro Indentation

Micro indentation (μI) is a microhardness measuring technique using an indenter probe, which causes a micro-sized depression on the sample for a given force value and dwell time. The indentation created by the probe causes plastic deformation and forms an impression on the sample. A computer-controlled microhardness tester having the flexibility of changing forces, displacement, indenter, indentation time, and table movement make it convenient for texturing applications. Recently, some researchers have used the μI technique for texturing on tungsten carbide cutting tools and bronze samples [28, 29]. This is a simple and easy process for creating texture without delamination of the coating. Texture shape is limited to the shape of the indenter probe.

4.2.7 Chemical Etching

ChE is a subtractive machining process for engraving or creating complex shaped patterns on metal surfaces using chemical etchants. In this process, the workpiece surface is cleaned with an alkaline solution to remove dust, grease, and other depositions. A polymer-based maskants layer is pasted over it, and the desired geometry is scribed by removing the maskant from the selected portion. Thereafter, an etchant solution is applied to the work surface wherever material removal is required. The etchant reacts with the unprotected workpiece surface and causes its dissolution to produce the desired geometry [23]. After the completion of the etching, the workpiece is removed from the solution to remove the maskant and is finished for use. ChE is an easy, flexible, and economic process for creating textures and patterns. The details of the steps used to perform ChE to produce desired geometry are depicted in Figure 4.7.

4.2.8 Laser Surface Texturing

LST is a widely used process for improving tribological features, frictional behavior, wettability, process performance, adherence, and the thermal and electrical conductivity of industrial components. The laser generates high-energy pulses on the work surface, which causes heating and vaporization of work materials [30]. The accuracy and quality of the texture produced on the work surface depend on the method of projection of the laser beam. Revolving disc, patterned mask, and galvano scanning are the three most frequently used methods for texturing. In the *revolving disc method*, the unfocused laser beam is chopped to create a pattern of texture. A *patterned mask*

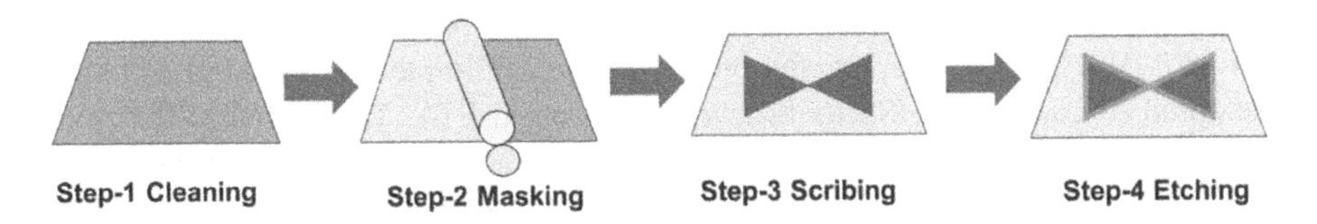

FIGURE 4.7 Schematics of different steps performed in chemical etching.

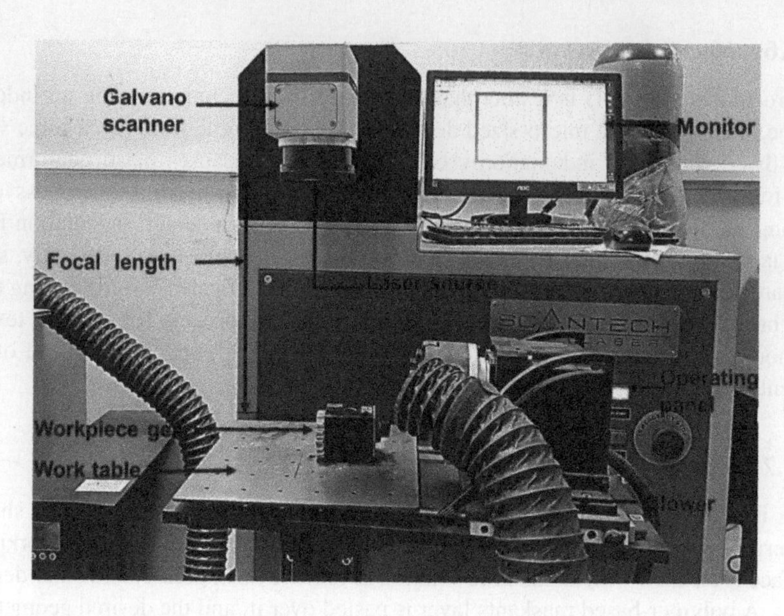

FIGURE 4.8 Photograph of continuous fiber laser machine used for laser texturing on gears.

is applied over the work surface to distinguish textured and nontextured areas, and the laser beam is applied over it. In the *galvano scanning method*, a laser beam is attached with a computer-controlled galvano scanner and a software interface to provide flexibility in creating simple, as well as complex, texture patterns. The control system associated with the galvano scanner moves the laser beam over the work surface to produce the desired textures. Figure 4.8 depicts a photograph of a computer numerical control (CNC) continuous fiber laser with a galvanometric scanner used for texturing on gears at IIT Indore. The formation of textures depends on (a) microstructure, chemical composition, absorption mechanism, and other properties of the work materials and (b) laser process parameters, that is, wavelength, laser power, scan speed, number of passes, focal length, and type of textures. It is very fast and accurate process for texturing a large surface area. It is frequently used for texturing mechanical seals, piston, cylinder liner, piston rings, bearings, camshaft, bushes, pins, cutting tools, knee and hip implants, sutures, and gears [31–35].

4.3 ADVANCES IN THE MACHINING PERFORMANCE USING TEXTURED CUTTING TOOLS

Sawant et al. [36] used the μ-PTA process to create dimple textures and spot textures on the rake face of the T42 HSS cutting tools for turning the titanium alloy Ti-6Al-4V. Figure 4.9 depicts the optical image of dimple (Figure 4.9a) and spot-textured (Figure 4.9b) HSS tools developed using μ-PTA. They revealed that spot-textured tool produced a low value of cutting forces, cutting temperature, flank wear, surface roughness as compared to dimple textured tool than that of nontextured tool. Also, the spot textured tools behaved as fins and enhanced heat dissipation into the surroundings, which helps increase tool life.

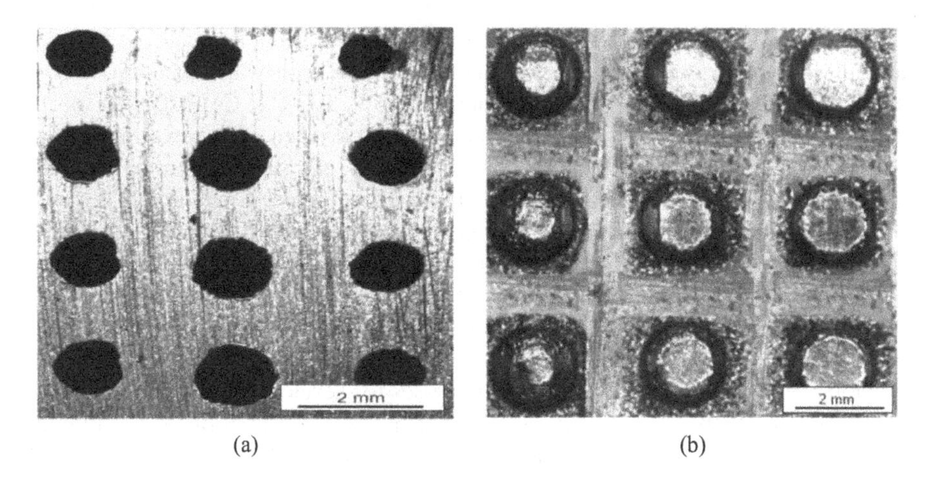

FIGURE 4.9 The optical image of the rake face of T42 high-speed steel tool with (a) dimple textures and (b) spot textures.

(Reprinted with permission from Elsevier © 2018.)

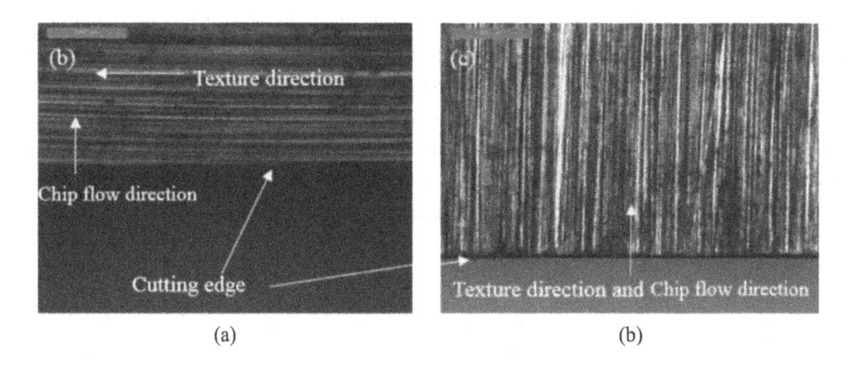

FIGURE 4.10 Surface micrograph of homothetic textures created on rake face tool insert using micro grinding (μG) in (a) a parallel direction and (b) a perpendicular direction.

(Reprinted with permission from Elsevier © 2019.)

Xie et al. [17] deployed a bronze-bonded diamond micro grinding (μG) wheel to create an array of micro-grooved textures on the top surface (rake area) of carbide tool for dry turning the titanium alloy Ti-6Al-4V. They found that both the diagonal and orthogonal micro-grooved textured tools produced fewer sparks during cutting, and less tool wear as compared to the nontextured tool. Also, the diagonal micro-grooved textured tool showed 6.7% less tool wear and 37.3% less surface roughness than the orthogonal micro-grooved textured tool. Dhage et al. [18] developed parallel and perpendicular homothetic texture on cemented tungsten carbide tools, as shown in Figure 4.10, using μG for machining of AISI-1045 steel bar. They reported that texturing on a tool-rack face develops microcapillaries, which enhanced colling action and reduced tool wear and friction.

Li et al. [37] used μEDM-assisted high-frequency vibration to create micro-hole and linear grooved textures on cemented carbide cutting tools with chemical composition as WC-8%, Co (MA-type tool), and WC-15%, TiC-6%, Co (MB-type tool). They reported no adhesion of tool material on electrode seen while texturing on the MB-type tool compared with the MA-type tool. Also, the textured formed on both the tools were free from cracks, heat-affected zone, and burrs. Singh et al. [38] created micro-dimple textures using μEDM on the rake face of an HHS tool for machining aluminum alloy (Al 7075–T6). They reported that a textured cutting tool reduced cutting forces by 5% to 13% compared to machining by a nontextured tool. Furthermore, less adhesion of chip particles and built-up-edge (BUE) formation on the textured tool was seen than on the nontextured tool.

Pratap et al. [39] used μ-EDM based processes for developing micro dimples and micro-grooved textures on a polycrystalline diamond grinding tool for machining BK7 glass. Figure 4.11 depicts a photograph of (a) scheme of texturing on tools, (b) μEDM-textured micro-groove, and (c) μEDM-textured and (d) wire-electro-discharge grinding (WEDG) scanned tool protruding abrasive particles. They revealed that texture tools showed a 35% to 40% reduction in cutting force compared to the nontextured tool. Textured tools had a low value of surface roughness compared to nontextured tools.

Pratap and Patra [40] performed experiments on polycrystalline textured micro-grinding tools for machining BK7 glass under minimum quantity lubrication (MQL). They found a 35% decrease in surface roughness, cutting forces, and adhesion of chips on the work–tool interface using a textured grinding tool. Also, no damage in the cutting edges of textured tools was seen whereas cutting edges of nontextured tools fractured in many places. The droplet size of cutting fluid in MQL equal to the width of the textured tool produced a maximum cooling effect.

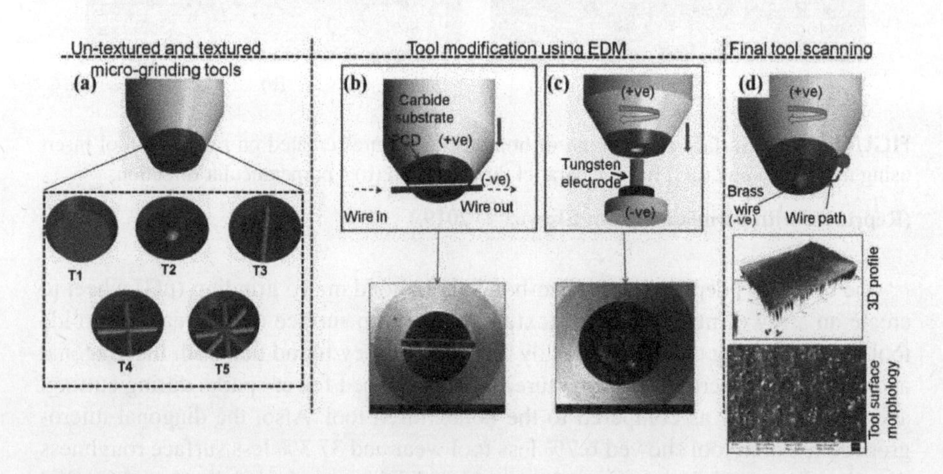

FIGURE 4.11 (a) Scheme of texturing on tools; (b) μEDM textured micro groove; (c) μEDM-textured and (d) wire-electro-discharge grinding (WEDG) scanned tool protruding abrasive particles.

(Reprinted with permission from Elsevier © 2019.)

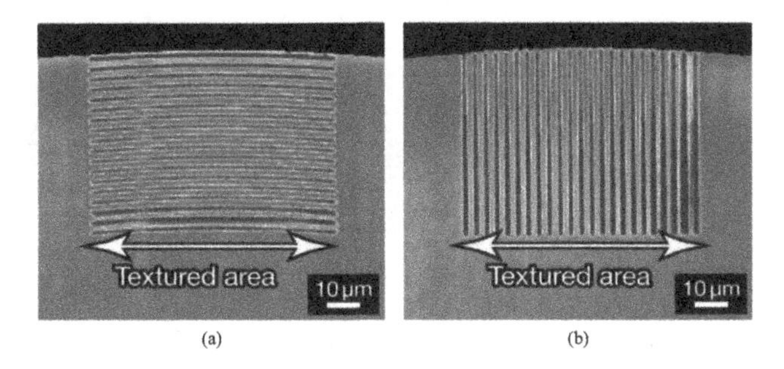

FIGURE 4.12 Surface micrograph of an FIB-textured tool rake face with (a) a parallel micro-groove and (b) a perpendicular micro-groove.

(Reprinted with permission from Elsevier © 2017.)

Kawasegi et al. [41] fabricated parallel and perpendicular micro-groove texture on the rake face of a diamond tool using the FIB process followed by heat treatment for machining an aluminum alloy (A5052) and nickel-phosphorus-layered stainless steel (NiP). Figure 4.12 depicts the surface micrograph of the FIB textured tool rake face with (a) a parallel micro-groove and (b) a perpendicular micro-groove. They revealed that texturing on the diamond tool helped reduce cutting forces and the coefficient of friction for both the alloys without affecting their surface quality compared to the nontextured tool.

Elias et al. [42] used the μI approach to create pyramid-shaped textures on a tungsten carbide tool for MQL-assisted turning of titanium alloy grade 5. They found that texturing on the tool decreased the cutting force, surface roughness, cutting temperature, and tool wear by 21%, 6%, 7%, and 19%, respectively, compared to the nontextured tool. Furthermore, no delamination of the coating was observed as like other texturing processes.

Prasad and Syed [43] developed micro-dimple and protruded textures over the T42-grade HSS tool using inkjet printing followed by ChE for machining aluminum alloy (6063) T6. They stated that (a) cutting and thrust forces were reduced by 23–7.3%, respectively, using dimple textured tool and 13–4.8%, respectively, using a protruded textured tool as compared to a nontextured tool and (b) the protruded textured tool showed less adhesion of chip compared to the dimple textured and nontextured tools.

Kawasegi et al. [15] developed micro- and nano-straight grooved parallel and perpendicular textures and cross-patterned textures in direction of chip flow on cemented carbide tool (WC-Co) using s femtosecond laser for machining aluminum alloy A5052. Figure 4.13 represents schematics of direction textures developed on the rack face of a tool in (a) the direction perpendicular to chip flow, (b) the direction parallel to chip flow, and (c) cross-patterned textures. They reported that nanotextured tool was more effective in reducing the cutting force compared to micro-texture tool and nontextured tool. Furthermore, the perpendicular textured tool offered more reduction in cutting force compared to nontextured tool than that of the parallel textured tool.

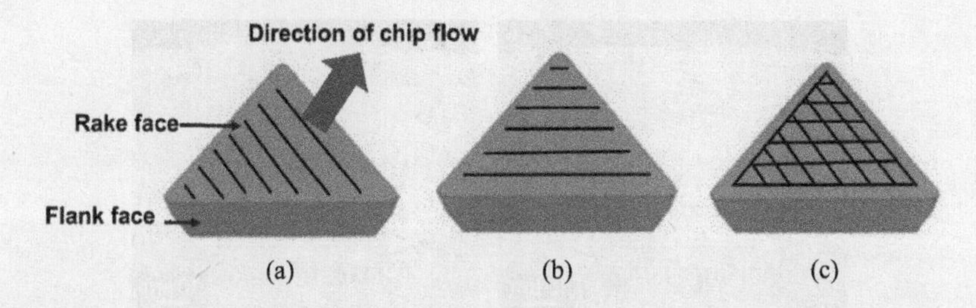

FIGURE 4.13 Schematics of direction textures developed on the rack face of tool in (a) the direction perpendicular to chip flow, (b) the direction parallel to chip flow, and (c) cross-patterned textures.

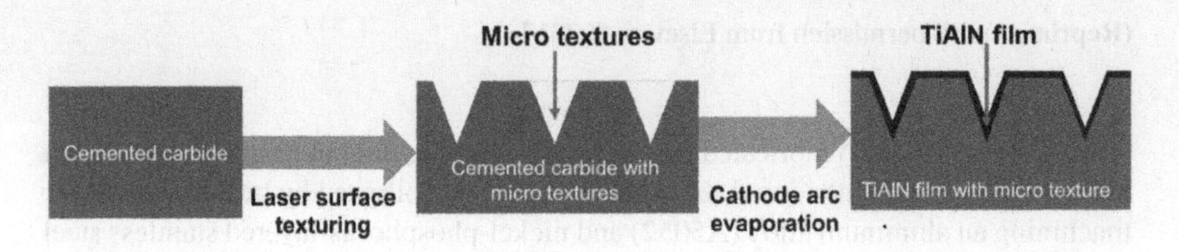

FIGURE 4.14 Schematic of steps followed during laser texturing and coating of TiAlN on the tool.

Zhang et al. [44] developed a microscale texture on cemented carbide turning insert using pulsed Nd:YAG laser followed by a coating of TiAlN film using sputtering deposition for machining of hardened steel (AISI 4045). The steps followed during texturing and coating on the tool are shown schematically in Figure 4.14. They reported that the textured tool reduced the cutting force by 21–34.7%, respectively, and tool wear by 59–76%, respectively, with low chip adhesion compared to the nontextured tool. Sun et al. [45] compared the performance of micro-grooved, micro pits, and hybrid Nd:YAG laser textured cemented carbide tool for machining pure iron and AISI 1045 steel. They observed a reduction of cutting forces by 7–33%, surface roughness by 43–69%, and cutting temperature by 7–21% compared to the nontextured tool. Furthermore, the hybrid texture tool performed more efficiently compared to other types of textured tools. Duan et al. [46] assessed the performance of Nd:YAG laser micro-grooved textured cemented carbide tool for the machining of medium carbon steel (AISI 1045). They revealed that textured tool offered a reduction of 16–25% in cutting forces and 7–9% in chip–tool contact length compared to the nontextured tool with a higher microhardness and smoother surface. Sugihara and Enemata [47] evaluated performance of femtosecond laser dimple textured and non-textured cemented carbide cutting tools for machining medium carbon steel. They revealed that the cutting tool reduced crater wear by 66% compared to the non-textured tool with a low-value coefficient of friction.

Sasi et al. [32] evaluated the performance of pulsed Nd:YAG laser micro-dimple textured HSS tool for turning Al7075-T6 aluminum alloy on a lathe machine. They observed that the textured cutting tool reduced the thrust force by 17–19% and the cutting forces by 9%, respectively, at a high machining speed during dry cutting compared to nontextured tool. Niketh and Samuel [33] investigated the effect of

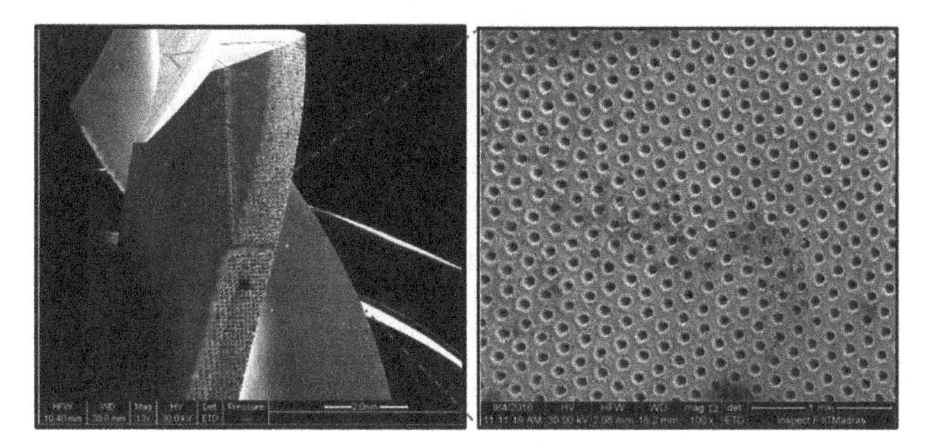

FIGURE 4.15 Scanning electron micrograph of micro-dimple textured carbide drill. **(Reprinted with permission from Elsevier © 2017.)**

pulsed Nd:YAG laser micro-dimples and homothetic textured carbide drill tool on its tribological performance while drilling in the titanium alloy Ti-6Al-4V. Figure 4.15 depicts a scanning electron microscope (SEM) micrograph of micro-dimple textured carbide drill. They reported that texturing a drill tool improved the tribological performance by reducing torque by 12% and thrust force by 10%, with less smearing of chips, BUE, and burrs formation. Ye et al. [48] investigated the effect of femtosecond laser textured micro-parallel and micro-perpendicular grooves on a cemented carbide tool for turning of C45 steel. They revealed that micro-parallel grooved textured tool showed more reduction in cutting forces and a lower coefficient of friction compared to the micro-parallel grooved textured tool than that of the nontextured tool.

Hao et al. [49] compared the performance of pulsed fiber laser–produced hybrid and homothetic texture on carbide tools for machining titanium alloy (Ti-6Al-4V) and revealed that the coefficient of friction was reduced by 9% using a hybrid textured tool and 5% by using a homothetic textured tool compared to a nontextured tool. Alagan et al. [50] developed micro-dimple, square pyramidal, and hybrid textures on an uncoated sintered carbide tool by using a pulsed Nd:YAG laser for facing alloy 718 with a high-pressure cooling system. Their experimental investigations showed that hybrid textures reduced tool wear by 30% compared to nontextured tools without any significant effect on reduction in chip–tool contact area. Vignesh et al. [51] compared the performance of a fiber laser dimple textured tungsten carbide tool followed by a coating of MoS_2 with a nontextured tungsten carbide tool for the machining of tool steel (AISI H-13). They observed a reduction of the cutting forces by 33–42% and in tool wear by 10% using a textured tool compared to a nontextured cutting tool. Sugihara et al. [52] studied the effect of the location of micro-grooved femtosecond laser textures on WC-Co alloy for machining aluminum (Al-1100). They observed that textures near the sliding area and the sticking area of the cutting edge of the tool offer 30% and 60% reduction, respectively, in the friction force compared to the nontextured tool. Furthermore, the location of textures on the rake face of the tool significantly affects the cutting forces. Xu et al. [53] studied the performance of a fiber laser micro-hole textured cemented carbide tool for machining aluminum and SiC composites (SiCp/Al). They observed a 4–17% reduction in cutting forces compared to the nontextured tool. Devaraj et al. [54] analyzed a laser micro-textured tungsten carbide tool for machining an aluminum metal matrix composite using tungsten disulfide (WS_2) lubricant. They varied the diameter, depth, and pitch of the micro-holes and reported that (a) increasing the diameter of textured hole decreases surface roughness by 21%, power consumption by 19%, and flank wear by 23%; (b) increasing the textured hole depth decreases the surface roughness by 23% and power consumption by 7% and increases flank wear by 43%; and (c) increasing the textured hole pitch decreases power consumption by 9%. Lu et al. [55] evaluated the performance of pulsed Nd:YAG laser micro-textured circular cutting blades for cutting stainless steel (SS-304) pipe. They revealed that the cutting force, temperature, and surface roughness were reduced by 3–12%, 6–14%, and 4–5%, respectively, using a textured cutting blade compared to a nontextured cutting blade. Wang et al. [56] studied the frictional and cutting performance of a femtosecond laser textured diamond tool and observed that the textured cutting tool reduced cutting forces, tool wear area, and depth by 15–25%, 48–92%, and 32–98%, respectively, compared to a nontextured tool.

4.4 SUMMARY AND CONCLUSION

Following conclusions can be drawn from the review of texturing of cutting tools:

- Comparing all the texturing processes, it is seen that laser surface texturing has a widely used texturing process by researchers. Flexibility and easy control of process parameters make it a very popular process for developing a variety of textures, such as micro-dimple, micro-groove, cross-pattern, hybrid, homothetic, ring, concentric, and square pyramid-shaped textures on HHS, cemented carbide, sintered carbide, tungsten carbide, and diamond tools.
- Textures on the cutting tool act as a fin to improve heat transfer and as a reservoir for storing cutting fluid to absorb heat developed in the tool. While machining, an increase in the tool temperature causes an upward flow of cutting fluid, and it gets to enter in the gap between tool and workpiece. This causes a weakening of the friction bond between the tool and the workpiece in the stickiness zone. Furthermore, texturing reduces chip–tool contact length and acts as a chip breaker. These all reasons contributed to improving the topological performance of the tool.
- Tool wear, cutting forces, surface roughness, power requirement, tool chip contact length, and chip adhesion were highly influenced by the size, type, pitch, and location of textures. Hybrid textures with variable shapes and densities were reported to be more effective in reducing cutting forces, the coefficient of friction, and tool wear compared to other types of textures due to more storage space for the coolant to remove the heat at tool-tip workpiece interface. A micro-dimple texture helped in reducing the BUE by storing worn debris and reducing tool–chip contact length. Spot textures on the tool rake surface offered a higher reduction in cutting temperature by increasing convective heat transfer to the surrounding and cutting fluid.

- FIB machining followed by heat treatment was reported to be a more effective method than LST for developing textures on diamond tools due to an effective removal-induced nondiamond phase.
- μG was reported to be a very effective process for texturing cutting tools to reduce tool wear, friction, and tool–chip contact length, but it causes an increase in surface roughness.
- μI was reported to be a simple and fast process for producing textures on cutting tools without delamination of the coating as in LST and μEDM.
- The use of solid lubricants and the MQL technique enhanced the machining performance of a textured tool by maximizing the flow of coolant at the tool–workpiece interaction and frequently removing chips from the cutting zone.

REFERENCES

1. Mills, B. 1996. "Recent developments in cutting tool materials". *Journal of Materials Processing Technology*, 56(1), 16–23. doi: 10.1016/0924-0136(95)01816-6
2. Shokrani, A., Dhokia, V., and Newman, S. T. 2012. "Environmentally conscious machining of difficult-to-machine materials with regard to cutting fluids". *International Journal of Machine Tools and Manufacture*, 57, 83–101. doi: 10.1016/j.ijmachtools.2012.02.002
3. Jedrzejewski, J., and Kwasny, W. 2017. "Development of machine tools design and operational properties". *The International Journal of Advanced Manufacturing Technology*, 93(1), 1051–1068. doi: 10.1007/s00170-017-0560-2
4. Deshpande, S., and Deshpande, Y. 2019. "A review on cooling systems used in machining processes". *Materials Today: Proceedings*, 18, 5019–5031. doi: 10.1016/j.matpr.2019.07.496
5. Gajrani, K. K., and Sankar, M. R. 2017. "State of the art on micro to nano textured cutting tools". *Materials Today: Proceedings*, 4(2, Part A), 3776–3785. doi: 10.1016/j.matpr.2017.02.274
6. Astakhov, V. P. 2011. "Machining of Hard Materials – Definitions and Industrial Applications". In *Machining of Hard Materials*. ed. J. P. Davim. 1–32. London: Springer London.
7. Kligerman, Y., Etsion, I., and Shinkarenko, A. 2005. "Improving tribological performance of piston rings by partial surface texturing". *Journal of Tribology*, 127(3), 632–638. doi: 10.1115/1.1866171
8. Tala-Ighil, N., Maspeyrot, P., Fillon, M., and Bounif, A. 2007. "Effects of surface texture on journal-bearing characteristics under steady-state operating conditions". *Proceedings of the Institution of Mechanical Engineers, Part J: Journal of Engineering Tribology*, 221(6), 623–633. doi: 10.1243/13506501JET287
9. Pou, P., Riveiro, A., del Val, J., Comesaña, R., Penide, J., Arias-González, F., and Pou, J. 2017. "Laser surface texturing of Titanium for bioengineering applications". *Procedia Manufacturing*, 13, 694–701. doi: 10.1016/j.promfg.2017.09.102
10. Ribeiro, F. S. F., Lopes, J. C., Bianchi, E. C., and de Angelo Sanchez, L. E. 2020. "Applications of texturization techniques on cutting tools surfaces—A survey". *The International Journal of Advanced Manufacturing Technology*, 109(3), 1117–1135. doi: 10.1007/s00170-020-05669-0
11. Sawant, M. S., Jain, N. K., and Palani, I. A. 2018. *Investigations on Additive Manufacturing of Metallic Materials by Micro-Plasma Transferred Arc Powder Deposition Process.* Doctoral dissertation, Department of Mechanical Engineering, IIT Indore.

12. Jhavar, S., Jain, N. K., and Paul, C. P. 2014. "Development of micro-plasma transferred arc (μ-PTA) wire deposition process for additive layer manufacturing applications". *Journal of Materials Processing Technology*, 214(5), 1102–1110. doi: 10.1016/j.jmatprotec.2013.12.016
13. Pratap, A., Patra, K., and Dyakonov, A. A. 2019a. "A comprehensive review of micro-grinding: emphasis on toolings, performance analysis, modeling techniques, and future research directions". *The International Journal of Advanced Manufacturing Technology*, 104(1), 63–102. doi: 10.1007/s00170-019-03831-x
14. 4M Association. 2008. Metals processing: Microgrinding. http://www.4m-association.org/content/Microgrinding.
15. Pratap, A., Patra, K., and Dyakonov, A. A. 2016. "Manufacturing miniature products by micro-grinding: A review". *Procedia Engineering*, 150, 969–974. doi: 10.1016/j.proeng.2016.07.072
16. Zhang, B., Liu, X., Brown, C. A., and Bergstrom, T. S. 2002. "Microgrinding of nanostructured material coatings". *CIRP Annals*, 51(1), 251–254. doi: 10.1016/S0007-8506(07)61510-8
17. Xie, J., Luo, M.-J., He, J.-L., Liu, X.-R., and Tan, T.-W. 2012. "Micro-grinding of micro-groove array on tool rake surface for dry cutting of titanium alloy". *International Journal of Precision Engineering and Manufacturing*, 13(10), 1845–1852. doi: 10.1007/s12541-012-0242-9
18. Dhage, S., Jayal, A. D., and Sarkar, P. 2019. "Effects of surface texture parameters of cutting tools on friction conditions at tool-chip interface during dry machining of AISI 1045 steel". *Procedia Manufacturing*, 33, 794–801. doi: 10.1016/j.promfg.2019.04.100
19. Jain, N. K., and Chaubey, S. K. 2017. Review of miniature gear manufacturing. In *Comprehensive Materials Finishing*, ed. M. S. J. Hashmi, 504–538. Oxford: Elsevier, Oxford.
20. Schubert, A., Zeidler, H., Kühn, R., and Hackert-Oschätzchen, M. 2015. "Micro-electrical discharge machining: a suitable process for machining ceramics". *Journal of Ceramics*, 2015, 470801. doi: 10.1155/2015/470801
21. Joshi, S. 2014. "Micro-Electrical Discharge Micromachining". In *Introduction to Micromachining*, ed. V. K. Jain, 11.1–11.29. New Delhi, India: Narosa Publishing House, ISBN: 978-81-8487-361-0
22. Lindquist, J. M., Young, R. J., and Jaehnig, M. C. 1993. Recent advances in application of focused ion beam technology. *Microelectronic Engineering*, 21(1), 179–185. doi: 10.1016/0167-9317(93)90051-6
23. Jain, V. K. 2014. *Advanced Machining Processes* (2nd ed.). Allied Publishers Private Limited, New Delhi, India
24. Nanophysics. 2021. *Focused Ion Beam Technology Capabilities and Applications*. https://www.nanophys.kth.se/nanolab/fei-nova200/FejFIB1.pdf
25. Orso, S. 2005. Application of a FIB system to biological samples. https://elib.uni-stuttgart.de › 02_Orso_Bio_FIB
26. Jain, N. K., and Jain, V. K. 2001. Modeling of material removal in mechanical type advanced machining processes: a state-of-art review. *International Journal of Machine Tools and Manufacture*, 41(11), 1573–1635. doi: 10.1016/S0890-6955(01)00010-4
27. Xu, S., Shimada, K., Mizutani, M., and Kuriyagawa, T. 2014. Fabrication of hybrid micro/nano-textured surfaces using rotary ultrasonic machining with one-point diamond tool. *International Journal of Machine Tools and Manufacture*, 86, 12–17. doi: 10.1016/j.ijmachtools.2014.06.005
28. Elias, J. V., Venkatesh N. P., Lawrence, K. D., and Mathew, J. 2021. Tool texturing for micro-turning applications – an approach using mechanical micro indentation. *Materials and Manufacturing Processes*, 36(1), 84–93. doi: 10.1080/10426914.2020.1813899

29. Wu, J., Yu, A., Chen, Q., Wu, M., Sun, L., and Yuan, J. 2020. Tribological properties of bronze surface with dimple textures fabricated by the indentation method. *Proceedings of the Institution of Mechanical Engineers, Part J: Journal of Engineering Tribology*, 234(10), 1680–1694. doi: 10.1177/1350650120940126
30. Arslan, A., Masjuki, H. H., Kalam, M. A., Varman, M., Mufti, R. A., Mosarof, M. H., and Quazi, M. M. 2016. "Surface texture manufacturing techniques and tribological effect of surface texturing on cutting tool performance: A review". *Critical Reviews in Solid State and Materials Sciences*, 41(6), 447–481. doi: 10.1080/10408436.2016.1186597
31. Kawasegi, N., Sugimori, H., Morimoto, H., Morita, N., and Hori, I. 2009. "Development of cutting tools with microscale and nanoscale textures to improve frictional behavior". *Precision Engineering*, 33(3), 248–254. doi: 10.1016/j.precisioneng.2008.07.005
32. Sasi, R., Kanmani Subbu, S., and Palani, I. A. 2017. "Performance of laser surface textured high speed steel cutting tool in machining of Al7075-T6 aerospace alloy". *Surface and Coatings Technology*, 313, 337–346. doi: 10.1016/j.surfcoat.2017.01.118
33. Niketh, S., and Samuel, G. L. 2017. "Surface texturing for tribology enhancement and its application on drill tool for the sustainable machining of titanium alloy". *Journal of Cleaner Production*, 167, 253–270. doi: 10.1016/j.jclepro.2017.08.178
34. Riveiro, A., Maçon, A. L. B., del Val, J., Comesaña, R., and Pou, J. 2018. "Laser surface texturing of polymers for biomedical applications". *Frontiers in Physics*, 6(16). doi: 10.3389/fphy.2018.00016
35. Petare, A. C., Mishra, A., Palani, I. A., and Jain, N. K. 2018. "Study of laser texturing assisted abrasive flow finishing for enhancing surface quality and microgeometry of spur gears". *The International Journal of Advanced Manufacturing Technology*. doi: 10.1007/s00170-018-2944-3
36. Sawant, M. S., Jain, N. K., and Palani, I. A. 2018. "Influence of dimple and spot-texturing of HSS cutting tool on machining of Ti-6Al-4V". *Journal of Materials Processing Technology*, 261, 1–11. doi: 10.1016/j.jmatprotec.2018.05.032
37. Li, Y., Deng, J., Chai, Y., and Fan, W. 2016. "Surface textures on cemented carbide cutting tools by micro EDM assisted with high-frequency vibration". *The International Journal of Advanced Manufacturing Technology*, 82(9), 2157–2165. doi: 10.1007/s00170-015-7544-x
38. Singh, B., Sasi, R., Kanmani Subbu, S., and Muralidharan, B. 2019. Electric discharge texturing of HSS cutting tool and its performance in dry machining of aerospace alloy. *Journal of the Brazilian Society of Mechanical Sciences and Engineering*, 41(3), 152. doi: 10.1007/s40430-019-1654-6
39. Pratap, A., Patra, K., and Dyakonov, A. A. 2019b. On-machine texturing of PCD micro-tools for dry micro-slot grinding of BK7 glass. *Precision Engineering*, 55, 491–502. doi: 10.1016/j.precisioneng.2018.11.004
40. Pratap, A., and Patra, K. 2020. Combined effects of tool surface texturing, cutting parameters and minimum quantity lubrication (MQL) pressure on micro-grinding of BK7 glass. *Journal of Manufacturing Processes*, 54, 374–392. doi: 10.1016/j.jmapro.2020.03.024
41. Kawasegi, N., Ozaki, K., Morita, N., Nishimura, K., and Yamaguchi, M. 2017. Development and machining performance of a textured diamond cutting tool fabricated with a focused ion beam and heat treatment. *Precision Engineering*, 47, 311–320. doi: 10.1016/j.precisioneng.2016.09.005
42. Elias, J. V., Venkatesh, N. P., Lawrence, K. D., and Mathew, J. 2021. Tool texturing for micro-turning applications – an approach using mechanical micro indentation. *Materials and Manufacturing Processes*, 36(1), 84–93. doi: 10.1080/10426914.2020.1813899

43. Wu, J., Yu, A., Chen, Q., Wu, M., Sun, L., and Yuan, J. 2020. Tribological properties of bronze surface with dimple textures fabricated by the indentation method. *Proceedings of the Institution of Mechanical Engineers, Part J: Journal of Engineering Tribology*, 234(10), 1680–1694. doi: 10.1177/1350650120940126
44. Zhang, K., Deng, J., Xing, Y., Li, S., and Gao, H. 2015. "Effect of microscale texture on cutting performance of WC/Co-based TiAlN coated tools under different lubrication conditions". *Applied Surface Science*, 326, 107–118. doi: 10.1016/j.apsusc.2014.11.059
45. Sun, J., Zhou, Y., Deng, J., and Zhao, J. 2016. "Effect of hybrid texture combining micro-pits and micro-grooves on cutting performance of WC/Co-based tools". *The International Journal of Advanced Manufacturing Technology*, 86(9), 3383–3394. doi: 10.1007/s00170-016-8452-4
46. Duan, R., Deng, J., Ai, X., Liu, Y., and Chen, H. 2017. "Experimental assessment of derivative cutting of micro-textured tools in dry cutting of medium carbon steels". *The International Journal of Advanced Manufacturing Technology*, 92(9), 3531–3540. doi: 10.1007/s00170-017-0360-8
47. Sugihara, T., and Enomoto, T. 2017. "Performance of cutting tools with dimple textured surfaces: A comparative study of different texture patterns". *Precision Engineering*, 49, 52–60. doi: 10.1016/j.precisioneng.2017.01.009
48. Ye, D., Lijun, Y., Bai, C., Xiaoli, W., Yang, W., and Hui, X. 2018. "Investigations on femtosecond laser-modified microgroove-textured cemented carbide YT15 turning tool with promotion in cutting performance". *The International Journal of Advanced Manufacturing Technology*, 96(9), 4367–4379. doi: 10.1007/s00170-018-1906-0
49. Hao, X., Chen, X., Xiao, S., Li, L., and He, N. 2018. "Cutting performance of carbide tools with hybrid texture". *The International Journal of Advanced Manufacturing Technology*, 97(9), 3547–3556. doi: 10.1007/s00170-018-2188-2
50. Alagan, N. T., Zeman, P., Hoier, P., Beno, T., and Klement, U. 2019. "Investigation of micro-textured cutting tools used for face turning of alloy 718 with high-pressure cooling". *Journal of Manufacturing Processes*, 37, 606–616. doi: 10.1016/j.jmapro.2018.12.023
51. Vignesh, G., Barik, D., Aravind, S., Ragupathi, P., and Arun, M. 2021. "An experimental study on machining of AISI H – 13 steel using dimple-textured and non-textured tungsten carbide cutting tools". *IOP Conference Series: Materials Science and Engineering*, 1017, 012021. doi: 10.1088/1757-899x/1017/1/012021
52. Sugihara, T., Kobayashi, R., and Enomoto, T. 2021. "Direct observations of tribological behavior in cutting with textured cutting tools". *International Journal of Machine Tools and Manufacture*, 168, 103726. doi: 10.1016/j.ijmachtools.2021.103726
53. Wang, X., Popov, V. L., Yu, Z., Li, Y., Xu, J. and Yu, H. 2021. Study on cutting performance of SiCp/Al composite using textured tool. *Research Square*. doi: 10.21203/rs.3.rs-302368/v1
54. Devaraj, S., Malkapuram, R., and Singaravel, B. 2021. Performance analysis of micro textured cutting insert design parameters on machining of Al-MMC in turning process. *International Journal of Lightweight Materials and Manufacture*, 4(2), 210–217. doi: 10.1016/j.ijlmm.2020.11.003
55. Lu, Y., Deng, J., Sun, Q., Ge, D., Wu, J., and Zhang, Z. 2021. Effect of micro textures on the cutting performance of circular saw blade. *The International Journal of Advanced Manufacturing Technology*, 115(9), 2889–2903. doi: 10.1007/s00170-021-07348-0
56. Wang, Q., Yang, Y., Yao, P., Zhang, Z., Yu, S., Zhu, H., and Huang, C. 2021. Friction and cutting characteristics of micro-textured diamond tools fabricated with femtosecond laser. *Tribology International*, 154, 106720. doi: 10.1016/j.triboint.2020.106720

5 Advances in MQL Machining

Milon Selvam Dennison
Kampala International University, Western Campus, Uganda

Kirubanidhi Jebabalan
Technical University of Liberec, Liberec, Czech Republic

M. Neelavannan
Dr. Mahalingam College of Engineering and Technology, Pollachi, India

M. Abisha Meji
Kampala International University, Western Campus, Uganda

CONTENTS

DOI: 10.1201/9781003284574-6

5.1 INTRODUCTION

With the advent of the Industrial Revolution in England from 1760 to 1820, there was a big transition from hand manufacturing to machine manufacturing, leading to a mechanized manufacturing system [1]. The manufacturing process is the most vital stage for the production of a good/product, and it is the most influential factor in the global economy. Any idea/concept gets the final shape after the manufacturing process [2]. At present, there has been a continuous requirement for producing finished components with ecological awareness and this requires a technology known as sustainable manufacturing [3]. According to the U.S. Environmental Protection Act, the definition of this technology is cost-effectively producing goods with minimal environmental degradation while conserving energy and natural resources, which, in turn, enhances product safety, thereby increasing its global competitiveness.

The United Nations industrial development organization released a report in 2018 that says that, in the global manufacturing economy, China is the major contributor followed by the United States and Japan [4]. However, manufacturing is also the major reason for greenhouse emissions (GHEs) because producing a final product often involves the emission of carbon and its oxides to the atmosphere. A research report states that carbon dioxide emissions have increased up to 1.5 times from 1995 to 2013. Moreover, the increasing demand for manufactured products calls for renewable resources that causes less harm to the ecological system. In this Industry 4.0 era, the rise of worldwide eco-awareness and strict directions has been imposed to carry out the sustainable manufacturing process, and this is the process of using natural resources efficiently and to generate minimal wastage in manufacturing [1, 5, 6].

Nowadays the researchers and scientists are putting their efforts to establish eco-friendly manufacturing for meeting the pollution control strategies of the government. Many territories have made their regulations rigorous in the environmental aspects and formed regulatory bodies'/organizations such as the European Environment Agency, the U.S. Environmental Protection Agency, the United Nations Environment Programme, the Global Alliance on Health and Pollution, the International Union for Conservation of Nature, the Earth System Governance Project, and others. India is also committed to protecting its natural resources and framed many important legislations for environmental protection, such as the National Green Tribunal Act, Environmental Protection Act, the Water/Air Prevention and Control of Pollution Act, and others [7].

In any manufacturing process, "machining" a part/component is inevitable because machining is the most significant manufacturing process. *Machining* is defined as a versatile process extensively used in the manufacturing industry to convert the given raw material to the desired shape. Sometimes it is used as a secondary operation wherein the workpiece will undergo processes such as casting or forging.

Fundamentally, machining is used to remove the excess material in the form of a burr to give the desired shape and surface of the intended object [2, 8].

The complex shapes/design of the part/component can be easily made with the aid of different machining processes, such as turning, milling, drilling, grinding, and so on. In the machining process, the instrument that facilitates the removal of excess material through direct mechanical contact is known as a "cutting tool" and the machine which delivers direct mechanical action via a cutting tool is known as the "machine tool." Based on the operations the machine tools are classified into many types such as turning center (lathe), milling center, grinder, and others [8].

5.1.1 Heat Generation in Machining

Machining a part/component can be achieved when the cutting tool comes in direct contact with the workpiece. To achieve the desired shape/design, the excess material on the workpiece is removed in the form of a burr, and this could be achieved when the cutting tool continuously rubs the workpiece. When there is continuous interaction (friction) between the workpiece and cutting tool, a huge sum of heat will be generated at the cutting tool–workpiece interface, and this heat tremendously damages the cutting tool by causing tool wear and damages the workpiece surface by resulting in microstructural distortion and poor surface quality. In any form of metal machining, nearly a loss of 95% of the gross energy was wasted in the form of heat at the tool–work interface. Although high-speed machining is desirable for bigger productivity, certain limitations, such as rapid tool wear due to excessive friction between the cutting tool and workpiece, make the process sluggish. So this form of excess heat at the tool–work interface should be abolished, and the remedy is applying cutting fluid or metalworking fluid (MWF) at the tool–work interface [9].

Figure 5.1 depicts the heat generation in the machining process. In any kind of machining operation, due to friction, heat would generate at various contact zones of the cutting tool and workpiece. This generated heat ultimately affects the machined product in context to surface topography cutting tool stability and end-product accuracy. Currently, various cooling methods are available to suppress the heat generated at the tool-work contact zones. Because water was the first coolant applied to the

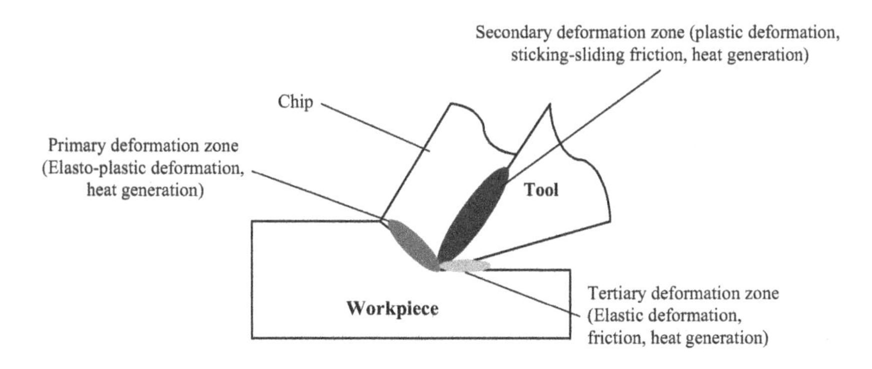

FIGURE 5.1 Heat generation in machining [10].

tool–work contact zones, which is the cheapest and widely available coolant, it has disadvantages, such as having a poor lubricating property and creating corrosion in the machined components. The technological development led to the use of mineral-based MWF, which provides both cooling and lubrication, for any machining operations. The global lubricant demand was 39.4 million tons in 2015, and it is expected to reach 43.9 million tons in 2022. The industrial lubricant market is divided into a few classes based on the applications. The widely used lubricants are gear oils, lubricants used in hydraulic applications, oil for engines, and MWF for machining. Among that above 80% contribution of the global lubricant market goes for MWF purposes [9].

5.1.2 Role of MWF in Machining

In today's manufacturing industry, the demand is increasing for low cost, high profitability, and better product quality. High productivity is naturally connected with high cutting velocity, feed rate, and depth of cut, which fundamentally leads to heat generation and the increase of the temperature at the cutting zone. Subsequently, the dimensional accuracy, tool life, and surface quality of the component will be getting affected. Nowadays, MWFs are considered a key element in any machining process to enhance the machining efficiency. The improved lubrication system in the machining process will produce better quality products.

The use of MWF in machining operation was first detailed in 1883 by F. W. Taylor, who observed that the cutting velocity was increased up to 33% from the traditional cutting velocity by applying the MWF at the tool–work interface. MWFs are specifically used to cool and lubricate moving parts in the metalworking process, particularly in stamping and machining operations commonly known as flooded lubrication/cooling systems [10]. In this topic, we discuss the necessity of cutting fluid why is needed and what type of cutting fluid is needed keeping in mind sustainable manufacturing purposes. It has been already explained in the previous section that when the temperature at the tool–work interface is quite high, it is destructive for both the cutting tools and the machined parts/components, and hence, it has to be controlled, without compromising on productivity and product quality. The following methods are generally employed for controlling machining temperature [10]:

- Superior selection of cutting tools, material, and geometry
- Selection of optimized machining parameters such as cutting velocity, feed, and depth of cut
- Appropriate selection of MWF

Among the earlier listed methods, the most important factor that forms the basis of this chapter is the MWF. In this chapter, we move toward the concept of sustainable manufacturing by explaining how flooded machining causes pollution to land and water, thereby causing environmental degradation. The next hazard of MWF is related to the degradation of the machine tool operator's health due to more exposure to mineral-based MWF leads to skin, digestive, and respiratory cancer. At the same time, workers who were exposed to oil-based MWF did not report chronic

TABLE 5.1
Advantages and Limitations of MWFs in Machining Operations [11]

Advantages	Limitations
• Increase in tool life. • Enhances workpiece quality. • Auxiliary chip removal by the MWF stream. • Power requirement and cutting force required are drastically reduced. • Increased cutting speeds and feed rates.	• The cost of mineral-based MWF is very high and also the storage and maintenance of the lubricant are quite cumbersome. • Improper maintenance of emulsion-based MWF over a prolonged period causes microbial formations. • Leads to environmental hazard when fed on land. • Causes health hazards to operators

diseases, but workers who were exposed to synthetic-based MWF reported respiratory diseases such as cough and phlegm. This has encouraged researchers to look into alternative solutions that will minimize the risk of workers who interact with the MWFs [11, 12]. The advantages and limitations of MWFs in machining operations are given in Table 5.1.

5.1.3 Flood Lubrication/Cooling System

The improved surface quality of the machined component/part and the economics of the manufacturing operation are becoming very significant in manufacturing. A flooded lubrication/cooling system is a predominant cooling practice in machining processes. The flooded lubrication/cooling system plays a vital role in planning and controlling the machining parameters. In spite of the advantages of applying a flooded lubrication/cooling system in the machining process the improper selection and usage of MWF is a great threat to the environment and the health of the machine tool operator. In a flooded lubrication/cooling system enormous quantity of MWFs has to be used (30–60 liters/hr). MWFs contribute to a significant part of manufacturing costs. It accounts for almost 8 to 16% of the total manufacturing cost in the production sector [9, 11]. The breakup of machining costs is depicted in Figure 5.2.

In order to overcome these disadvantages of flooded lubrication/cooling systems, many researchers and practitioners are working on developing alternate methods that enable less usage of MWFs without compromising on the machining performance. The increasing costs associated with the application and disposal of MWFs led engineers to explore new technology that uses minimal lubricant or relatively no lubricant.

5.1.4 Dry Machining System

A dry machining system means that no MWF is used in the cutting process, which completely avoids the negative influence of MWF, and is an ideal green cutting method. In 2003, more than 20% of the German manufacturing industries adopted a dry machining system. In terms of dry machining research and application, Germany

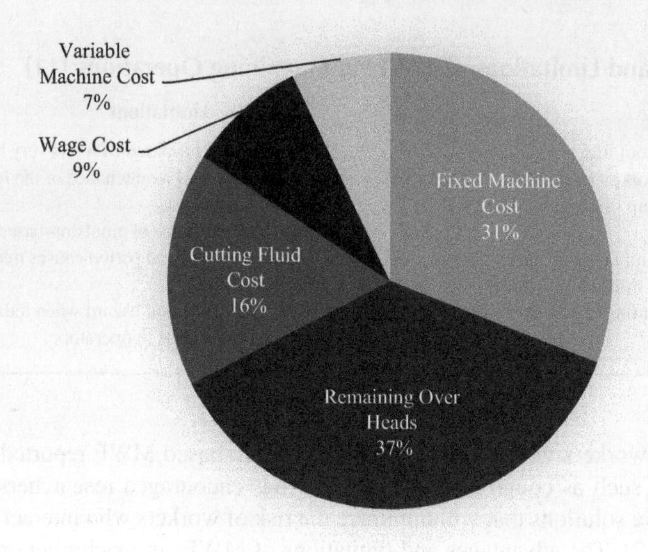

FIGURE 5.2 Machining cost breakup.

is an international leader. Dry machining is a new process to achieve clean and efficient machining, and it is the general trend of manufacturing technology development [5, 9, 12, 13]. The benefits of a dry machining system are depicted in Figure 5.3.

The dry machining system has the following characteristics:

- The chips formed are clean, pollution-free, easy to recycle and process.
- Dry machining eliminates the transmission, recovery, filtration, other related equipment, and costs related to MWF, which simplifies the production system.
- It also saves production costs by eliminating the costs related to chip and MWF treatment.

However, in the dry machining process, due to the lack of lubrication, cooling, and auxiliary chip removal by the MWF stream, the cutting tool bears a large load, a strong heat source, and a high cutting temperature. The tool and the workpiece are prone to thermal deformation; the tool's durability and the quality of the machined surface would reduce. Therefore, a blended technique that gives the merits of traditional flooded machining and dry machining would be the remedy for the previously addressed problems.

5.1.5 Need for Alternative System

The International Academy for Production Engineering advocates the use of green manufacturing technologies that are conducive to environmental protection, saving capital, increasing efficiency, and reducing waste/emissions. Unlike traditional

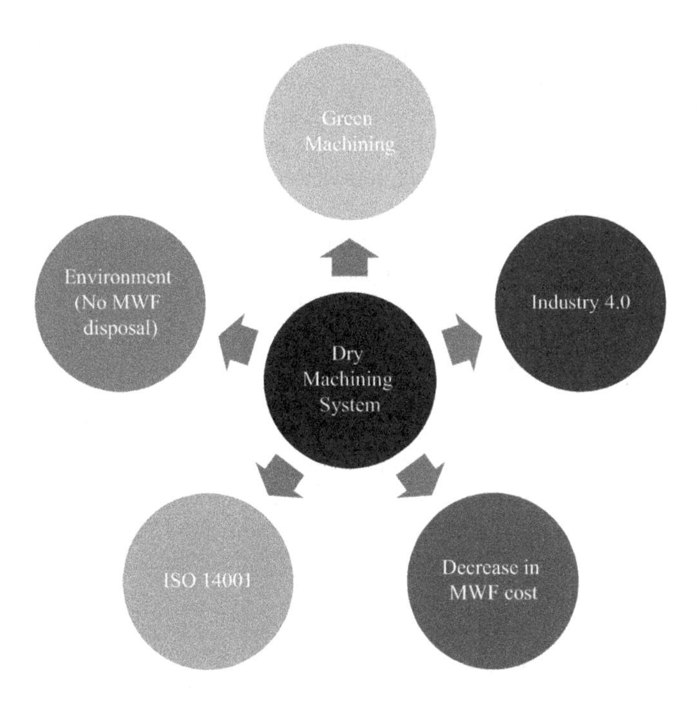

FIGURE 5.3 Benefits of dry machining system [14].

machining technologies, a dry machining system is an environmentally friendly cutting technology. Traditional dry machining has its advantages, such as being simple to employ with no cooling or lubricating devices. Also, its disadvantages include high temperatures resulting in quicker tool wear, leading to poor surface finish, which leads to the requirement of a novel lubricating technique. It also contributes to the plastic deformation of the workpiece as the temperature increases. Thus, flooded lubrication/cooling was introduced to overcome these disadvantages. Using coolant lubricants in abundance, in the form of flood constantly flowing at the tool-work interface proves to be very efficient in reducing the temperatures almost to room temperature and increasing the tool life. However, using coolant lubricants in flooded form creates an unclean environment with a lot of spillages and bad odor. Flooded lubrication is also not friendly to the environment because a huge volume of lubricants that cannot be recycled are scrapped into the surroundings. The preceding conditions and their consequences provide the scope of an innovative and new method of cooling and lubrication, bringing together the advantages of both dry machining and flooded machining.

In the initial years' industries, governments and researchers focused only on application and end use, but with time in the current scenario, there has to be a perfect balance between demands coupled with sustainability. This is because governments are signatories of climate change agreements and industries are obliged to follow ISO 14001 regulations and with the increasing awareness of the ecosystem, there has been a demand to provide an eco-friendly manufacturing process. This search has led us to a sustainable machining option known as minimum quantity lubrication (MQL). MQL, a method of using the coolant in the form of mist, along with a continuous supply of pressurized air, is likely to possess the advantages of dry and flooded machining. Researchers in the last decade explored this technique on how minute quantities of cutting fluids can be used for the various machining processes. The MQL system is a highly significant technology from an economic and ecological point of view.

A developing number of organizations are making the transition to the MQL systems to reduce costs, environmental impact, and health hazards. Moreover, the MQL system enhances tool life, and chips are discharged in a practically dry condition, thereby avoiding MWF recycling costs. However, if costs related to health and environments are not considered, many manufacturers consider the expenses for innovation high.

MQL has proved to be a reasonable alternative method for applying cutting fluid since it satisfies the necessities of green manufacturing. Water treatment costs, fewer ecological impacts and health problems of mineral-based cutting fluids deliberate the requirement for sustainable and biodegradable lubricants. Vegetable-based cutting fluids can be considered environmentally friendly since these fluids are sustainable and highly biodegradable.

5.1.6 MQL and Its Advantages

MQL is a technique that utilizes a very little quantity of MWF that is mixed with compressed air to generate a coolant mist, which is less visible to the human eyes and hence almost perceived as a dry machining system [9, 15]. Presently, many researchers and practitioners are working in the field of MQL for improving its effectiveness in machining. The advantages of MQL [9, 14, 16, 17] have been reported in the literature arena. Some of the benefits of MQL follow:

- A dry machining system allows the cutting zone to reach high peak temperatures, but MQL reduces the heat to a greater extent, as the compressed air performs half the work and a little amount of MWF is sufficient to enhance the machining performance.
- Because the MWF spillage does not occur; the whole workplace is prevented from becoming wet.
- Machining takes place like a dry machining system because the operator is free from any contact with the MWFs.
- In a flooded lubrication/cooling system, the MWF is allowed to be recycled within the machine tool for a prolonged duration that leads to the development of microorganisms (resulting in the foul smell) and it is harmful to the environment at its disposal, whereas MQL eliminates the recycling of coolants, which prevents the existence of microorganisms within the workplace, thereby eliminating any health risks to the operator.
- In a flooded lubrication/cooling system, the cyclic reusing of MWF leads to heating, cooling, interaction with rust, prolonged interaction with chips, and so on, all contributing to the deterioration of the MWF itself. The MWF gradually loses its lubricating ability as it loses its viscosity. In MQL, every

droplet of fresh and unused MWF treats the tool–work interface. Thus, the MQL machining can provide peak performance throughout the machining process.
- In a dry machining system, there is a possibility of red-hot metal chips flying toward the operator, whereas in the MQL system, the mist jet directs the metal chips away from the operator, thereby protecting the operator and having the pressure of the jet aiding the chip removal process on the whole.
- One of the key performance factors of MQL is that the MWF is used in very small quantities, thus eliminating wastage.

5.1.7 MQL: A Comparison with Other Systems

From the literature, it is evident that the MQL system perform better than the traditional machining system. The key observations from various literature are presented in this section.

- MQL gives a better product quality, and it is a possible contrasting option to the dry machining system. Cutting performance during MQL is superior to that of a dry machining system [6].
- The use of MQL in different machining operations costs less than that of traditional cooling techniques [6].
- MQL is an intense near-dry application of MWF and offers a favorable techno-economical machining condition [9].
- MQL aims at sustainable and green manufacturing in a modern machine shop [9].
- The cutting performance of MQL is superior to that of the conventional flooded machining, and MQL decreases the cutting temperature resulting in a better surface finish of the product [15].
- MQL is an effort toward utilizing fewer lubricants and coolants [14].
- MQL is one such development that turned out to be exceptionally valuable in promoting a greener manufacturing process [14].

5.2 SUSTAINABLE MANUFACTURING AND CLEAN MACHINING

A sustainable development strategy is a major strategy for mankind to get rid of the dilemma of lack of resources and environmental degradation and to realize the transition from industrial civilization to an ecological civilization [17, 18]. It contains at least the following two meanings: first is development, and then is the sustainability of development. Specific to the sustainable development of the machining field, it is not only to continue to develop in the direction of high efficiency, high precision, high flexibility, and automation but also to adopt clear-cutting methods to achieve "sustainable manufacturing" or "green manufacturing" [3, 5]. The concept of green manufacturing was first proposed by the Organisation for Economic Co-operation and Development. It is a modern manufacturing model that comprehensively considers the environmental impact and resource efficiency. The goal is to make the product throughout the product life cycle, from design to manufacturing to transportation,

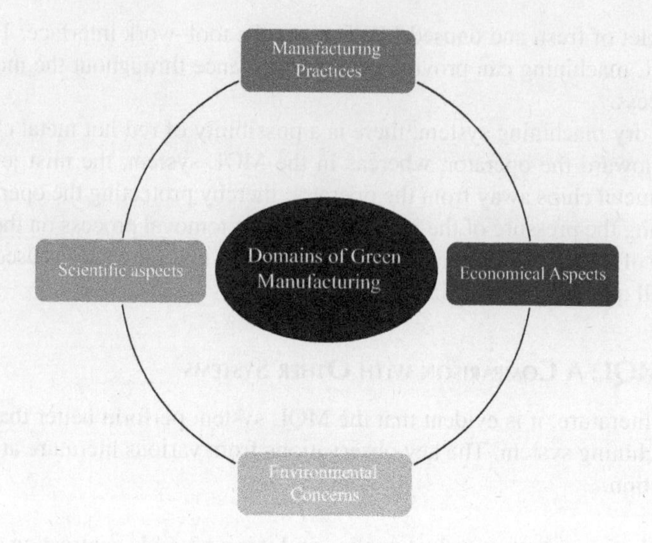

FIGURE 5.4 Domains of green manufacturing [3].

with an optimum utilization of resources. Environment, resources, and population are the three main issues of the universe. It is imperative to implement green manufacturing in the manufacturing industry in the 21st century. The domains of green manufacturing are depicted in Figure 5.4. In the manufacturing sector, traditional machining technology is facing challenges due to its serious pollution to the environment. Therefore, green machining has emerged as the times require.

5.2.1 Challenges in Sustainability and MQL

In this subtopic, we divulge toward understanding sustainability as a whole and challenges faced by MQL in particular. *Sustainable manufacturing* or *green manufacturing* or *clean manufacturing* or *environmentally sensible manufacturing* is defined as the process of creating products through an economically sound process, thereby minimizing environmental impact leading to conservation of energy and natural resources. Sustainable manufacturing leads to a better environment that positively enhances our economy and society [19]. Sustainable manufacturing practices are of different types and in the machining process; 'MQL' is the sustainable practice [19, 20]. The following are the challenges of sustainability:

- Global warming arises due to carbon dioxide (CO_2) emissions
- Contamination that occurs in air and water
- Population explosion and improper wastewater management
- Overexploitation of natural resources

Environmental concerns and strict parliamentary regulatory acts, along with the rise in global warming, are making governments move toward sustainable manufacturing [21]. MQL-enabled machining studies through various research articles appear to give a promising solution and generate better surface quality, lowered cutting forces, and extended cutting tool life in various machining processes; however, the following are several areas of concern based on various findings in the literature.

5.2.1.1 Cooling Effect

Researcher Benedicto et al. [22] have clearly stated in their research finding that MQL offers superior lubrication at the tool–work interface whereas the cooling system is not accommodated properly; hence, the machining process doesn't stabilize thermally. It is because water droplets present in the MQL mist absorb the generated heat from the tool–work interface that leads to tool breakage and distortion.

5.2.1.2 Workpiece that is Difficult to Machine

Benedicto et al. [22] have stated in their finding that the research related to the MQL machining of difficult-to-machine materials, such as nickel-based superalloys, titanium, and others, are minimal that will enable the future researchers to think that the MQL systems are inadequate against these metals.

5.2.1.3 Formation of Chips

In machining processes such as deep-hole drilling and milling, the clogging of chips happens at the tool and workpiece interface. The flooded lubrication/cooling system washes away those pieces whereas MQL systems prove to be inadequate in this case, and that leads to temperature buildup on the cutting tool, which, in turn, results in workpiece distortion and tool damage [23].

5.2.1.4 Selection of Optimized Parameters

In the MQL machining system, the dedicated data of the MQL operating conditions, such as flow rate, pressure, a mixture of air and oil, and nozzle design, along with essential machining parameters, such as feed, speed, depth of cut, and cutting tool selection, for different machining processes, are not readily available in the research fraternity that leads to the improper selection of parameters at the factory end [23, 24].

5.2.1.5 Economic Factors

The cost associated with shifting from the traditional machining process to MQL is said to be high due to the factors like the purchase of equipment, implementation, and subsequent maintenance [22, 23].

5.2.1.6 Machining and High-Speed Machining

In machining processes such as drilling and milling, MWFs experience difficulty in reaching the tool–work interface, resulting in improper cooling. High-speed machining is an area that faces issues in MQL; due to the high cutting force, the lubrication does not reach the tool–workpiece interface zone [24, 25].

5.2.1.7 Formation of Mist

In the MQL machining system, the formation of mist is unavoidable, and the machine tool operator is prone to generate oil mist, which leads to respiratory issues in them [22].

5.2.1.8 Lack of Numerically Simulated Data

The important parameters/factor in the MQL technique are flow rate and behavior of droplets on the contact surface. The investigation related to the numerical studies on the behavior of these parameters and their interaction with material that they come into contact with to overcome issues like landing, penetration, cohesion, and so on are insufficient [24].

By understanding the preceding challenges, more researchers are encouraged to develop and understand better lubrication for removing heat generation, a mechanism for flushing the generated chips, including a wider range of materials and cutting tools, and provide optimal parameters for machining various materials along with the required MQL parameters.

5.3 ADVANCES IN MQL

MQL machining does have certain challenges, but it comes with certain advantages in the aspects of the economy, ecology, sustainability, and process capability. Hamran et al. (2020) has clearly explained that there has been an increase in the number of articles published on MQL between 2014 and 2019; specifically in 2015, there was a spike of about 123.1%, and since then, there has been an increase in publications every year [5].

MQL machining systems have advanced in the various section right from understanding industrial challenges and aligning itself with Industry 4.0 standards, improvement in lubrication, advancement in the supply system, inclusion of several machining processes, cutting tools, and so on.

The alternative techniques to the traditional MQL machining system include MQL with solid lubrication (SL), ionic liquid (iL), biodegradable MWFs (vegetable-based), cryogenic cooling (CO_2 and LN_2), and cold compressed air (CCA). More efforts in the analysis of these alternative techniques from the fiscal point of view and environmental perspectives are desperately essential.

A hybrid MQL (MQL + biodegradable oil) can be used as an alternative to the already used mineral oil–based MWF. The use of nanoparticles, high-pressure fluid jet (gaseous based fluid) sprayed at the cutting zone would bring down the temperature [3, 26]. The use of subcritical pressure gases like CO_2 has been an MQL lubricant with a limitation of being a contributor to greenhouse gas emissions, and these lubricants can be used to process difficult-to-machine materials [3]. Some of the alternatives and advances in the MQL machining systems are presented in Figure 5.5.

5.3.1 Advancement of MQL Concerning Industry 4.0 Standards

With the advancement in technology, we are moving toward industry 4.0 standards, which involve artificial intelligence coupled with automation, which enhances productivity through autonomous feedback systems [3]. In this highly sophisticated manufacturing environment, we will see how MQL has fit into this system.

The 'Smart MQL' is an extension of the already-existing conventional MQL system but with certain modifications to suit the Industry 4.0 standards under

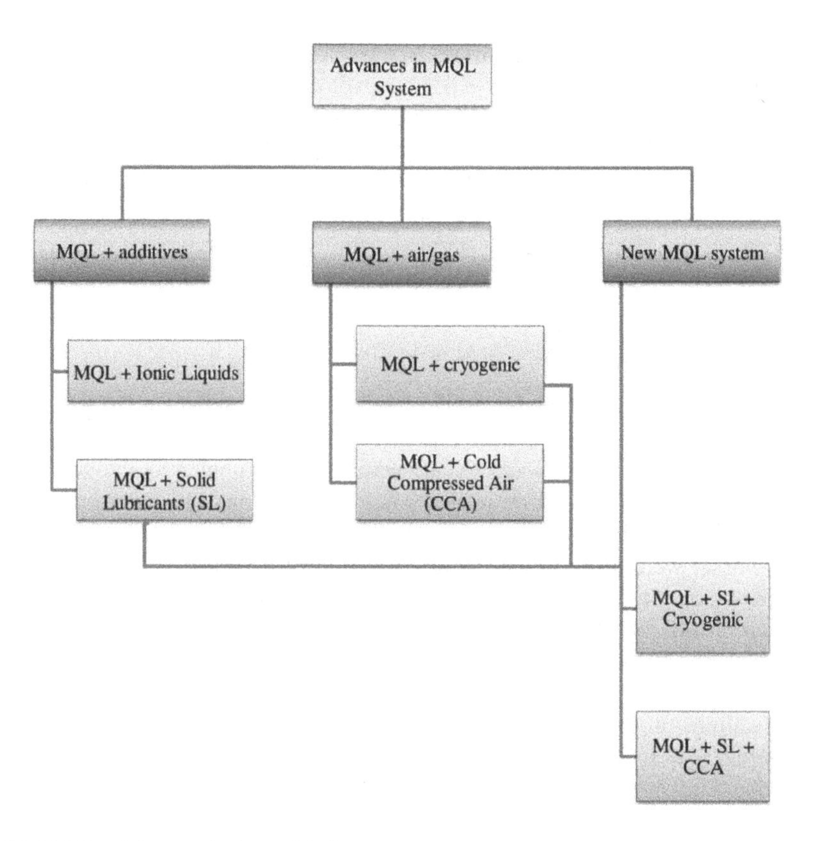

FIGURE 5.5 Advances in the MQL System

cyber-physical production systems (CPPSs), which come within the domain of smart factories that steer toward an ecologically sustainable manufacturing process. Here the MQL system is designed to deliver the required quantity of measured lubricant, in this case aerosols, to machining zones experiencing higher temperatures; the smart system comes into play by creating a wired or wireless feedback system that is feed to the master computer to avoid the formation of mist [3, 5]. Based on this smart system, Zheng et al. [27] performed ultrasonic assisted grinding. The experimental results of this research proved that the smart MQL parameters implemented in the grinding process resulted in reduced friction along with a better penetration rate.

5.3.2 Awareness among Researchers

As discussed earlier, there has been a significant increase in the number of publications made in the area of MQL. A simple search in ScienceDirect catalogue (https://www.sciencedirect.com/search?qs=MQL) can prove our point; for example, the number of articles published in 2019 is about 367 whereas in 2021, even before our year ends, there about 475 research articles have been published, which shows that there is a jump of about 29.42%. This shows that the researchers are interested to explore a new area in MQL. The development of research in MQL lies mostly in the area of sustainable manufacturing. Boswell et al. [25] have suggested in their research article that a few areas have been unexplored, such as MQL machining with a hybrid of cryogenic MQL system coupled with SL and CCA, with a few research papers being published in which researchers have predicted a new potential area of research. The new area of research that has to be taken into consideration is testing machining with an MQL system with SLs. The types of solid lubricants used in the advancement area are micro-fluids and nano-fluids while the type of cryogenic liquids used is carbon dioxide (CO_2) and liquid nitrogen (LN_2).

5.3.3 SMEET Framework

The prime goal of the MQL machining system is the implementation of sustainability without compromising on the profit-making ability of the company this is done through a framework, Social, Manufacturing, Environmental, Economical and Technology (SMEET), which emphasizes better utilization of resources through the manufacturing process. Conventionally three elements such as environment, social and economic were taken into account for the manufacturing process for effective implementation. More than three factors should be considered; therefore, five factors are taken into account in the present situation [28]. Valase et al. [28] have clearly stated in their research that the SMEET framework moves to sustainable modeling through manufacturing sustainability. The domains such as technology and manufacturing facilitate profit-making domains such as social, environmental, and economic aspects that lead to a human–environmental balance known as ecological balance. These five domain strategies provide a critical analysis and do exhibit better correlations with the "Environmental" and "Sustainable Manufacturing" domains. Here the implementation of MQL techniques does improve the ecosystem while maintaining profitability; this is seen as an advancement of MQL within the SMEET framework.

During machining of high-strength materials, such as titanium, nickel, magnesium alloy, structural ceramics, and hardened ferrous alloy, the advanced cooling and machining operation could not meet the SMEET framework to meet its standard cryogenic cooling as well as hybrid cooling should be used so that challenges arising due to economical, ecological, and social sustainability domain can be met.

5.3.4 MQL Supply System

The traditional MQL system consists of five parts, namely, a spray nozzle, a storage tank for storing MWF, necessary valves, a compressor, and a tube to carry fluid. MQL uses a technique of spraying a small amount of a mixture of oil and pressurized air at a flow rate of 100 ml/hr and sprays at the cutting zone directly; this is nearly 10^5 less than the volume of cutting fluid compared with the traditional flooding technique. Thus, the modified delivery system prevents flooding at the cutting zone [29]. The restructured delivery system has attracted the interest of researchers and in the past six years, the number of publications has increased up to 25.5% [3]. Most researchers concentrated on restructuring the nozzle or adding an extra nozzle, others tried changing the diameter of the nozzle, and some tried integrating MQL with a cleaning jet, as well as technological assistance such as electrostatic and ultrasonic

TABLE 5.2
Some of the Recent Papers in MQL Advancements

Advancement	References	Process	Workpiece	Tool	Environment	MWF & Additives	Responses
MQL + Solid Lubricants (SL)	Khanafer et al. [30]	Micro-drilling	Inconel 718	Solid tungsten carbide drill	MQL+ MQL nozzle was located at 45^0	ECOLUBRIC E200 vegetable oil (rapeseed oil)	Delamination factor
	Virdi et al. [31]	Grinding	Inconel 718	Alumina AA80 K5 V8 Carborundum Grinding wheels	Dry grinding and Minimum Quantity Lubrication (MQL)	Biodegradable oil-based Nanofluid MQL	Surface roughness and Grinding Energy
MQL + ionic Liquid (iL)	Singh et al. [32]	Grinding	AISI 52100 steel	White Aluminium oxide (Al_2O_3) grinding wheels	MQL	Coconut oil + iL	Surface roughness, grinding forces, specific grinding energy, grinding temperature
	Babu et al. [33]	Turning	Inconel 825	Coated carbide insert	MQL	iL	surface roughness, tool wear, cutting temperature, chip thickness
Hybrid MQL + new nozzle	Shokrani and Betts [34]	Milling	Titanium alloy (Ti6Al4V)	Coated solid carbide end mill	Hybrid MQL + new nozzle system	Tungsten disulphide (WS_2) nano particle + rapeseed oil	Tool life and cutting forces
MQL + biodegradable oil + Cryogenic	Shokrani et al. [35]	Milling	Titanium alloy (Ti6Al4V)	Coated solid carbide end mill	New hybrid MQL	Vegetable oil + Liquid Nitrogen (LN_2)	Tool wear and surface roughness
MQL + Cryogenic	Lai et al. [36]	Turning	17-4PH stainless steel	Coated carbide tool	MQL	Liquid Nitrogen (LN_2)	Surface roughness, flank wear
MQL + SL + Cryogenic	Yıldırım [37]	Turning	Steel alloy (AISI 420)	Coated carbide insert	New hybrid MQL	Graphene nano-platelet + LN_2	Tool–chip interface temperature, tool wear, and chip morphology
MQL + SL + Cold Compressed Air (CCA)	Sugiantoro [38]	Milling	ST60 steel and SS 304	HSS end mill	MQL+CCA	CuO/ Al_2O_3/ TiO_2 + water + Coconut oil and Palm oil	Tool wear and surface roughness
	Khalil et al. [39]	Turning	NiTi alloy	Coated carbide insert	MQL+CCA	Al_2O_3 + SolCut oil	Cutting force, tool wear, and surface roughness
Smart MQL	Bartolomeis and Shokrani [40]	Millling	Titanium alloy (Ti6Al4V)	Coated solid carbide tool	Electro Hydro-Dynamic Atomization (EHDA)-MQL	Pure rapeseed oil	Tool wear

vibrations, these advancements have been tested in the various machining process. Advancements in MQL have led to the machining of difficult-to-machine materials, such as nickel-based superalloys and titanium-based alloys [3, 5]. Some of the recent papers about MQL advancements are given in Table 5.2.

5.4 CONCLUSION

The International Society of Production Engineering has always paid full attention to ecological balance, environmental protection, and sustainable development and advocated the use of green manufacturing technologies that are conducive to environmental protection. Based on ensuring the comprehensive performance of MWFs, people are vigorously promoting the application of low-toxic, low-polluting green MWFs and striving to achieve "zero emission" of pollutants. By conducting an extensive literature survey based on factors/parameters like economical, ecological, and technological sustainability the following conclusions have been drawn.

Sustainable machining using MQL techniques, considering important factors such as the economic, ecological, and technical aspects of biodegradable oil, can be suggested as the superlative replacement. Coming to the technological front, "smart MQL" implementation helps in the proper implementation of industry 4.0 standards. Also, the technological improvement on the Industry 4.0 sector has been used to achieve the targets of sustainable machining through the "smart MQL" system by adjusting the internal delivery systems to deliver the required quantity. The quality of lubricants these modifications have been utilized in machining electrochemically assisted turning, ultrasonic-assisted milling, and laser-assisted milling for machining of harder grade, as well as difficult-to-machine materials. Taking into consideration the SMEET framework, very little research has reported on the aspects of social and technological factors in the domains of sustainability.

REFERENCES

1. Meisenzahl, R.R. and Mokyr, J., 2012. *9. The Rate and Direction of Invention in the British Industrial Revolution* (pp. 443–482). University of Chicago Press.
2. Selvam, M.D. and Senthil, P., 2016. Investigation on the effect of turning operation on surface roughness of hardened C45 carbon steel. *Australian Journal of Mechanical Engineering*, *14*(2), pp. 131–137.
3. Singh, G., Aggarwal, V. and Singh, S., 2020. Critical review on ecological, economical and technological aspects of minimum quantity lubrication towards sustainable machining. *Journal of Cleaner Production*, *271*(22), p. 122185.
4. Li, L., 2018. China's manufacturing locus in 2025: With a comparison of "Made-in-China 2025" and "Industry 4.0". *Technological Forecasting and Social Change*, *135*, pp. 66–74.
5. Hamran, N.N., Ghani, J.A., Ramli, R. and Haron, C.C., 2020. A review on recent development of minimum quantity lubrication for sustainable machining. *Journal of Cleaner Production*, *268*, p. 122165.
6. Selvam, M.D., Senthil, P. and Sivaram, N.M., 2017. Parametric optimisation for surface roughness of AISI 4340 steel during turning under near dry machining condition. *International Journal of Machining and Machinability of Materials*, *19*(6), pp. 554–569.

7. Naseem, M. and Naseem, S., 2020. *Environmental Law in India*. Kluwer Law International BV.
8. Dennison, M.S., Sivaram, N.M., Barik, D. and Ponnusamy, S., 2019. Turning operation of AISI 4340 steel in flooded, near-dry and dry conditions: a comparative study on tool-work interface temperature. *Mechanics and Mechanical Engineering*, *23*(1), pp. 172–182.
9. Selvam, M.D. and Sivaram, N.M., 2020. A comparative study on the surface finish achieved during turning operation of AISI 4340 steel in flooded, near-dry and dry conditions. *Australian Journal of Mechanical Engineering*, *18*(3), pp. 457–466.
10. Abukhshim, N.A., Mativenga, P.T. and Sheikh, M.A., 2006. Heat generation and temperature prediction in metal cutting: A review and implications for high speed machining. *International Journal of Machine Tools and Manufacture*, *46*(7–8), pp. 782–800.
11. Wickramasinghe, K.C., Sasahara, H., Abd Rahim, E. and Perera, G.I.P., 2020. Green Metalworking Fluids for sustainable machining applications: A review. *Journal of Cleaner Production*, *257*, p. 120552.
12. Sarhan, A.A., Sayuti, M. and Hamdi, M., 2012. Reduction of power and lubricant oil consumption in milling process using a new SiO2 nanolubrication system. *The International Journal of Advanced Manufacturing Technology*, *63*(5–8), pp. 505–512.
13. Dennison, M.S., Meji, M.A., Nelson, A.J.R., Balakumar, S. and Prasath, K., 2019. A comparative study on the surface finish achieved during face milling of AISI 1045 steel components using eco-friendly cutting fluids in near dry condition. *International Journal of Machining and Machinability of Materials*, *21*(5–6), pp. 337–356.
14. Krolczyk, G.M., Maruda, R.W., Krolczyk, J.B., Wojciechowski, S., Mia, M., Nieslony, P. and Budzik, G., 2019. Ecological trends in machining as a key factor in sustainable production–A review. *Journal of Cleaner Production*, *218*, pp. 601–615.
15. Rajarajan, S., Ramesh Kannan, C. and Dennison, M.S., 2020. A comparative study on the machining characteristics on turning AISI 52100 alloy steel in dry and microlubrication condition. *Australian Journal of Mechanical Engineering*, pp. 1–12. doi:10.1080/14484846.2019.1711019
16. Race, A., Zwierzak, I., Secker, J., Walsh, J., Carrell, J., Slatter, T. and Maurotto, A., 2021. Environmentally sustainable cooling strategies in milling of SA516: Effects on surface integrity of dry, flood and MQL machining. *Journal of Cleaner Production*, *288*, p. 125580.
17. Said, Z., Gupta, M., Hegab, H., Arora, N., Khan, A.M., Jamil, M. and Bellos, E., 2019. A comprehensive review on minimum quantity lubrication (MQL) in machining processes using nano-cutting fluids. *The International Journal of Advanced Manufacturing Technology*, *105*(5), pp. 2057–2086.
18. Dennison, M.S. and Umar, M.M., 2020. Data-set collected during turning operation of AISI 1045 alloy steel with green cutting fluids in near dry condition. *Data in Brief*, *32*, p. 106215.
19. https://www.epa.gov/sustainability/sustainable-manufacturing (as on 01/09/2021).
20. Ekinovic, S., Prcanovic, H. and Begovic, E., 2015. Investigation of influence of MQL machining parameters on cutting forces during MQL turning of carbon steel St52-3. *Procedia Engineering*, *132*, pp. 608–614.
21. Sangwan, K.S., Bhakar, V. and Digalwar, A.K., 2018. Sustainability assessment in manufacturing organizations: Development of assessment models. *Benchmarking: An International Journal*, *25*(3), pp. 994–1027.
22. Benedicto, E., Carou, D. and Rubio, E.M., 2017. Technical, economic and environmental review of the lubrication/cooling systems used in machining processes. *Procedia Engineering*, *184*, pp. 99–116.

23. Sharma, A.K., Tiwari, A.K. and Dixit, A.R., 2016. Effects of Minimum Quantity Lubrication (MQL) in machining processes using conventional and nanofluid based cutting fluids: A comprehensive review. *Journal of Cleaner Production, 127*, pp. 1–18.
24. Sharif, M.N., Pervaiz, S. and Deiab, I., 2017. Potential of alternative lubrication strategies for metal cutting processes: a review. *The International Journal of Advanced Manufacturing Technology, 89*(5–8), pp. 2447–2479.
25. Boswell, B., Islam, M.N., Davies, I.J., Ginting, Y.R. and Ong, A.K., 2017. A review identifying the effectiveness of minimum quantity lubrication (MQL) during conventional machining. *The International Journal of Advanced Manufacturing Technology, 92*(1), pp. 321–340.
26. Li, K., Aghazadeh, F., Hatipkarasulu, S. and Ray, T.G., 2003. Health risks from exposure to metal-working fluids in machining and grinding operations. *International Journal of Occupational Safety and Ergonomics, 9*(1), pp. 75–95.
27. Zheng, F., Kang, R., Dong, Z., Guo, J., Liu, J. and Zhang, J., 2018. A theoretical and experimental investigation on ultrasonic assisted grinding from the single-grain aspect. *International Journal of Mechanical Sciences, 148*, pp. 667–675.
28. Valase, K. and Raut, D.N., 2019. Mediation analysis of multiple constructs in the relationship between manufacturing and technology and environmental constructs in structural equation model for sustainable manufacturing. *The International Journal of Advanced Manufacturing Technology, 101*(5), pp. 1887–1901.
29. Osman, K.A., Ünver, H.Ö. and Şeker, U., 2019. Application of minimum quantity lubrication techniques in machining process of titanium alloy for sustainability: A review. *The International Journal of Advanced Manufacturing Technology, 100*(9), pp. 2311–2332.
30. Khanafer, K., Eltaggaz, A., Deiab, I., Agarwal, H. and Abdul-Latif, A., 2020. Toward sustainable micro-drilling of Inconel 718 superalloy using MQL-Nanofluid. *The International Journal of Advanced Manufacturing Technology, 107*(7), pp. 3459–3469.
31. Virdi, R.L., Chatha, S.S. and Singh, H., 2020. Machining performance of Inconel-718 alloy under the influence of nanoparticles based minimum quantity lubrication grinding. *Journal of Manufacturing Processes, 59*, pp. 355–365.
32. Singh, H., Singh, B. and Virdi, R.L., 2022. Exploration of Effectiveness of Ionic Liquid Adopted as an Additive to the Vegetable Oils. In *Sustainable Machining Strategies for Better Performance* (pp. 171–183). Springer, Singapore.
33. Babu, M.N., Anandan, V. and Babu, M.D., 2021. Performance of ionic liquid as a lubricant in turning inconel 825 via minimum quantity lubrication method. *Journal of Manufacturing Processes, 64*, pp. 793–804.
34. Shokrani, A. and Betts, J., 2020. A new hybrid minimum quantity lubrication system for machining difficult-to-cut materials. *CIRP Annals, 69*(1), pp. 73–76.
35. Shokrani, A., Al-Samarrai, I. and Newman, S.T., 2019. Hybrid cryogenic MQL for improving tool life in machining of Ti-6Al-4V titanium alloy. *Journal of Manufacturing Processes, 43*, pp. 229–243.
36. Lai, Z., Wang, C., Zheng, L., Lin, H., Yuan, Y., Yang, J. and Xiong, W., 2020. Effect of cryogenic oils-on-water compared with cryogenic minimum quantity lubrication in finishing turning of 17-4PH stainless steel. *Machining Science and Technology, 24*(6), pp. 1016–1036.
37. Yıldırım, Ç.V., 2020. Investigation of hard turning performance of eco-friendly cooling strategies: Cryogenic cooling and nanofluid based MQL. *Tribology International, 144*, p. 106127.

38. Sugiantoro, B., 2018. Performance evaluation of using water and bio oil-based nanocutting fluids under minimum quantity lubrication with compressed cold air during milling operations of steel. In *IOP Conference Series: Materials Science and Engineering* (Vol. 403, p. 012098). IOP Publishing.
39. Khalil, A.N.M., Azmi, A.I., Murad, M.N., Annuar, A.F. and Ali, M.A.M., 2019. Coupled effects of vortex tube hybrid cooling with minimal quantity reinforced nanoparticle lubricants in turning NiTi alloys. *The International Journal of Advanced Manufacturing Technology*, *105*(7), pp. 3007–3015.
40. De Bartolomeis, A. and Shokrani, A., 2020. Electrohydrodynamic atomization for minimum quantity lubrication (EHDA-MQL) in end milling Ti6Al4V titanium alloy. *Journal of Manufacturing and Materials Processing*, *4*(3), p. 70.

6 Nanofluids Application for Cutting Fluids

Mechiri Sandeep Kumar
B.V. Raju Institute of Technology, Narsapur, India

V. Vasu
National Institute of Technology, Warangal, India

CONTENTS

DOI: 10.1201/9781003284574-7

6.1 INTRODUCTION

Machining operations traditionally used conventional lubricants such as water and oil-based fluids to lubricate and cool the machining region. Conventional lubricants have environmental and human health–damaging effects. A well-known alternative to flood coolant and traditional lubricants is MQL with the addition of biodegradable oils. The solution is a comprehensive system that includes suitable guidelines for fluid use and increases heat transfer rates in metal-cutting operations. In recent years, scientists have developed new technologies in nanotechnology that have allowed for the characteristics of fluids to be increased with the use of nanoparticles. Due to the improved thermo-physical characteristics of the nanoparticles, the fluids' usefulness as coolants is mostly because of that. Many studies in the last decade have shown that the addition of ceramic-based microparticles results in increased thermophysical characteristics of cutting fluid. At the same time, however, further research efforts are still required to overcome the inherent instability and conductivity. Meanwhile, new hybrid nanofluids have been developed with improved thermal characteristics and stability to provide a greater heat transfer rate for machining.

6.2 MACHINING AND SUSTAINABILITY

The United Nations defines *sustainable development* as an activity that serves current needs without sacrificing the future abilities of the same group of people to satisfy their own requirements [1]. The U.S. Department of Commerce says that manufacturing firms may create sustainable goods by avoiding pollution, conserving energy and natural resources, and being profitable and safe for their workers, the community, and the customer [2].

Machining has seen significant development in developing nations, such as China, due to its use in manufacturing. As reported by the Freedonia Industry Research Institute, China's machine tool market has seen development in recent years. An analyst projected that the demand for machine tools in China would grow at a rate of 14.2% per year over the next 4 years [3]. Furthermore, a member of the European Committee for the Cooperation of Industrial Machine Tools also saw a significant rise in demand for industrial machinery in 2011 because of the 2009 economic crisis [4]. For instance, highly innovative, diverse, and high-precision industrial machinery accounted for about 33% of the worldwide industrial machinery market share. The rise in global demand for industrial machinery implies that the need for industrial machinery components is also rising rapidly.

There will certainly be economic advantages if this favorable trend continues. It would also raise usage, for instance, the demand for electricity and the use of cutting fluids both rose. Cutting fluids and energy supplies are mostly acquired via natural resources. Even though mineral-based cutting fluids have been utilized extensively, people are still largely dependent on them. As a result, the supply of natural resources has dwindled, as shown by this decrease in resource availability. Although there are limitations on natural resources, the increased use of energy and the usage of cutting fluids would generate environmental problems. Machining techniques typically use a lot more cutting fluid to enclose the cutting zones to prevent damage to the cutting tools and to lengthen the lifetime of the machine. In the usual course of events, fluid that has been removed is drifted back to a reservoir. However, the proportion that returns to a fluid reservoir has decreased because chips have vaporized, contaminants are embedded in machine tools, and the machined surface is covered with deposits. If workers breathe in the chemical vapor released by cutting fluid that has been vaporized, they may have health issues. To further enhance recyclability, the chips must be purified and processed again after exposure to the cutting fluid. Once you cut, the fluid must be disposed of correctly to avoid environmental damage and other creatures. A provision of extractors is provided, thus ensuring that the machine tool's workplace is safe, and treatment and purifying procedures of the fluid used in cutting are also included, which increases production costs. To produce the electricity needed to operate industrial machinery, the power is mainly produced by environmental assets such as fossil energy, coal, and other nonrenewable resources. The growth in demand for goods made by the machining operation means a rise in the demand for energy, and therefore an increase in the cost of energy. Not only that, but the manufacturing process for products also results in carbon emissions, which may further contribute to global warming. Because of this, a strong environmental policy is required. Additionally, the International Standard Organization (ISO) is one of the world's largest proponents of environmental management by introducing the ISO 14000 series. To be more competitive in the machining industries, machining processes had to be enhanced to meet all the issues associated with the machining process, and suitable business strategies had to be implemented to fulfill environmental regulations [5, 6].

Machining is made ecologically friendly without using cutting fluid. While the deployment of this technology has numerous benefits, there are also many difficulties that have yet to be solved. Efforts to further technical advancements are therefore continuing. So, as an alternative, the use of a backup technique, including certain near dry machining, is appropriate.

6.2.1 Dry Machining and Semi-Dry Machining

Some say that using cutting fluids to lower hot machining processes improves machining performance. Machining practitioners and scientists are becoming more concerned about the increased use of cutting fluids. This rising tendency impacts the economy and the environment. In addition to treatment expenses for disposal, a large amount of cutting fluid is utilized on the shop floor, resulting in higher purchase costs and additional treatment costs. At the same time, the usage of cutting fluid is quite aside from the economic concerns and creates environmentally damaging health problems [7–10]. Another adverse effect of using cutting fluid, as discussed earlier, is that it requires the use of alternative, new technologies that would be more economically sustainable feasible. Specifically, in industrialized nations with strict environmental regulations, the use of cutting fluid has forced the use of technological innovation that is far more economically sustainable feasible. The dry machining technique is well recognized as the most efficient way of metal removal. Dry machining does not utilize cutting fluids. Although the complete application is limited to specific processes, a dry machining technique is in the process of development. This is another technique that has been tried and researched extensively. Despite being less radical than dry machining, near-dry machining requires significantly lower cutting fluid quantities.

6.2.1.1 Dry Machining

Dry machining is a reference to machining techniques in which cutting fluids are not used. To decrease the friction, the temperature must be lowered in these sophisticated techniques at the tool chip and workpiece interface. With the assistance of a coating layer, friction minimization in this technique may be achieved. To avoid undesirable friction, the coating layer should have a low friction. Decisions regarding coating materials and coating technique affect the choice of coatings. To investigate the lubricating effect of various cutting fluid types that were combined using carbon tetrachloride as a lubricant improvement agent, Cassin and Boothroyd [11] employed this technique. They discovered that treating the workpiece material in CCL4 (i.e., carbon tetrachloride) causes the same cutting force magnitude as when the workpiece is dry machined. In this case, the use of dry machining was successfully used, as shown by the results of this experiment. When cutting speeds are low, the coating layer has a reduced coefficient of friction. However, as cutting speeds rise, the heat produced by the tool will reduce the impact of the coating layer. To further characterize the coating material, research was done to determine a suitable coating material and new approach. The investigation examines the properties of coating materials and how to apply coatings to the substrate materials to maximize performance. To this end, physical vapor deposition and chemical vapor deposition coating techniques have opened the door for shop-floor dry machining. One of the most notable shifts in the search for acceptable material for tool coating has been an increasingly steep upward trend for finding a suitable coating that may enhance the dry machining process [12–16]. Increasing the variety of cutting speeds suitable for dry machining extends the cutting versatility of the machine.

The researchers found that the selection of suitable tool coatings and coating technique may help achieve a higher surface finish on milling H13 tool steel, suggesting that tool coatings that are more effective will be able to serve as wear zone protection, therefore making dry machining a more viable option for greater cutting rates [17]. Additional, Dudzenski [18] explained the significance of the appropriate mix of tool material and coating material, particularly for dry machining performance, and pointed out that various cutting speeds may result in improved dry machining performance with the use of Inconel 718 alloy.

In order to enhance overall machining performance, it is essential to utilize precise cutting parameters. Researchers Diniz and Micaroni [19] discovered that careful parameter selection may enhance machining overall surface quality, tool life, and performance. Following the application on two AISI 1045 cylindrical shape workpieces that were each tempered to 55 HRC and 59 HRC, the principle in question was verified. The aforementioned research by Galanis et al. [20] shows that there is an optimal setting for dry machining performance utilizing workpiece material cut using AISI 422 stainless steel.

While workpiece and cutting tools are sheared, heat is generated. This additional heat production cannot be ignored when doing a dry cut analysis. Reducing temperature means reducing the speed of the cutting, which is a well-known concept. The heat produced at the contact zone will rise also as cutting speed increases. Due to the high heat produced, the heat will escape through the rigid surface that is part of the removal process, which includes the workpiece material. In this way, the microstructure changes take place in the solid body instead of the workpiece material. Consequently, it is feasible to increase burr size, and hence weaken workpiece material structural integrity [21]. Also, the residual stress produced by higher temperatures would also reduce the mechanical strength of machined parts and components [22].

Based on the studies listed earlier, it appears that a variety of dry machining applications (such as workpiece material, operating conditions, and cutting tools) could benefit from optimizing a superior friction characteristic with cutting tools so that cutting fluid completely takes over the job of cutting. For that reason, it must be able to do better than the level of cutting fluid (together with component quality and machining time) that may be reached when using fluid cutting [23, 24].

The adoption of the near-dry machining has proved to be very popular due to the use of the MQL process. This is further assisted by MQL systems that are commercially accessible. Because the use of the term *MQL* has spread throughout the academic and scientific communities, one study group retains its original lubrication [25], while another other uses machining in a near-dry state [26–28]. The name "MQL" is an unfortunate misnomer because it overlooks the feasibility of dry machining, and the lower threshold for MQL is dry machining. It would be more suitable to refer to it as low amount lubrication. Clear agreement on nomenclature is required. Based on this, MQL is employed in this research.

6.2.1.2 Semi-Dry Machining

Dry machining is considered to be the most sustainable kind of machining. Furthermore, owing to the technical limits of tooling, limitations on its implementation were introduced because of high cutting rates. A bridging technique is required to close the gap among wet machining as well as dry machining. The use of a recognized technology referred to as the near-dry machining has been suggested as a means for advancing sustainable machining practices. A near-dry machining technique is often referred to as using a minimum amount of liquid or replacing current cutting fluids with liquid or gas. Although the future of certain dry machining techniques has been investigated with new technology, past technologies remain the best options of near dry machining processes. Cryogenic machining, high-pressure jet machining (HPJ), and MQL are only a few of the machine cutting techniques used by machine shops. To fully mitigate the usage of cutting fluid, detailed study needs to be conducted on each technique. To cool the cutting zones as much as possible, liquid gas is used, such as nitrogen, helium, hydrogen, neon, air, and oxygen. These four techniques have recently been used in connection with cryogenic machining. Indirect cryogenic cooling [29], indirect cryogenic treatment of cutting tools [30], and precooling the workpiece [29] are techniques for this task.

Many studies have investigated cryogenic machining's effectiveness with various materials, such as low-hardenability materials [31–33] and difficult-to-cut materials. Due to its capacity for keeping temperatures at the cutting contact below the critical temperature of cutting tool materials, cryogenic machining is one of the more viable techniques besides cutting difficult-to-cut aerospace alloys as well as other hard-to-cut materials [34, 35]. Employing liquid nitrogen for cryogenic machining operations offers a substantial decrease in rake face heat production because of shorter tool–chip contact lengths [36]. It so increases the effectiveness of the machine's ability to cut hard-to-cut materials at greater cutting speeds. However, despite the ability of cryogenic machining to provide a substantial drop in the cutting-zone temperature and decrease the overall cutting zone length, lubrication has many challenges ahead of it. Hong and Ding [37] performed a sliding test to see whether cryogenic machining employing liquid nitrogen provides lubrication at the cutting contact. They discovered that an effect on the microscale of hydrostatic pressure, which varies with different temperatures, may be used to provide a lubricating action. Unlike conventional cutting fluids, liquid nitrogen doesn't have a lubricating effect, and it is readily vaporized.

In addition, because of the absence of lubrication for cryogenic machining, it is not only an unsustainable technique, but it is also an unnecessary way because of the advantages that may be obtained from its use. The positive aspects of operating a green business include improved environmental sustainability, increased material removal rates, better throughput, and improvements to machine component quality due to such a cleaner environment. Cryogenic machining, which requires significant investment and running expenses, making it unfeasible for modest production quantities.

HPJ machining is an environmentally friendly technology that has just begun to be of interest to machining scientists. This approach is somewhat similar to the conventional overhead technique, but the manner of delivery is different. Traditional machining methods need small amounts of low-pressure air to assist transport the cutting fluids to the cutting zone, while high-pressure jet machining requires high-pressure air to move the fluids to the cutting zone. In higher air pressure, the particles' speed goes up, which means that the particles' size decreases. Weaker cutting fluid forces combined with higher particle speeds and smaller particle sizes will improve the penetration capability of the fluids, and this means that the particles are more likely to be able to reach the operating region and form the lubricating action. Because evaporative cooling is more economical than convective cooling, cutting fluids with evaporative cooling capacity will have better cooling capabilities than those with just convective cooling capability [38]. While the primary advantage of high-pressure jet coolant in machining is improving cutting tool performance, a secondary benefit is high spindle speeds. A 250% increase in tool life was obtained by utilizing fluids that are water-soluble cutting fluids [39]. While conducting additional studies on the impact of high-pressure flushing via the rake face of tools, Wertheim et al. [41] discovered that the air pressure could be raised to 25 atm in this instance.

Rahman et al. [42] have discovered that high-pressure coolant may also provide a superior surface polish. Although HPJ machining is somewhat sustainable owing to the use of significant quantities of cutting fluids as a cooling and lubricating medium, it should be kept in mind that the technique has not been verified as sustainable. In order to decrease the quantity of cutting fluid used in HPJ machining, Kovacevic et al. [43] proposed the use of a narrow jet. Unfortunately, there is still a limit on how this technique may be implemented.

TABLE 6.1
Merits and Demerits of Different Environmentally Friendly Cooling Strategies Employed in the Machining of High Strength Alloys under Continuous Cutting Processes

Dry	Minimum Quantity Lubrication (MQL)	High Pressurized Cooling (HPC)	Cryogenic Cooling
No lubricant/ coolant required, no need for coolant disposal	Very small amount (ml/ min) of lubricant/ coolant is involved	Very high quantity of lubricant/ coolant is required	Generally, lubricants/ coolants are compressed air or liquid nitrogen
Poor temperature control in the cutting zone	For machining difficult to cut materials not effective due to low cooling capacity [46]	Superior temperature control in cutting zone	Superior temperature control in cutting zone [45]
Poor chip removal and dust formation	Better chip removal as compared to dry	Superior chip removal	Better chip removal as compared to dry
Poor tool life	Extended tool life at higher cutting speeds as compared to dry but highly dependent on arrangement of nozzle	Improved tool life [44]	Improved tool life [45]
Poor surface finish	Improved surface finish but highly dependent on arrangement of nozzle	Improved surface finish but highly dependent on arrangement of coolant delivery system and pressure involved	Inconsistency in surface integrity as it is highly dependent on the pairing of tool and workpiece materials [46]

Colloquially, MQL is another sustainable development bridge. This technology is a realistic option for shop-floor implementation. In other words, Ford began using the MQL system at the factory in May 2005. This kind of solution has the advantage of simplicity in its supply systems but is challenged by the fact that it improves machining performance with a little amount of cutting fluid. Table 6.1 outlines the many benefits and drawbacks associated with various machining techniques when working with high-strength alloys.

6.3 APPLICATION OF MQL IN MACHINING PROCESSES

In many machining operations, particularly primary processes like turning, drilling, milling, and grinding, the use of the MQL technique may be found. Furthermore, the MQL technique may also be used to cut a wide variety of workpiece material, like aluminum, steel, hardened material, and hard-to-cut material.

Sadeghi et al. found that surface quality characteristics (such as grindability) of AISI 4140 depended on surface grinding forces [47]. The researchers discovered that by using MQL, cutting forces were reduced and that a higher surface quality could be achieved. Separate research compared various workpiece materials and found that the MQL technique together with resin bond corundum provided the highest performance in grinding, which was found to be in contrast to wet and dry machining. These findings were corroborated by the earlier results that had shown Da Silva et al. [48] was correct. MQL is suitable for grinding operations since it has adequate surface integrity.

In addition, closed-type drilling operations have certain MQL application restrictions. The drill bit needs lubrication when drilling into the earth because of the extreme difficulty of airborne particles supplying cooling or lubrication to the drill tip. Internal cooling/lubricating channels delivering coolant are recommended for proper functioning. Compared to the use of external coolant, they decrease cutting temperature by about 50% [49].

Heinemann [50] concluded that changing the coolant delivery method from continuous to discontinuous may enhance the use of external twist drills. With this other technique, the cooling capacity will be raised, and tool wear will be decreased. Low-viscous cutting fluid also aids penetration while cooling and lubricating.

Machining high thermal conductivity materials such as aluminum should be done using MQL, rather than dry machining. It would impair the hole quality because of thermal expansion at high cutting temperatures. Thus, MQL is preferred, and flood-cooling efforts should be avoided when they're implemented. By the proper use of machine variables, the performance will be equivalent to that of flooding for drilling aluminum [51].

MQL is studied in terms of cutting temperature, chips, and dimensional accuracy in AISI 1040 steel turning by Dhar [33]. Using the MQL program, they found positive tool chip interaction. MQL performed better than wet and dry machining on hardened AISI 4340 steel, according to a study by Varadarajan [40].

6.4 MQL AND MACHINING PARAMETERS

The MQL equipment is easy to use, which is specifically developed to be compatible with many kinds of manufacturing machinery. On the other hand, appropriate MQL installation on machine tools is necessary. To better meet your requirements and financial limitations, the equipment must be selected.

Although the equipment is appropriate for the workplace, some of it has not been entirely sufficient since it was not prepared to meet the unique requirements of the workplace. This was done for advertising purposes as indicated by Astakhov [52]. Tests are therefore still necessary, even with all this information, to hunt for the best possible results for various needs and that is why tests are still required to search for the optimum MQL parameter set (i.e., flow rate and nozzle positions) as well as for the best possible cutting condition besides given workpiece material and chosen cutting tool.

6.5 IMPROVING MQL LUBRICATION

It was discovered that MQL provided superior results when used for machining, even though dry lubrication was preferred. The positive outcomes of these tests are the direct consequence of the microliter lubrication which was supplied as part of the MQL system to the cutting zone. Supplying lubricant to the cutting zone decreases the friction between the chip and the workpiece. To decrease the amount of heat that is generated during machining, lubricants with high thermal conductivity are used.

Nanotechnology is a rapidly expanding and wide-ranging field of study that has seen an increase in global research and development activities in the last few years.

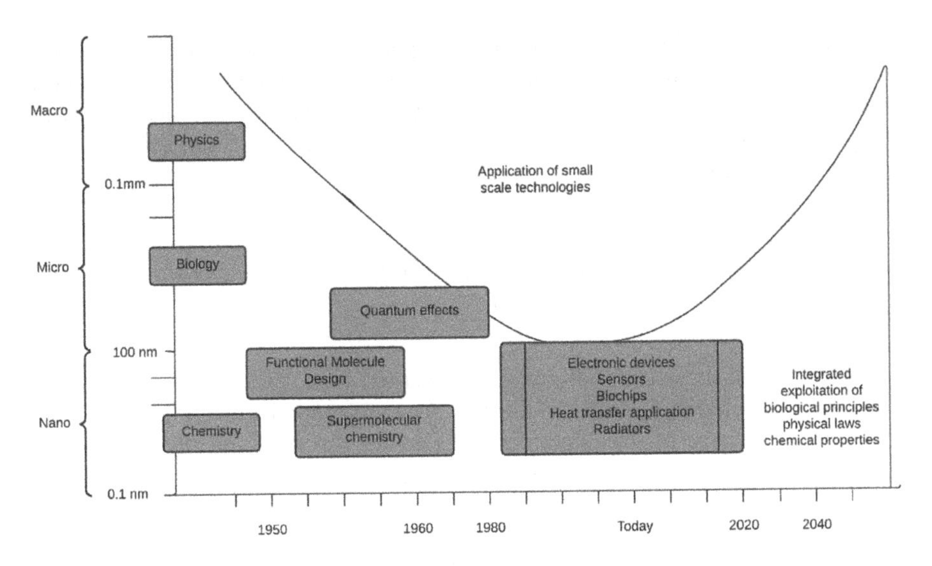

FIGURE 6.1 Small scale technologies and the convergence of sciences [53].

This has the capacity to fundamentally change the methods by which materials and products are produced, as well as the capability to access many functions. It's predicted that nanotechnology would have a major commercial effect in the future, as illustrated in Figure 6.1.

It became possible to do over 4 million multiplications per second on desktop PCs by the late 1980s. Major developments in technology have improved energy efficiency in a variety of sectors, including heavy-duty vehicle engine, as well as microelectronic cooling. Coolant fluids, such as water, oil, and ethylene glycol mixture, are poor heat-transfer fluids because of their conventional uses. Metal thermal conductivities are orders of magnitude greater than fluid conductivities. This is one good example of why copper has a thermal conductivity that is 700 times higher than that of water. Metallic liquids have a higher heat conductivity than nonmetallic liquids. Thus, substantially greater thermal conductivity may be anticipated to be achieved by using a fluid with a significantly high thermal conducting fluid that is a metallic liquid and a fluid that contains suspended solid metallic particles [54, 55].

Non-Newtonian fluid, invented by the Argonne National Laboratory in 1995, is a kind of working fluid used in heat transfer applications. It is made up of tiny particles of solid matter that are no more than 50 nanometers in diameter. Heat-transfer technologies like the micro-refrigerator [56], spray cooling [57], and heat pipes [58] are relatively new but very efficient and have excellent cooling rates. Although nanofluid may be utilized in microchannels, as well as large-scale cooling such as heavy-duty vehicle engines, it is better suited for microchannels because it requires fewer resources. The ability to create nanoparticles down to a few nanometers in diameter is owing, in part, to current manufacturing techniques that allow the creation of nanoparticles down to that size. Due to their appealing features, nanoparticles may give their support to the concept of a nanofluid. The first thing to remember is that nanoparticles are resistant to sedimentation. The sedimentation speed of a particle relies on the size of the particle, the viscosity of the base fluid, and the density differential between the particle and the base fluid (Equation 6.1). To prevent sedimentation, make sure particles are as small as possible, and the speed is eliminated with particles measuring a billionth of a meter.

$$U_s = \frac{2d_p^2}{9\mu_f}(\rho_p - \rho_f)g \tag{6.1}$$

Moreover, nanoparticles have an overall surface area that is 1,000 times greater than that of microparticles. This characteristic significantly increases the fluid's heat conduction because of the direct contact between the fluid and the surface. The higher the potential for increasing heat exchanges, the smaller the particle. This was discovered after experiments were conducted at Argonne National Laboratory, where nanoparticles and base fluids were used.

6.6 NATURE OF HEAT TRANSFER IN NANOPARTICLES

The mean free path of a phonon in an Al2O3 crystal, according to Keblinski Phillpot et al. [59], is 35 nm. Phonons may diffuse in 10-nm particles but must travel in a ballistic manner. High-packing fractions, soot-like particle assemblies, and Brownian motion of the particles will be required to maintain the spacing between nanoparticles small enough in order for the ballistic phonons started in one nanoparticle to remain in the liquid and reach another nanoparticle.

Xie et al. [60] discovered in their study on alumina nanofluids that, as the particle size approaches the mean free path of phonons, the thermal conductivity of the nanofluid decreases. The findings of this study, however, were not in accord with many of the previous experiments from other research organizations.

The researchers found that the large change in thermal conductivity, from ballistic heat conduction in nanotubes to diffusion heat conduction in liquid, occurs as the result of a significant limitation of ballistic heat conduction. Convection and diffusion contribute to nanotube dispersions' high thermal conductivity.

Despite the fact that ballistic heat conduction cannot be the only explanation for nanoparticle-enhanced thermal conductivity, sluggish heat diffusion in liquid creates a barrier, and barrier-based mechanisms must be involved in explaining nanoparticle-enhanced thermal conductivity. Without the use of other processes, we wouldn't comprehend the enhanced thermal conductivity of nanofluids. Nanotechnology has led to the emergence of fluids having very high thermal conductivity known as nanofluids. Ultrafine solid particles (10–100 nm) are a stable colloidal suspension of the base fluid in nanofluids. Ultrafine particles are added to the base fluid to improve fluid characteristics [61]. Due to the improved thermophysical characteristics of the nanoparticles, the fluids are widely applicable as coolants [62]. In general, the study has shown that either metal or metal oxide nanoparticle-based nanostructured materials may be suspended employing surfactants into the base fluid [63, 64].

6.7 NANOMATERIALS FOR NANOFLUIDS

However, the properties of ceramic nanoparticles are as awful as one would expect. They are very stable and chemically inert but have poor thermal conductivity when they are used in base fluids to enhance thermal conductivity. Because of their lower stability and reactivity, metallic nanoparticles, such as copper, have better heat conductivity. To help solve this issue, a copper-containing nanofluid was created by adding copper nanoparticles to an alumina matrix, which increased the thermal characteristics without compromising the nanofluid's stability [65]. The results in Table 6.2 show the different nanofluids that are utilized in metalworking fluids.

Nanofluids have been explored in the past and their use can boost the performance of MQL systems. Another property of nanofluids that plays a critical role in their use as cutting fluids is thermal conductivity. Particle volume fraction, particle size, temperature, and interfacial characteristics determine the conductivity increase. The following are the observations on the percentage increases in thermal conductivity, dependent on particle type and base fluid.

6.7.1 Nonmetallic Nanoparticle Dispersion

Prior work on nanofluids used nonmetallic nanoparticle dispersion without metallic nanoparticles. Nonmetallic nanoparticles (which are also called metal nanoparticles) include Al_2O_3, CuO, TiO_2, MgO, ZnO, SiC, and others. Unlike metallic nanoparticles, these nanoparticles have a lower thermal conductivity, which results in a lower density. It is easier to suspend nanoparticles in water and the nanofluids also have greater stability. Furthermore, Table 6.3 presents the summarized findings on the increase of nanofluid thermal conductivity as a result of ceramics nanoparticle dispersion.

TABLE 6.2
Report of Various Articles about Nanofluids as Metalworking Fluid Thermophysical Properties of Nanofluids

Type of Nanoparticles	Base Fluid	Workpiece	Investigated Parameters
Al_2O_3 [66]	Vegetable oil	Inconel 600 alloy	– Cutting temperature – Surface roughness, – Tool wear
Graphite [67]	Water soluble oil	AISI 1040	– Average chip–tool interface Temperature, – Tool wear
CuO [68]		AISI 1040	– Workpiece temperature
CNT [69]		AISI 1040	– Cutting temperature – Surface roughness, – Tool wear
Nano graphite [70]	Water soluble oil	AISI 1040	– Cutting temperature – Surface roughness, – Tool wear
MWCNT [71]	Coconut oil	Martensitic Stainless Steel	– Cutting temperature – Surface roughness
MWCNT [72]		AISI 1040	– Cutting temperature – Finite element analyses (FEAs)
Al_2O_3	Water	Nimonic 90	– Cutting forces – Surface roughness – Tool wear
Silver [73]	Water	Nimonic 90	– Cutting forces – Surface roughness – Tool wear

TABLE 6.3
Summary of Nanofluid Thermal Conductivity Enhancements over Base Fluids with Ceramics Nanoparticle Dispersion

Investigator	Nanofluid	Concentration (vol%)	Size (nm)	Observation Thermal Conductivity Enhancement in %
Masuda H [74]	Al_2O_3 + water	4.3%	13	30
Lee S [75]		4.3	33	15
Xie H [76]		5	68	21
Lee [77]		4.3	38.4	9
		0.3	30±5	1.44
Masuda H [74]		4.3	13	30
N. Putra [77]		4	38.4	9.4%–21°C/24.3%–51 °C
		4	131	24%–51 °C
C.H. Chon [78]		1	11	24.8%–70 °C
		1	47	10.2%–70 °C
		1	150	4.8%–60 °C
		4	47	28.8%–70 °C
X. Wang [79]		5.5	28	16
C.H. Lie [80]		6	36	28.2
		6	47	26.1
X. Zhang [81]		5	20	15
E.V. Timofeeva [82]		5	11	8
		5	20	7
		5	40	10
Lee [80]	Al_2O_3 + EG	4.3	38.4	18
Wei Yu [83]		5		28.2
Zhou L P[84]		0.4	50	16
Das [85]		4	28.6	13
Masuda H [74]		3.4	24	11
Wang et [79]		10	23	30
Masuda H [74]	CuO + EG	4	24	20
Wang [79]		15	23	50
Murshed [86]	TiO_2 + water	5	15	30
Duanfthongsuk et al. [87]		2	21	7
Murshed [86]		5	10	32.8
Wei Yu [83]	TiO_2 + EG	5		27.2
	MgO + EG	5		40.6
	ZnO + EG	5		26.8
	SiO_2 + EG	5		25.3
Xie H. [88]	SiC + water	4.2	26	16
	SiC + EG	4	600	23

6.7.2 Metallic Nanoparticle Dispersion

Due to having a greater degree of thermal conductivity than base fluids, metallic nanoparticles have been discovered to increase the thermal conductivity of fluids. Gold, copper, iron, and aluminum are widely used metallic nanoparticles. Nanofluid thermal conductivity improvements, in addition to base fluids with dispersion of metallic nanoparticles, are shown in Table 6.4.

TABLE 6.4
Summary of Nanofluid Thermal Conductivity Enhancements over Base Fluids with Metal Nanoparticle Dispersions

Investigator	Nanofluid	Concentration (vol%)	Size (nm)	Observation Thermal Conductivity Enhancement in %
Choi SUS [89]	Cu+EG	0.3	10	40
Assael [90]		0.48	120	3
Eastman [91]		0.3	<10	40
Xuan Y [92]	Cu + water	7.5	100	78
Hong T K [93]	Fe + EG	0.55	10	18
Patel H E [94]	Au + water	0.026	10–20	21
Putnam [95]	Au + Ethanol	0.6	4	1.3 ± 0.8
	Au + Tolonene	–	1.65	–
Patel H E [94]	Ag + water	0.001	60–80	17
Patel H E [94]	Ag + Citrate	0.001	10–20	3
Murshed [96]	Al + water	5	80	>40

6.7.3 Carbon Nanotube Dispersion

Carbon nanotubes are structures composed of tubes with a length of about 1 nanometer and a diameter of about 1 nanometer. They have hundreds of concentric carbon shells with the intervening carbon shell network intimately associated with honeycomb-like structure. They have the greatest heat conductivity, but they have much less density. Table 6.5 details the effects of carbon nanotubes' dispersion on nanofluid thermal conductivity.

TABLE 6.5
Summary of Nanofluid Thermal Conductivity Enhancements over Base Fluids with Carbon Nanotubes Incorporation

Investigator	Nanofluid	Concentration (vol%)	Size (nm)	Observation Thermal Conductivity Enhancement in %
Choi SUS [89]	MWCNT+ water	0.6	D = 100 nm, L = 70µm	38
Xie H. [60]		1	D = 15 nm, L = 30µm	7
Assael MJ [97]		0.6	D = 15–130 nm, L = 50µm	38
Xie H. [60]	MWCNT+EG	1	D = 15 nm, L = 30µm	13
Liu [98]		1	D = 20–30 nm, L = ---	12.4
Xie H. [60]	MWCNT+ decene	1	D = 15 nm, L = 30µm	20
Assael MJ [97]	MWCNT + oil	1	D = 25 nm, L = 50µm	150
Liu [98]		2	D = 20–30 nm, L= -----	30
Biercuk [100]	CWCNT + Epoxy	1	D = 3–30	175
Wei Yu [83]	GNS	5		61

TABLE 6.6
Summary of Nanofluid Thermal Conductivity Enhancements over Base Fluids with Alloy Nanoparticle Incorporation

Investigator	Nanofluid	Concentration (vol%)	Size (nm)	Observation Thermal Conductivity Enhancement in %
Chopkar [100]	Al_2Cu	2	30–40	100
	Ag_2Al	2	30	100
S. Suresh[101]	Al_2O_3–Cu	0.1–2	17	12
D. Madhesh [102]	Cu–TiO_2	0.1–2	55	52
L. Syam Sundar, [103]	MWCNT–Fe_3O_4/ water	0.1–0.3		31
Mechiri Sandeep Kumar [104]	Cu, Zn, Cu-Zn	0.1–0.5	60	48
Mechiri Sandeep Kumar [105]	Cu-Zn (50:50)	01–0.5	20–28	53

6.8 HYBRID NANOFLUIDS

Nanofluids having a mix of two or more different nanoparticles are called "hybrid nanofluids." Over the years, numerous scientists have focused on the transfer of heat and rheological characteristics of hybrid nanofluid that incorporates hybrid nanoparticles. As shown in Table 6.6, nanoparticle dispersion, combined with hybrid nanoparticles, improves thermal conductivity in nanofluid heat-transfer applications when that nanoparticle dispersion is used in a certain size range.

In contrast to single-particle-based nanofluids, the heat-transfer coefficient rose as hybrid nanoparticles grew in volume concentration. Hybrid nanostructures have a stronger effect on the viscosity of the base fluid at high concentrations [99]. Single-particle nanoparticles have a lower density compared to hybrid nanoparticles, allowing them to be suspended in nanofluid for a longer period. Additionally, throughout the measured temperature range, the nanofluid acted as a Newtonian fluid for different concentrations of nanoparticles.

6.9 MACHINE TOOLS APPLICATION

The principles behind nanofluid research that have been studied up to this point in the article are presented in this section, as well as the effects of employing nanofluid on various machining parameters, which are assessed in the next section.

6.9.1 Grinding

Cooling and lubricating characteristics have led to nanofluids being seen as a viable option for machine applications. Water-based Al2O3 and diamond nanofluids were utilized in the MQL grinding process developed by Shen et al. [106]. This has been done using wheels, tribological properties, and nanoparticles in wet (pure water), dry, and MQL, respectively. An improvement in surface roughness, lower grinding temperature, and reduced grinding forces were found when using MQL nanofluids. Also, Malkin et al. [107] examined the grinding efficiency for MQL, a generic quantitative and narrative description of market analysis software, using ester oil and nanoparticle-embedded molybdenum disulfide nanoparticles. The researchers found that in comparison to ordinary ester oil, they had superior findings regarding the reduction of thermal distortion, owing to greater lubrication and a lower specific energy. In an experiment described in "The Nano-Diamond Revolution in the Micro-Drilling Process on Aluminum 6061" by Jung et al. [108], nano-diamond particles were used in the micro-drilling process on aluminum 6061. A substantial increase in the number of drilled holes and a decrease in torque and thrust forces have been discovered via experimental data. In a miniaturized machine tool system, an MQL mesoscale process in which nano-diamond particles are used to make a nanofluid was studied by Lee et al. [109], who saw a notable decrease in cutting forces and surface roughness when compared to a dry and pure MQL system.

As the last machining operation for components with smooth surfaces and exact tolerances, grinding is frequently used [110]. Another issue with grinding studies is managing the high workpiece–tool temperature. In certain cases, surface quality may be adversely affected by the oxidation and metallurgical transformation produced heat [111]. Protecting workpiece and wheel against burning, phase transitions, residual tensile stresses, distortion, and errors all need cooling and lubrication. Operator and environmental risks may be greatly reduced by only using fluids with appropriate cooling qualities (such as hypothermia and chills) [112]. About 85% of the lubricants being used worldwide are based on petroleum. It is estimated that approximately 80% of all occupational illnesses of operators are caused by contact with cutting fluids. It has been shown that nanoparticle fluids may bring about thermal conductivity and convection heat-transfer factors that are greater than those of smaller volumes of fluid and lower environmental and operator hazard [113].

6.9.2 Turning

In industrial sectors, turning is one of the most essential operations. This is used to produce workpieces like shafts, pins, and threads that have a round form. Movement of stationary tools axially along spinning workpieces and the removal of surplus material in the form of chips [114] are common practices in this process. Generated heat in this process is high because the tool and the workpiece are continuously being sheared; therefore, the heat must be dissipated by using a suitable coolant. As a result, it is essential to constantly provide coolant and lubricant to the tool–workpiece contact. Lubrication is normally required at low cutting speeds, because of the high amount of friction, while cooling is critical at higher cutting speeds, because the machine's temperature rises dramatically during milling. Although variables such as feed rate, cutting speed, depth of cut, and workpiece and tool material all affect the generated part and machine tool energy consumption, cutting factors such as these have little or no impact on the created part [115]. When cutting at optimal condition, which has both fast cutting speed, low depth of cut, and feed rate, as well as the appropriate cutting fluid, you get the combination of good surface quality and reduced power consumption for turning. Nanofluids are being used extensively in investigations and studies, and several metalworking fluids have been examined.

6.9.3 Milling

A chip removal operation using rotating and multi-teeth tools on a stationary workpiece is part of the milling process. Milling is used in molding for a broad range of applications. Regarding motion, tool movement occurs in a rotating manner whereas tool movement occurs in a forward manner. Teeth touch the workpiece for a brief period, following which they chip away and leave. Horizontal, vertical, and universal milling machines may be defined as having three different groupings [116]. There is a need to implement coolants in milling [117, 118]. According to many studies, the MQL provides the optimal lubrication for milling and delivers superior surface quality and a longer tool life [114, 119].

6.9.4 Drilling

A spiral fluted tool, along with the simplest machine tool, is the basic process used for drilling. The material is being pushed out of the hole in a circular motion while chips are being removed through spiral flutes [120]. These two characteristics, heat production and high temperature, go hand in hand when creating the tool tip. When drilling at a high temperature, the drill bit may burn, and therefore, the surface quality may be reduced.

6.10 NANOFLUIDS: EFFECT ON MACHINING PARAMETERS

Numerous tests to test various nanoparticles in water-based or other solvent solutions have been carried out in order to enhance fluid cooling and lubricating capabilities. In recent years, some of these studies have been described in Table 6.7.

6.10.1 Cutting Force

In experiment results, researchers observed a reduction in cutting forces, due to the usage of nanofluid in grinding processes, turning, milling, and drilling [110–125] due to the depletion of lubrication and cooling characteristics [121]. However, the cutting forces decrease with the increase in the concentration of nanoparticles up to a certain threshold, and furthermore, cutting forces grow at a modest level.

6.10.2 Surface Roughness

When nanofluid is used as a coolant, machined surfaces always have higher quality than dry or those treated with other conventional fluids. Because there is less force being used, the cutters will move with more comfort, and you will experience less microcracking, resulting in a workpiece temperature drop [110]. In general,

TABLE 6.7
Report of Various Articles about Nanofluids as Metalworking Fluid

No.	Operation	Cooling Method*	Type of Nps	Base Fluid	Tool	Workpiece	Investigated Parameters	Ref.
1	Facing operation	Wet	CuO	– **	Uncoated carbide tool	AISI 1040	– Tool wear – Cutting temperature	[117]
2	Milling	MQL	MoS_2	ECOCUT HSG 905S neat cutting oil	Tungsten carbide	Al 6061-T6 alloy	– Cutting temperature – Surface roughness – Cutting forces	[118]
3	Micro milling	Wet	Graphite Nano platelet	Distilled water	Uncoated micro-grain carbide	H13 Steel	– Cutting forces – Surface roughness	[123]
4	Milling	Wet	Copper Nano particle	Water soluble oil	Carbide tool	St37	– Cutting temperature – Surface roughness	[126]
5	Micro drilling	MQL	Diamond	– Paraffin oil – Vegetable oil	Uncoated carbide twist drill	Aluminum6061	– Number of drilled holes – Drilling torques – Thrust forces	[124]
6	Grinding	MQL	Al_2O_3 and Diamond	Water	Al_2O_3 grinding wheel	Cast iron	– Grinding temperature – Surface roughness – Grinding force – G-ratio	[110]
7	Grinding	Wet	Al_2O_3	Emulsifier TRIM E709	Al_2O_3 grinding wheel	EN-31 Steel	– Grinding temperature – Surface roughness	[112]
8	Grinding	MQL	Al_2O_3	Water	SiC grinding wheel	Ti-6Al-4V alloy	– Surface roughness – Grinding force	[22]
9	Grinding	– Nano coolant, – Conventional coolant	Zinc Oxide	Water	Al_2O_3 grinding wheel	Ductile Cast Iron	– Optimizing cutting parameters such as table speed, depth of cut – Tool wear, – G-ratio	[113]
10	Micro Grinding	MQL	Al_2O_3 and Diamond	Paraffin oil	Vitrified CBN	SK-41C tool steel	– Surface roughness – Grinding force	[127]
11	Grinding	MQL	Al_2O_3	Water	White aluminum oxide grinding wheel	AISI 52100	– Grinding temperature – Surface roughness – Grinding forces – Surface morphology	[125]

(Continued)

TABLE 6.7 (Continued)

No.	Operation	Cooling Method*	Type of Nps	Base Fluid	Tool	Workpiece	Investigated Parameters	Ref.
12	Grinding	MQL	Al_2O_3	Deionized water	White aluminum oxide grinding wheel	hardened AISI 52100	– Grinding temperature – Surface roughness – Grinding force	[128]
13	Grinding	Wet and ELID method	CNT	Water-soluble oil	CBN Diamond bonded	glass	– Surface roughness – Morphology and micro crack observation	[129]
14	Grinding	Wet	CNT	SAE20W40 oil	Vitrified alumina	AISI D2 tool steel	– Surface roughness – The effect of ultrasonic vibration	[130]
15	Grinding	MQL	Al_2O_3	Water	–	–	– Concentration of fluid and pH on stability of nanofluid	[131]
16	Grinding	MQL	Al_2O_3	Water	White aluminum oxide grinding wheel	Hardened AISI 52100	– Grinding temperature – Surface roughness – Grinding force – Grinding force ratio	[132]
17	Grinding	MQL	Al_2O_3	Water	Ceramic bond aluminum oxide	AISI 52100	– Heat transfer coefficient analysis – Grinding temperature	[133]
18	Grinding	MQL	Nano-Diamond	Paraffin oil	Vitrified CBN	SK-41C tool steel	– Surface roughness – Grinding force	[134]
19	Grinding	Wet	SiO_2	Water	-	Ductile cast iron	– Material removal rate (MRR) – Surface roughness	[135]
20	Turning	Wet	Nano boric acid	Coconut oil	Carbide tool (SNMG)	AISI 304 Austenitic Stainless Steel	– Cutting temperature – Surface roughness – Tool wear	[136]
21	Turning	MQL	Al_2O_3	Vegetable oil	Coated carbide cutting tool	Inconel 600 alloy	– Cutting temperature – Surface roughness – Tool wear – Average chip–tool interface	[121]
22	Turning	MQL	Graphite	Water soluble oil	– HSS – carbide tool	AISI 1040	Temperature – Tool wear – Cutting forces	[122]

No.	Operation	Cooling Method*	Type of Nps	Base Fluid	Tool	Workpiece	Investigated Parameters	Ref.
23	Turning	Wet	CuO	Water soluble oil	DNMG 150604-QM	AISI 4340	– Cutting forces – Surface roughness	[137]
24	Turning	MQL	CuO	—	– Cemented carbide insert– HSS	AISI 1040	– Workpiece temperature	[138]
25	Turning	Wet	Nano Boric acid	– SAE40 oil – Coconut oil	Cemented carbide tool (SNMG)	AISI 1040	– Cutting temperature – Surface roughness – Tool flank wear	[139]
26	Turning	Wet	$AgNO_3$	Sodium borohydride	HSS	Mild steel	– Cutting force – Surface roughness – Tool temperature	[140]
27	Turning	MQL	CNT	—	– HSS – Cemented carbide tools	AISI 1040	– Cutting temperature – Surface roughness – Tool wear – Cutting force	[141]
28	Turning	MQL	Nano graphite	Water soluble oil	– HSS – Cemented carbide tools	AISI 1040	– Cutting temperature – Surface roughness – Tool wear – Cutting force	[142]
29	Turning	MQL	MWCNT	Coconut oil	Carbide tool insert	Martensitic Stainless Steel	– Cutting temperature – Surface roughness	[143]
30	Turning	MQL	MWCNT	–	Multi-layered TiN top-coated insert	AISI 1040	– Cutting temperature – Finite element analyses (FEAs)	[144]
31	Hobbing	Conventional lubrication	Al_2O_3	25W-50 oil	P6M5K5 steel with TiN top coated	DIN1.7131	– Surface roughness	[145]

nanofluids will help decrease surface roughness when they are applied correctly. While these findings are consistent across other studies conducted with various nanofluids and machining procedures, one or two findings may vary.

6.10.3 Machining Temperature

Using nanofluid is significantly different from the other variables while turning. Like nanofluids, metal fluids exhibit both good heat transmission properties (particularly high thermal conductivity) and impact on the tool and workpiece temperature. Lower heat generation also results from nanofluids that reduce the amount of friction when the tool and workpiece come into contact [139, 140].

6.10.4 Tool Wear

A machine tool's wear is also critical, because it affects how well the machine runs and its tool life [146]. Wear of tools, both due to cutting and due to friction, results in tool failure during cutting, cutting tools and their frictional wear frequently impede the overall performance of cutting operations [147]. While the main causes for tool wear are abrasion, adhesion, and diffusion, there are many other factors that have a considerable impact. When both adhesion and diffusion occur at the tool–chip contact, then they are occurring at the optimal temperature [148]. Early wear of tools will produce uneconomical machining operation. Various scientific studies have shown that nanoparticles improve cooling and lubricity to provide substantial reductions in tool wear [142].

6.10.5 Environmental Aspects

Machine workers are harmed by corrosive chemicals, such as chlorinated paraffin, which may cause skin and respiratory issues as well as cancer when present in high pressures (especially in engineering practice) and other common cutting fluid additives [149, 150]. Used oil often contains more halogens, particularly chlorine, than treated wastewater; thus, it is considered special waste with very high disposal costs [151]. And this is only the beginning of many greater advances in cooling techniques. Additionally, because gas pressure distributes a certain amount of fluid into the air, using MQL may expose workers to respiratory issues [152]. Polar nanofluid MQL has a variety of applications, including the reduction of environmental and operator risks [129]. Nonetheless, very few tests have been done on the environmental effects of nanoparticle application in cutting fluids, although preliminary evidence shows that nanoparticles have an antibacterial property that combats environmental issues [134].

6.11 DIFFICULTIES OF APPLYING NANOFLUIDS IN MACHINING

Simple solid–liquid combinations are also used for nanofluids. Nanofluids must meet specific criteria in order to be useful, such as homogeneous and stable suspension, reduced particle clustering, and others. There are a variety of methods to do this to get these specific circumstances. Preparing of a nanofluid is difficult because it clumps. Conversely, altering the pH of solution suspension, the use of surfactants, dispersers, or vibrators may all accomplish these things [153]. In addition, the cost of nanoparticles and how to keep them from settling are important concerns to keep in mind.

6.12 CONCLUSION

Many different machining techniques and nanofluids were used initially, followed by an assessment of their impact on cutting parameters (cutting forces, surface roughness, tool-tooling temperature, tool wear, environmental issues). The use of nanofluids influences machine parameters, but so do nanoparticle type and base fluid concentration, as well as nanoparticle size and concentration. Viscosity, thermal conductivity, density, and thermal capacity are important physical characteristics associated with heat transfer. If they were to apply proper lubrication, there will be an improvement in heat transfer because of the change in thermal characteristics (viscosity, thermal conductivity, density, thermal capacity). Nanofluids are used as coolant and lubricant, which lowers the temperature of tools, lessens tool wear, and increases surface quality. But the expense of nanoparticles, the need of a specific equipment, the presence of aggregates, and silt are all problems associated with nanofluids use in metalworking.

REFERENCES

1. United Nation, World Commission on Environment and Development "Our Common Future, Chapter 2: Towards Sustainable Development," 1987. Available at www.un-documents.net/ocf-02.html.
2. Department of Commerce, United States of America, how does Commerce define Sustainable Manufacturing. Available at http://www.trade.gov/competitiveness/sustainablemanufacturing/how_doc_defines-SM.asp.
3. Machine Tools in China to 2014-Demand and Sales Forecasts, Market Share, Market Size, Market Leader, 2011. Available at www.freedoniagroup.com/Machine-Tools-In-hina.html
4. Last Trends in the European Machine Tool Industry, 2011. Available at www.cecimo.eu/machine-tools/datastatistics/latesttrend.html.
5. Sutherland J., Gunter K., Allen D., Bauer D., Bras B., Gutowski T., Murphy C., Piwonka T., Sheng P., Thuston D., and Wolff E., A global perspective on the environmental challenges facing the automotive industry: state-of-the-art and directions for the future, *Int. J. Veh. Des.*, 2004, 35(1), 86–110
6. Quinn M.M., Kriebel D., Geiser K., and Moure-Eraso R., Sustainable production: a proposed strategy for the work environment, *Am. J. Ind. Med.*, 1998, 34, 297–304
7. Hewstone R.K., Environmental health aspects of lubricant additives, *Sci. Total Environ.*, 1994, 156, 243–254
8. Greaves I.A., Eisen E.A., Smith T.J., Pothier L.J., Kriebel D., Woskie S.R., Kennedy S.M., Shalat S., and Monson R.R., Respiratory health of automobile workers exposed to metal-working fluid aerosols: respiratory symptoms, *Am. J. Ind. Med.*, 1997, 32, 450–459

9. Lim C.-H., Yu I.J., Kim H.-Y., Lee S.-B., Kang M.-G., Marshak D.R., and Moon C.-K. Respiratory effect of acute and sub-acute exposure to endotoxincontaminated metal working fluid (MWF) aerosols on Sprague-Dawley rats *Arch. Toxicol.*, 2005, 79, 321–329
10. Godderis L., Deschuyffeleer T., Roelandt H., Veulemans H., and Moens G., Exposure to metalworking Xuids and respiratory and dermatological complaints in a secondary aluminium plant, *Int. Arch. Occup. Environ. Health*, 2008, 81, 845–853
11. Cassin C., and Boothroyd G., Lubricating action of cutting fluids, *J. Mech. Eng. Sci.*, 1965, 1(7), 67–81
12. D'Errico G.E., Guglielmi E., and Rutelli G., A study of coatings for end mills in high speed metal cutting, *J. Mater. Process. Technol.*, 1999, 92–93, 251–256
13. Guleryuz C.G., Krzanowski J.E., Veldhuis S.C., and Fox-Rabinovich G.S., Machining performance of TiN coatings incorporating indium as solid lubricant, *Surf. Coat. Technol.*, 2009, 203, 3370–3376
14. Hanyu H., Murakami Y., Kamiya S., and Saka M., New diamond coating with finely crystallized smooth surface for the tools to achieve fine surface finish of non-ferrous metals, *Proc. Front. Surf. Eng.*, 2003, 169–170, 258–261
15. Renevier N.M., Oosterling H., Konig U., Dautzenberg H., Kim B.J., Geppert L., Koopmans F.G.M., and Leopold J., Performance and limitations of MoS_2/Ti composite coated inserts, *Surf. Coat. Technol.*, 2003, 172(1), 13–23
16. Dos Santos G.R., Da Costa D.D., Amorim F.L., and Torres R.D., Characterization of DLC thin film and evaluation of machining forces using coated inserts in turning of Al-Si alloys, *Surf. Coat. Technol.*, 2007, 202, 1029–1033
17. Mativenga P.T., and Hon K.K., A study of cutting forces and surface finish in high-speed machining of AISI H13 tool steel using carbide tools with TiAlN based coatings, *Proc. Inst. o Mech. Eng.*, 2003, 217, 143–151
18. Dudzenski D., Devillez A., Moufki A., Larrouquere D., Zerrouki V., and Vigneau J., A review of developments toward dry and high speed machining of Inconel 718 alloy, *Int. J. Mach. Tools Manuf.*, 2004, 44, 439–456
19. Diniz A.E., and Micaroni R., Cutting conditions for finish turning process aiming: the use of dry cutting, *Int. J. Mach. Tools Manuf.*, 2002, 42, 899–904
20. Galanis N.I., Study of the performance of the turning and drilling of austenitic stainless steels using two coolant techniques. *Int. J. Mach. Mach. Mater.*, 2008, 3, 1–2
21. Shefelbine W., and Dornfeld D., The effect of dry machining on burr size, 2004. Available at http://repositories.edlib.org/lma/eodef/wendy_1_03.
22. Outeiro J.C., Pina J.C., M'Saoubi R., Pusavec F., and Jawahir I.S., Analysis of residual stressses induced by dry turning of difficult-to-machine materials, *CIRP Ann. Manuf. Technol.*, 2008, 57, 77–80
23. Klocke F., and Eisenblatter G., Dry cutting, *Ann. CIRP Manuf. Technol.*, 1997, 46(2), 519–526
24. Sreejith P.S., and Ngoi B.K.A., Dry machining: machining of the future. *J. Mater. Process. Technol.*, 2000, 101, 287–291
25. Obikawa T., Kamata Y., Asano Y., Nakayama K., and Otieno A.W., Microliter lubrication machining of Inconel 718. *Int. J. Mach. Tools Manuf.*, 2008, 48(15), 1605–1612
26. Stoll A., and Furness R., Near-dry machining (MQL) is a key technology for driving paradigm shift in machining operations, *Mach. Technol., Soc. Manuf. Eng., Fourth Quarter*, 2006, 17(4), 1–22
27. Astakhov V.P., Ecological Machining: Near-Dry Machining, in P.J. Davim (Ed.): *Machining Fundamentals and Recent Advances*, Springer, London, 2008, pp. 195–223

28. Min S., Inasaki I., Fujimura S., Wakabayashi T., and Suda S., Investigation of adsorption behaviour of lubricants in near-dry machining, *Proc. Inst. Mech. Eng. Part B: J. Eng. Manuf.*, 2005, 219, 665–671
29. Ding Y., and Hong S.Y., Improvement in chip breaking in machining a low carbon steel by cryogenically precooling the workpiece, *Trans. ASME, J. Manuf. Sci. Eng.*, 1997, 120(1), 76–83
30. Yildiz Y., and Nalbant M., A review of cryogenic cooling in machining processes, *Int. J. Mach. Tools Manuf.*, 2008, 48, 947–964
31. Dhananchezian M., and Kumar M.P., Experimental investigation of cryogenic cooling by liquid nitrogen in the orthogonal machining of Aluminium 6061-T6 alloy, *Int. J. Mach. Mach. Mater.*, 2010, 7(3–4), 274–285
32. Paul S., Dhar N.R., and Chattopadhyay A.B., Beneficial effects of cryogenic cooling over dry and wet machining on tool wear and surface finish in turning AISI 1060 steel, *J. Mater. Process. Technol.*, 2001, 116, 44–48
33. Dhar N.R., Paul S., and Chattopadhyay A.B., The influence of cryogenic cooling on tool wear, dimensional accuracy and surface finish in turning AISI 1040 and E4340C steel, *Wear*, 2002, 249, 932–942
34. Ezuqwu E.O. Key improvements in the machining of difficult-to-cut aerospace superalloys, *Int J Mach Tool Manu*, 2005, 45, 1353–1367
35. Wang Z.Y., and Rajurkar K.P., Cryogenic machining of hard-to-cut materials, *Wear*, 2000, 239, 168–175
36. Bermingham M.J., Kirsch J., Sun S., Palanisamy S., and Dargusch M.S., New observations on tool life, cutting forces, and chip morphology in cryogenic machining Ti-6Al-4V, *Int. J. Mach. Tools Manuf.*, 2011, 51, 500–511
37. Hong S.Y., and Ding Y., Cooling approaches and cutting temperatures in cryogenic machining of Ti-6Al-4V, *Int. J. Mach. Tools Manuf.*, 2001, 41, 1417–1437
38. Kopac J., Achievements of sustainable manufacturing by machining, *J. Achiev. Mater. Manuf. Eng.*, 2009, 34(2), 180–187
39. Pusavec F., Krajnik P., and Kopac J., Transitioning to sustainable production – Part 1: Application on machining technologies, *J. Clean. Prod.*, 2010, 18, 174–184
40. Varadarajan A.S., Philip P.K., and Ramamoorthy B., Investigations on hard turning with minimal cutting fluid application (HTMF) and its comparison with dry and wet turning, *Int. J. Mach. Tools Manuf.*, 2002, 42, 193–200
41. Wertheim R., Rotbery J., and Ber A., Influence of high-pressure flushing through the rake face of the cutting tool, *Annals of the CIRP*, 1992, 4(1), 101–106
42. Rahman M., Kumar A.S., and Choudhury M.R., Identification of effective zones for high pressure coolant in milling, *Ann. CIRP*, 2000, 49(1), 47–52
43. Kovacevic R., Cherukuthota C., and Mazurkiewicz M., High pressure waterjet cooling/lubrication to improve machining efficiency in milling, *Int. J. Mach. Tools Manuf.*, 1995, 35(10), 1459–1473
44. Klocke F., Sangermann H., Kramer A., and Lung D., Influence of a high-pressure lubricoolant supply on thermo-mechanical tool load and tool wear behavior in the turning of aerospace materials, *Proc. Inst. Mech. Eng. B J. Eng. Manuf.*, 2011, 225, 52
45. Timmerhaus K.D., and Reed R.P., *Cryogenic Engineering: Fifty Years of Progress*, Springer Verlag, Berlin, 2007.
46. Shokrani A., Dhokia V., and Newman S.T., Environmentally conscious machining of difficult-to-machine materials with regard to cutting fluids, *Int. J. Mach. Tools Manuf.*, 2012, 57, 83–101

47. Sadeghi M.H., Hadad M.J., Tawakoli T., Vesali A., and Emami M., An investigation on surface grinding of AISI 4140 hardened steel using minimum quantity lubrication-MQL technique, *Int. J. Mater. Form.*, 2010, 3(4), 241–251
48. Da Silva L.R., Bianchi E.C., Fusse R.Y., Catai R.E., Franc T.V., and Aguiar P.R., Analysis of surface integrity for minimum quantity lubricant—MQL in grinding, *Int. J. Mach. Tools Manuf.*, 2007, 47, 412–418
49. Zeilmann R.P., and Weingaertner W.L., Analysis of temperature during drilling of Ti6Al4V with minimal quantity of lubricant, *J. Mater. Process. Technol.*, 2006, 179, 124–127
50. Heinemann R., Hinduja S., Barrow G., and Petuelli G., Effect of MQL on the tool life of small twist drills in deep-hole drilling, *Int. J. Mach. Tools Manuf.*, 2006, 46, 1–6
51. Davim J.P., Sreejith P.S., Gomes R., and Peixoto C., Experimental studies on drilling of aluminium (AA1050) under dry, minimum quantity of lubricant, and flood-lubricated conditions, *Proc. IMechE Part B: J. Eng. Manuf.*, 2006, 220, 1605–1611
52. Astakhov V.P., Ecological Machining: Near-Dry Machining, in P.J. Davim (Ed.): *Machining Fundamentals and Recent Advances*, Springer, London, 2008, pp. 195–223
53. Rohrer H., The nanoworld: Chances and challenges. *Microelectron. Eng.*, 1996, 32, 5–14
54. Eastman J.A., Phillpot S.R., Choi S.U.S., and Keblinski P., Thermal transport in nanofluids, *Annu. Rev. Mater. Res.*, 2004, 34, 219–246
55. Lienhard I.V., and John H. *A Heat Transfer Text Book*, 3rd edn., Phlogiston Press, Cambridge, 2004.
56. Zhang Y., Zeng G., Piprek J., Bar-Cohen A., and Shakouri A., Superlattice microrefrigerators flip-clip bonded with optoelectronic devices, *Journal IEEE Trans. Computer Packing Tech.*, 28(4), 2005, 658–666
57. Vanam K., Junghans J., Barlow F., Selvam R.P., Balda J.C., and Elshabini A., A novel packaging methodology for spray cooling of power semiconductor devices using dielectric liquids, in *Applied Power Electronics Conf. Exposition, APEC2005*, Vol. 3, 2005, pp. 2014–2018
58. Peterson G.P., *An Introduction to Heat Pipes Modeling, Testing, and Applications*, John Wiley & Sons, New York, 1994.
59. Kumar D.H., Patel H.E., Kumar V.R., Sundararajan T., Pradeep T., and Das S.K., Model for heat conduction in nanofluids, *Phys. Rev. Lett.*, 2004, 93(14), 144301.
60. Xie H., Wang T., Xi T., and Ai F., Thermal conductivity enhancement of suspensions containing Nanosized Alumina particles, *J. Appl. Phys.*, 2002, 91, 4568–4572
61. Koo J., and Kleinstreuer C., Impact analysis of nanoparticles motion mechanisms on the thermal conductivity of nanofluids, *Int. Commun. Heat Mass Transfer*, 2005, 32(9), 1111–1118
62. Prasher R., Bhattacharya P., and Phelan P.E., Thermal conductivity of Nanoscale colloidal solutions (Nanofluids), *Phys. Rev. Lett.*, 2005, 94, 025901.
63. Nan C.W., Liu G., Lin Y., and Li M., Interface effective thermal conductivity of carbon nanotube composites, *Appl. Phys. Lett.*, 2004, 85, 3549–3551
64. Prasher R., *Applications of Nanotechnology in Electronics Cooling*, Seminar, Troy, New York, March 17, 2006.
65. Suresh S., Venkitaraj K.P., Selvakumar P., and Chandrasekar M., Synthesis of Al_2O_3-Cu/water hybrid nanofluids using two step method and its thermo physical properties, *Colloids Surf. A: Physicochem. Eng. Asp.*, 2011, 388, 41–48
66. Vasu V., and Reddy G.P.K., Effect of minimum quantity lubrication with Al_2O_3 nanoparticles on surface roughness, tool wear and temperature dissipation in machining Inconel 600 alloy, *Proc. Inst. Mech. Eng. Part N: J. Nanoeng. Nanosys.*, 2011, 225, 3–16

67. Amrita M., Srikant R., Sitaramaraju A., Prasad M., and Krishna P.V., Experimental investigations on influence of mist cooling using nanofluids on machining parameters in turning AISI 1040 steel, *Proc. Ins. Mech. Eng., Part J: J. Eng. Trib.* 2013, 227, 1334–1346
68. Rao D.N., Srikant R., Krishna P.V., and Subrahmanyam M., Nanocutting fluids in minimum quantity lubrication, in *International Multi-Conference on Engineering and Technological Innovation (IMETI 2008)*, Orlando, FL, USA, 2008.
69. Rao S.N., Satyanarayana B., and Venkatasubbaiah K., Experimental estimation of tool wear and cutting temperatures in MQL using cutting fluids with CNT inclusion, *Int. J. Eng. Sci. Technol.*, 2011, 3, 2928–2931
70. Prasad M., and Srikant R., Performance evaluation of nano graphite inclusions in cutting fluids with MQL technique in turning of AISI 1040 Steel, *Int. J. Res. Eng. Tech.*, 2013, 2, 381–393
71. Kumar M.S., and Kumar K.S., An investigation multi walled carbon nanotubes based nano cutting fluids in turning of martensitic stainless steel by using Taguchi and Anova analysis, *IOSR J. Mech. Civil Eng.*, 2010, 7, 8–15
72. Roy S., and Ghosh A., High-speed turning of AISI 4140 steel by multi-layered TiN top-coated insert with minimum quantity lubrication technology and assessment of near tool-tip temperature using infrared thermography, *Proc. Ins. Mech. Eng., Part B: J. Eng. Man.*, 2014, 228(9), 1058–1067
73. Behera B.C., Ghosh S., and Rao P.V., Application of nanofluids during minimum quantity lubrication: A case study in turning process, *Tribol. Int.*, 2016, 101, 234–246
74. Masuda H., Ebata A., Teramae K., and Hishinuma N., Alteration of thermal conductivity and viscosity of liquid by dispersing ultra-fine particles dispersion of y-$A1_2O_3$, SiO_2, and TiO_2 ultra-fine particles, *Netsu. Bussei.*, 1993, 4, 227–233
75. Lee S.P., Choi S.U.S., Li S., and Eastman J.A., Measuring thermal conductivity of fluids containing oxide nanoparticles, *ASME J. Heat Transfer*, 1999, 121, 280–289
76. Xie H., Wang J., Xi T., and Liu Y., Thermal conductivity of suspensions containing nanosized SiC particles, *Int. J. Thermophys.*, 2002, 23(2), 571–580
77. Putra N., Roetzel W., and Das S.K., Natural convection of nanofluids, *Heat Mass Transf.*, 2003, 39, 775–784
78. Chon C.H., Kihm K.D., Lee S.P., and Choi S.U.S., Empirical correlation finding the role of temperature and particle size for nanofluid (Al2O3) thermal conductivity enhancement, *Appl. Phys. Lett.*, 2005, 87, 153107.
79. Wang X., Xu, X., and Choi, S.U.S., Thermal conductivity of nanoparticle-fluid mixture, *J. Thermophys. Heat Transf.*, 1999, 13, 474–480
80. Lee J.H., Hwang K.S., Jang S.P., Lee B.H., Kim J.H., Choi S.U.S., and Choi C.J., Effective viscosities and thermal conductivities of aqueous nanofluids containing low volume concentrations of Al_2O_3 nanoparticles, *Int. J. Heat Mass Transf.*, 2008, 51, 2651–2656
81. Zhang X., Gu H., and Fujii M., Effective thermal conductivity and thermal diffusivity of nanofluids containing spherical and cylindrical nanoparticles, *Exp. Thermal Fluid Sci.*, 2006, 31, 593–599
82. Timofeeva E.V., Gavrilov A.N., McCloskey J.M., and Tolmachev Y.V., Thermal conductivity and particle agglomeration in alumina nanofluids: experiment and theory, *Phys. Rev. E*, 2007, 76, 061203
83. Wei Y., and Xie H., A review on nanofluids: preparation, stability mechanisms, and applications, *J. Nanomater.*, 2012, 1–17
84. Zhou L.P., and Wang B.X., Experimental research on the thermophysical properties of nanoparticle suspensions using the quasi-steady method, *Annu. Proc. Chin. Eng. Thermophysics*, 2002, 889–892

85. Das S.K., Putra N., Thiesen P., and Roetzel W., Temperature dependence of thermal conductivity enhancement for nanofluids, *ASME J. Heat Transfer*, 2003, 125, 567–574
86. Murshed S.M.S., Leong K.C., and Yang C., Enhanced thermal conductivity of TiO_2-water based nanofluids, *Int. J. Therm. Sci.*, 2005, 44, 367–373
87. Duangthongsuk W., and Wongwises S., Measurement of temperature-dependent thermal conductivity and viscosity of TiO_2-water nanofluids, *Exp. Thermal Fluid Sci.*, 2009, 33, 706–714
88. Xie H., Wang J., Xi T., and Liu Y., Thermal conductivity of suspensions containing nanosized SiC particles, *Int. J. Thermophys.*, 2002, 23(2), 571–580
89. Choi S.U.S., Yu W., Hull J.R., Zhang Z.G., and Lockwood F.E.. Nanofluids for vehicle thermal management, *SAE Technical Paper 2001-01-1706*; 2001, 139–144
90. Assael M.J., Metaxa I.N., Kakosimos K., and Constantinou D., Thermal conductivity of nanofluids—Experimental and theoretical, *Int. J. Thermophys.*, 2006, 27, 999–1017
91. Eastman J.A., Choi S.U.S., Li S., Yu W., and Thompson L.J., Anomalously increased effective thermal conductivities of ethylene glycol-based nanofluids containing copper nanoparticles, *Appl. Phys. Lett.*, 2001, 78, 718–720
92. Xuan Y., and Li Q., Heat transfer enhancement of nanofluids, *Int. J. Heat Fluid Flow*, 2005, 2, 58–64
93. Hong T.K., Yang H.S., and Choi C.J., Study of the enhanced thermal conductivity of Fe nanofluids, *J. Appl. Phys.*, 2005, 97, 064311
94. Patel H.E., Das S.K., and Sundararajan T., Thermal conductivities of naked and monolayer protected metal nanoparticle based nanofluids: manifestation of anomalous enhancement and chemical effects, *Appl. Phys. Lett.*, 2003, 83, 2931–2933
95. Putnam S.A., Cahill D.G., and Braun P.V., Thermal conductivity of nanoparticle suspensions, *J. Appl. Phys.*, 2006, 99, 084308
96. Murshed S.M.S., Leong K.C., and Yang C., Investigations of thermal conductivity and viscosity of nanofluids, *Int. J. Therm. Sci.*, 2008, 47, 560–568
97. Assael M.J., Chen C.F., Metaxa I., and Wakeham W.A., Thermal conductivity of suspensions of carbon nanotubes in water, *Int. J. Thermophys.*, 2005, 25(4), 971–985
98. Liu M.S., Lin M.C.C., Huang I.T., and Wang C.C., Enhancement of thermal conductivity with carbon nanotube for nanofluids, *Int. Commun. Heat Mass Transfer*, 2005, 32, 1202–1210
99. Biercuk M.J., Llaguno M.C., Radosavljevic M., Hyun J.K., Johnson A.T., and Fischer J.E., Carbon nanotube composites for thermal management, *Appl. Phys. Lett.*, 2002, 80, 2767–2769
100. Chopkar M., Sudarshan S., Das P.K., and Manna I., Effect of particle size on thermal conductivity of nanofluids, *Metall. Mater. Trans. A*, 2008, 39, 1535–1542
101. Suresh S., Venkitaraj K.P., Selvakumar P., and Chandrasekar M., Synthesis of Al_2O_3-Cu/water hybrid nanofluids using two step method and its thermo physical properties, *Colloids and Surfaces A: Physicochemical Engineering Aspects*, 2011, 388, 41–48
102. Madhesh D., Parameshwaran R., and Kalaiselvam S., Experimental investigation on convective heat transfer and rheological characteristics of Cu–TiO_2 hybrid nanofluid, *Exp. Thermal Fluid Sci.*, 2014, 52, 104–115
103. Syam Sundar L., Singh M.K., and Sousa A.C.M., Enhanced heat transfer and friction factor of MWCNT–Fe_3O_4/water hybrid nanofluids, *Int. Commun. Heat Mass Transfer*, 2014, 52, 73–83
104. Mechiri S.K., Vasu V., and Venu Gopal A., Thermal conductivity and viscosity of vegetable oil–based Cu, Zn, and Cu–Zn hybrid nanofluids, *J. Test. Eval.*, 2015, 44, 1–7, doi:10.1520/JTE20140286

105. Mechiri S.K., Vasu V., and Venu Gopal A., Investigation of thermal conductivity and rheological properties of vegetable oil based hybrid nanofluids containing Cu-Zn hybrid nanoparticles, *Exp. Heat Transf.*, 2016, 30, 205–217, doi:10.1080/08916152.2016.1233147
106. Shen B., Shih A.J., Tung S.C., and Hunter M., Application of nanofluids in minimum quantity lubrication grinding, Tribol. Trans., 2008, 51(6), 730–737
107. Malkin S., and Sridharan U., Effect of minimum quantity lubrication (MQL) with nanolfuids on grinding behavior and thermal distortion, *Trans. NAMRI/SME*, 2009, 37, 629–636
108. Nam J.S., Lee P.H., and Lee S.W., Experimental characterization of micro-drilling process using nanofluid minimum quantity lubrication, *Int. J. Mach. Tools Manuf.*, 2011, 51(7–8), 649–652
109. Lee P.-H., Nam T.-S., Li C., and Lee S.W., Experimental study on meso-scale milling process using nanofluid minimum quantity lubrication and compressed chilly air, in *Proceedings of the International Conference on Mechanical, Inductrial, and Manufacturing Technologies (MIMT 2010)*, Sanya, China, 2010, pp. 457–463
110. Shen B., Shih A.J., and Tung S.C., Application of nanofluids in minimum quantity lubrication grinding, *Tribol. Trans.*, 2008, 51, 730–737
111. Sanchez J., Pombo I., Alberdi R., Izquierdo B., Ortega N., Plaza S., and Martinez Toledano J., Machining evaluation of a hybrid MQL-CO_2 grinding technology, *J. Clean. Prod.*, 2010, 18, 1840–1849
112. Vasu V., and Kumar K.M., Analysis of nanofluids as cutting fluid in grinding EN-31 steel, *Nano-Micro Lett.*, 2011, 3, 209–214
113. Setti D., Ghosh S., and Rao P.V., Application of nano cutting fluid under minimum quantity lubrication (MQL) technique to improve grinding of Ti–6Al–4V alloy, *Proc. World Acad. Sci. Eng. Tech.*, 2012, 70, 512–516
114. Byers J.P., *Metalworking Fluids*, CRC Press, USA, 2012
115. Ghani J., Choudhury I., and Hassan H., Application of Taguchi method in the optimization of end milling parameters, *J. Mater. Process. Technol.*, 2004, 145, 84–92
116. Razfar M.R., *Principles of Machining*, Amirkabir University of Technology Press, Tehran, Iran, 2010
117. Srikant R., Rao D., Subrahmanyam M., and Krishna V.P., Applicability of cutting fluids with nanoparticle inclusion as coolants in machining, *Proc. Ins. Mech. Eng., Part J: J. Eng. Trib.*, 2009, 223, 221–225
118. Rahmati B., Sarhan A.A., and Sayuti M., Investigating the optimum molybdenum disulfide (MoS_2) nanolubrication parameters in CNC milling of AL6061-T6 alloy, *Int. J. Adv. Manuf. Technol.*, 2014, 70, 1143–1155
119. Lawal S., Choudhury I., and Nukman Y., Application of vegetable oil-based metalworking fluids in machining ferrous metals—A review, *Int. J. Mach. Tools Manuf.*, 2012, 52, 1–12
120. Alinejad G., and Ghafari M., *Principles of Universal Machining Machine Tool*, Babol University Technology Press, Babol, Iran, 2011
121. Vasu V., and Reddy G.P.K., Effect of minimum quantity lubrication with Al_2O_3 nanoparticles on surface roughness, tool wear and temperature dissipation in machining Inconel 600 alloy, *Proc. Inst. Mech. Eng., Part N: J. Nanoeng. Nanosys.*, 2011, 225, 3–16
122. Amrita M., Srikant R., Sitaramaraju A., Prasad M., and Krishna P.V., Experimental investigations on influence of mist cooling using nanofluids on machining parameters in turning AISI 1040 steel, *Proc. Ins. Mech. Eng., Part J: J. Eng. Trib.*, 2013, 227, 1334–1346

123. Marcon A., Melkote S., Kalaitzidou K., and Debra D., An experimental evaluation of graphite nanoplatelet based lubricant in micro-milling, *CIRP Ann. Man. Tech.*, 2010, 59, 141–144
124. Nam J.S., Lee P.H., and Lee S.W., Experimental characterization of micro-drilling process using nanofluid minimum quantity lubrication, *Int. J. Mach. Tools Manuf.*, 2011, 51, 649–652
125. Mao C., Tang X., Zou H., Huang X., and Zhou Z., Investigation of grinding characteristic using nanofluid minimum quantity lubrication, *Int. J. Precision Eng. Man.*, 2012, 13, 1745–1752
126. Dahmus J.B., and Gutowski T.G., An environmental analysis of machining, in *ASME 2004 International Mechanical Engineering Congress and Exposition, American Society of Mechanical Engineers*, 2004. (BOOK or JOURNAL (give VOLUME and ISSUE)
127. Lee P.H., Nam J.S., Li C., and Lee S.W., An experimental study on micro-grinding process with nanofluid minimum quantity lubrication (MQL), *Int. J. Precision Eng. Man.*, 2012, 13, 331–338
128. Mao C., Zou H., Huang X., Zhang J., and Zhou Z., The influence of spraying parameters on grinding performance for nanofluid minimum quantity lubrication, *Int. J. Adv. Manuf. Technol.*, 2013 64, 1791–1799
129. Prabhu S., and Vinayagam B., Analysis of surface characteristics by electrolytic inprocess dressing (ELID) technique for grinding process using single wall carbon nano tube-based nanofluids, *Arab. J. Sci. Eng.*, 2013, 38, 1169–1178
130. Prabhu S., and Vinayagam B., Nano surface generation of grinding process using carbon nano tubes, *Sadhana*, 2010, 35, 747–760
131. Mao C., Zou H., Zhou X., Huang Y., Gan H., and Zhou Z., Analysis of suspension stability for nanofluid applied in minimum quantity lubricant grinding, *Int. J. Adv. Manuf. Technol.*, 2014, 71, 2073–2081
132. Mao C., Zhang J., Huang Y., Zou H., Huang X., and Zhou Z., Investigation on the effect of nanofluid parameters on MQL grinding, *Mater. Manuf. Process.*, 2013, 28, 436–442
133. Mao C., Zou H., Huang Y., Li Y., and Zhou Z., Analysis of heat transfer coefficient on workpiece surface during minimum quantity lubricant grinding, *Int. J. Adv. Manuf. Technol.*, 2013, 66, 363–370
134. Lee P.H., Nam T.S., Li C., and Lee S.W., Environmentally-friendly nano-fluid minimum quantity lubrication (MQL) meso-scale grinding process using nanodiamond particles, in *Manufacturing Automation (ICMA), International Conference on IEEE*, The University of Hong Kong, China, 2010
135. Rahman M.M., Kadirgama K., and Aziz A., Artificial neural network modeling of grinding of ductile cast iron using water based SiO_2 Nanocoolant, *Int. J. Autom. Mech. Eng.*, 2013, 9, 1649–1661
136. Sodavadia K., and Makwana A., Experimental investigation on the performance of coconut oil based nano fluid as lubricants during turning of AISI 304 austenitic stainless steel, *Int. J. Adv. Mech. Eng.*, 2014, 4, 55–60
137. Shabgard M.R., Jafarian Zenjanab M., and Azarafza R., Experimental study on the influence of CuO nanofluid on surface roughness and machining force in turning of AISI 4340 steel, *Modares Mech. Eng.*, 2014, 14, 27–33 138. Rao D.N., Srikant R., Krishna P.V., and Subrahmanyam M., Nano cutting fluids in minimum quantity lubrication, in *International Multi-Conference on Engineering and Technological Innovation (IMETI 2008)*, Orlando, FL, USA, 2008

139. Vamsi Krishna P., Srikant R., and Nageswara Rao D., Experimental investigation on the performance of nanoboric acid suspensions in SAE-40 and coconut oil during turning of AISI 1040 steel, *Int. J. Mach. Tools Manuf.*, 2010, 50, 911–916
140. Saravanakumar N., Prabu L., Karthik M., and Rajamanickam A., Experimental analysis on cutting fluid dispersed with silver nano particles, *J. Mech. Sci. Technol.*, 2014, 28, 645–651
141. Rao S.N., Satyanarayana B., and Venkatasubbaiah B., Experimental estimation of tool wear and cutting temperatures in MQL using cutting fluids with CNT inclusion, *Int. J. Eng. Sci. Technol.*, 2011, 3, 2928–2931
142. Prasad M., and Srikant R., Performance evaluation of nano graphite inclusions in cutting fluids with MQL technique in turning of AISI 1040 steel, *Int. J. Res. Eng. Tech.*, 2013, 2, 381–393
143. Kumar M.S., and Kumar K.S., An investigation multi walled carbon nanotubes based nano cutting fluids in turning of martensitic stainless steel by using Taguchi and Anova analysis, *IOSR J. Mech. Civil Eng.*, 2010, 7, 8–15
144. Roy S., and Ghosh A., High-speed turning of AISI 4140 steel by multi-layered TiN top-coated insert with minimum quantity lubrication technology and assessment of near tool-tip temperature using infrared thermography, *Proc. Ins. Mech. Eng., Part B: J. Eng. Man.*, 2014, 228(9), 1058–1067
145. Poorazari S., Effect of alumina nanoparticles on the surface finish of the gear lubricant oil contains the simple production process, *Hobbing Mech. Eng.*, 2014, 93, 38–47
146. Rahman M., and Kadirgama K., Performance of water-based zinc oxide nanoparticle coolant during abrasive grinding of ductile cast iron, *J. Nanoparticles*, 2014, 2014, 1–7
147. Weinert K., Inasaki I., Sutherland J., and Wakabayashi T., Dry machining and minimum quantity lubrication, *CIRP Annals-Manuf. Tech.*, 2004, 53, 511–537
148. Nouari M., List G., Girot F., and Coupard D., Experimental analysis and optimization of tool wear in dry machining of aluminum alloys, *Wear*, 2003, 255, 1359–1368
149. Kuram E., Ozcelik B., Cetin M.H., Demirbas E., and Askin S., Effects of blended vegetable-based cutting fluids with extreme pressure on tool wear and force components in turning of Al 7075-T6, *Lubr. Sci.*, 2013, 25, 39–52
150. Cetin M.H., Ozcelik B., Kuram E., and Demirbas E., Evaluation of vegetable based cutting fluids with extreme pressure and cutting parameters in turning of AISI 304L by Taguchi method, *J. Clean. Prod.*, 2011, 19, 2049–2056
151. Bartz W.J., Ecological and environmental aspects of cutting fluids, *Lubrication Eng. Illinois*, 2001, 57, 13–16
152. Shokoohi Y., Khosrojerdi E., and Rassolian Shiadhi B.H., Machining and ecological effects of a new developed cutting fluid in combination with different cooling techniques on turning operation, *J. Clean. Prod.*, 2015, 94, 330–339
153. Shekarian E., Tarighaleslami A.H., and Khodaverdi F., Review of effective parameters on the nanofluid thermal conductivity, *J. Middle East App. Sci. Tech.*, 2014, 15, 776–780

145. Poursaeri S., Effect of adding nanoparticles on the surface finish of the gear lubricant [illegible] compared the [illegible] process, [illegible] *Mech. Eng.*, 2014, 9, 28–33.

7 Nanofluids for Machining in the Era of Industry 4.0

Wei Li

Hunan University, Changsha, China

Ahmed Mohamed Mahmoud Ibrahim

Hunan University, Changsha, China

Minia University, Minia, Egypt

Yinghui Ren

Hunan University, Changsha, China

Ahmed Moustafa Abd-El Nabi

Minia University, Minia, Egypt

CONTENTS

7.1 INTRODUCTION

The rapid progress in modern industries requires new generations of manufacturing methods which are considered the fourth era of industry. However, the energy savings and hygenic impact, besides the ultra-high accuracy requirements, and recent

DOI: 10.1201/9781003284574-8

global attitudes as well as, are challenges. The traditional metalwork fluids cannot fulfill the recent global standards in terms of environmental and health impacts. In spite of the numerous advantages of metalworking fluids, like the extended tool life span and better surface finish than dry cutting, the traditional metalworking fluids (MWFs) are strictly used due to the health and environmental implications (Klocke & Eisenblätter, 1997). Meanwhile, costly and complicated treatments are required to dispose of MWFs after use. The researchers proved that there is substantial evidence connecting some types of cancers (pancreas, larynx, scrotum, rectum, skin) to the subjection to the MWFs (Calvert et al., 1998). Dry cutting was presented as having less environmental impact and being an inexpensive alternative to machining operations to solve difficulties related to the environmental obstacles inherent in flood-cooling modes. Despite the numerous environmental and economic rewards that dry cutting provides, surface burning and heat degradation are apparent. In addition, dry cutting changes significantly the metallurgical phase, mechanical, and physical properties of machined surfaces (Klocke & Eisenblätter, 1997; Lee et al., 2015). Nanofluids have been suggested as an eco-friendly alternative to the MWFs with a lower impact on workers' health. Meanwhile, the nanofluids could overcome the problems associated with dry cutting (Ibrahim et al., 2020). This chapter discusses the types, preparation, applications, and sustainability of nanofluids for machining in the era of industry.

7.2 PREPARATION OF THE NANOFLUIDS

The operation performance of the nanofluids is governed by the stability and preparing routes of these nanofluids. Sedimentation ends when the particle size falls below critical as gravity forces by Brownian forces are counterbalanced, according to the colloidal hypothesis. However, smaller sizes of nanoparticles with higher surface energies may suffer from agglomeration rather than the uniform dispersion in the nanofluids (Ilyas, Pendyala, & Marneni, 2014; Jailani, Franks, & Healy, 2008; Witharana, Palabiyik, Musina, & Ding, 2013). Therefore, the preparation route of nanofluids performs an important function in the stability of the dispersion of the nanoparticles.

As shown in Figure 7.1, There are two routes to prepare the nanofluids. The first route is the single-step method in which the nanoparticle production and the dispersion process are performed at the same time. This method ensures less agglomeration and better dispersion stability of the nanofluids (Li, Tung, Schneider, & Xi, 2009). One example of the single-step method is proposed by Zhu et al. (Zhu, Lin, & Yin,

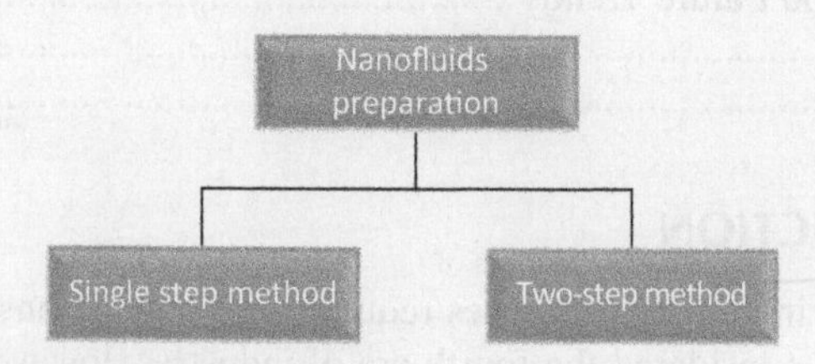

FIGURE 7.1 Nanofluids preparation methods.

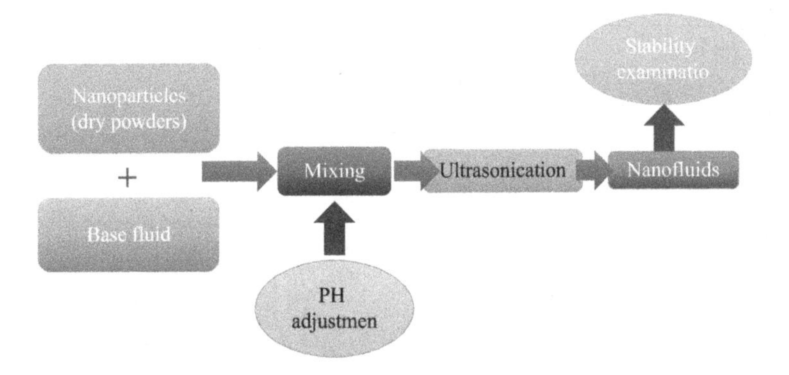

FIGURE 7.2 Two-step nanofluid production method.

2004) in which a stable and uniform copper dispersion is obtained by reducing $CuSO_4$-$5H_2O$ with NaH_2PO_2-H_2O in ethylene glycol under microwave irradiation.

The two-step method is based on producing the nano additives by chemical or physical powder in the form of powders. In the second processing stage, the nanopowder is dispersed into a fluid by intensive magnetic agitation, ultrasonic agitation, high screen mixing, and the like. As shown in Figure 7.2, the dry nanoparticle powder is mixed with the base fluid mechanically with consideration of adjusting the pH values, which have a substantial influence on the dispersion stability of the nanofluids. The mixture is then sonicated to provide excellent dispersion stability by creating adequate shear stress and cavitation. (Ibrahim et al., 2020; Johnson, Dobson, & Coleman, 2015). This method showed many benefits, such as simplicity, in that it doesn't require any complicated chemical or physical steps. In addition, the two-step method is an economical method in which the nanopowders are produced at an industrial level, which lessens costs effectively (W. Yu & Xie, 2012).

7.3 TYPES OF NANOFLUIDS

7.3.1 Graphene-Based Nanofluids

Graphene is a two-dimensional (2D) hexagonal carbon lattice that is the building component for 3D graphite. In several technical applications, outstanding graphene performance has received considerable interest in recent years (Young, Kinloch, Gong, & Novoselov, 2012). Graphene as a solid additive demonstrated an improvement in friction and wear strength. In addition, graphene not only produces minimal wear but also enables simple shearing (Berman, Erdemir, & Sumant, 2013; Ibrahim, Shi, Zhai, & Yang, 2015; Ibrahim, Shi, Zhang, Yang, & Zhai, 2015; Ibrahim et al., 2014). Concerning thermal conductivity, thanks to a high thermal conductivity in the planned temperatures, graphene exceeds most carbon-based materials because of its covalent sp2 bonding between carbon atoms (Balandin, 2011; Pop, Varshney, & Roy, 2012). The aforementioned unique properties of graphene give it the high potential to be used as a nano-additive for machining applications. Graphene nano-additives can be dispersed in palm oil with different content percentages of 0, 0.03, 0.1, 0.2,

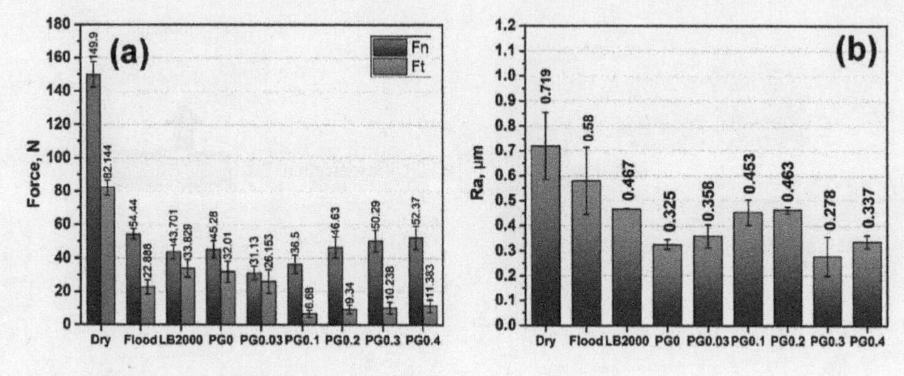

FIGURE 7.3 Cutting forces (a), and surface roughness (b); during the grinding operation at different lubrication conditions (Ibrahim et al., 2020).

0.3, and 0.4 wt.% to be used as a cutting fluid in the grinding operations. As shown in Figure 7.3, the graphene-based nanofluid outperforms the flood, dry, and high-performance LB2000 lubrication modes. Nevertheless, the palm oil graphene-based nanofluid lessened the cutting forces sharply with the comparison to the dry and wet lubrication modes. The minimum value of normal forces was recorded under palm oil 0.03 wt.% graphene nanoplatelets-based nanofluid, while the minimum tangential force was obtained when using palm oil 0.1 wt.% graphene nanoplatelets-based nanofluid. Besides, the palm-graphene nanofluid had a noticeable beneficial effect on the surface roughness.

Another attempt was done by Singh et al. (Kumar Singh et al., 2018) when they dispersed the graphene nanoplatelets in water and they used it as a cooling/lubrication medium for turning AISI 304. The experimental results proved that utilizing graphene-based nanofluids could decrease sharply the resulting surface roughness. Meanwhile, the higher percentages of graphene exhibited a better surface finish. As graphene has unique thermal properties, the graphene-based nanofluids transferred the cutting temperature effectively from the cutting zone to the surrounding medium through the graphene channels. Therefore, the reasonable heat dissipation lowered the thermal damages of the newly machined surfaces.

Canola oils are also used as a base fluid for graphene-based nanofluids in many machining operations. The results showed that the canola/graphene nanofluids extended the life span of the turning tool by 178–190%, while the cutting temperature and cutting forces decreased by 36–40% and 31–42%, respectively, in the comparison with dry conditions during machining of Ti-6Al-4V alloy (Singh, Dureja, Dogra, Gupta, & Mia, 2019). However, the friction coefficient decreases by 16–39%. Meanwhile, the graphene raised the thermal conductivity of canola oil, which enhances the heat dissipation from the cutting zone (Singh et al., 2020). The graphene presence is the main reason for the improvement effect of the nanofluids based on vegetable or biodegradable oils. Referring to the distinguished thermal conductivity of graphene, it can dissipate the heat generated from the cutting zone out to the surrounding; hence, it provides the required protection to the newly machined surfaces from thermal damages. Using the graphene-based nanofluids reduces the cutting energy sharply. This significant reduction in the specific cutting energy is attributed to the distinctive lubrication characteristics of graphene. Graphene eases the slipping of the cutting tool over the machined surfaces as it forms with the dispersing oil a thin film that has lower shear. Thus, lower friction can be obtained leading to a remarkable reduction in the heat generation in the cutting zone and lower specific cutting energy (Ibrahim et al., 2020).

7.3.2 Carbon Nanotube–Based Nanofluids

Another additive that attracted the attention of researchers due to its excellent lubrication properties is the carbon nanotube (CNT). As a carbon material, carbon nanotubes exhibit remarkable strength, high stiffness, high thermal conductivity, high chemical inertness, and distinctive thermal conductivity (Wong, Sheehan, & Lieber, 1997). Therefore, carbon nanotubes have attained much attention as eco-friendly additives. CNTs were dispersed in water to form an eco-friendly nanofluid combining the advantages of the high heat capacity of the water and the high thermal conductivity of CNTs. The main problem that faces the production of water-based CNTs nanofluids is the lack of solubility of CNTs in water (Chen & Collier, 2005). There are many ways to overcome such problems, including the chemical surface modification of CNTs. Multiwalled carbon nanotubes were covalently grafted polyacrylamide by redox polymerization of acrylamide with ceric ammonium nitrate being the initiator (Pei, Hu, Liu, & Hao, 2008). One more suggested method is to add humic acid to improve the dispersion of the CNTs in water (Kristiansen, Zeng, Wang, & Israelachvili, 2011). The addition of sodium dodecyl sulfate (SDS) as a surfactant is also suggested as an approach to boost the dispersion stability of CNTs in water (Sahu, Andhare, & Raju, 2018). Relative higher dispersion stability of CNTs is noticed in coconut oil. However, coconut oil-based CNTs nanofluid improved the machining performance of turning of Inconel 718 and Ti-6AL-4V alloys. It is noticed that coconut oil-based nanofluid lowered the machining temperature and extended the tool life (Sahu et al., 2018; Sarkar & Datta, 2021).

7.3.3 Al_2O_3-Based Nanofluids

As most carbon materials suffer from the lack of dispersion stability in different fluids, noncarbon materials were suggested to replace these carbon materials as a trial to produce stable nanofluids. Alumina nanoparticles were suggested as a possible substitution. The alumina nanoparticle is the most used nano additive due to the relatively lower costs with the enhanced lubrication properties. The lubricating action of the alumina particles is different from the lamellar materials, such as MoS_2 and graphite, whereas the alumina particles utilize the rolling to facilitate the sliding between the rubbing surfaces. Furthermore, alumina showed better dispersion properties than all carbon materials, especially when dispersed in water-based mediums. Sunflower oil/alumina nanofluid showed much better cooling/lubrication properties than dry and traditional cutting fluids in terms of force, torque, surface quality, and tool temperature (Pal, Chatha, & Sidhu, 2021). Furthermore, Al_2O_3 nanofluid helps effectively flush chips from the grinding zone and therefore solves the major difficulty during Ti–6Al–4V grinding (Setti, Sinha, Ghosh, & Rao, 2015). Because the alumina exhibits good thermal conductivity, it is used to prepare different kinds of nanofluids for a wide range of applications. Dispersing alumina nanoparticles with a percentage of 1 wt.% in water increases the thermal conductivity and viscosity by 19.74% and 29.77%, respectively (Chaudhari, Chakule, & Talmale, 2019). The beneficial effect of the alumina-based nanofluids extends to producing a better surface finish as it prevents thermal damages, eases the slipping between the tool and the machined surface, increases the wet area, and helps flush out the chips (Wang et al., 2017).

7.3.4 MoS_2-Based Nanofluids

The use of molybdenum disulfide (MoS_2) as a solid lubricant back to several years (Winer, 1967). The outstanding lubrication properties of MoS_2 were attributed to the crystal structure, in which the layers slip over with a relatively lower shear force. Regarding the remarkable lubricating characteristics of MoS_2 give this kind of solid lubricant the potential to be a reasonable additive to the nanofluids in the field of machining and manufacturing. MoS_2 can be suspended in soybean oil, palm oil, rapeseed oil, and liquid paraffin to be used during grinding of a 45 steel workpiece (Zhang, Li, Jia, Zhang, & Zhang, 2015). The presence of MoS_2 showed great benefits in terms of the friction coefficients and surface roughness. Nonetheless, 40 ml/hr of discharge of MoS_2-based nanofluid was able to noticeably decrease the surface roughness and cutting zone temperature in the comparison with conventional cooling/lubrication modes (Uysal, Demiren, & Altan, 2015). However, the MoS_2 additive doesn't show a considerable improvement in thermal conductivity. It can reduce the heat generated from the fiction significantly. Cottonseed oil is also used to disperse the MoS_2 nanoparticles to form eco-friendly lubrication oil. It has been proved the cottonseed/MoS_2-based nanofluid can improve the conditions of machining, however; other kinds of nanofluids outperformed the MoS_2-based nanofluid (Bai, Li, Dong, & Yin, 2019).

7.3.5 Pentaerythritol Rosin Ester–Based Nanofluids

Pentaerythritol rosin ester (PRE) has been chosen as a water/vegetable oil additive owing to its biodegradable and reasonable tribological properties. The hygienic and environmental impact have been investigated by US Food and Drug Administration that approved it in food industries (Yao et al., 2011). Moreover, PRE is widely used in varnishing, adhesives, painting, and drug microcapsules (Comyn, 1995; Moyano, París, & Martín-Martínez, 2016; Pathak & Dorle, 1987; Sheorey & Dorle, 1994; Yu, Chen, Gong, & Zhang, 2012). PRE is mainly produced by esterifying abietic acid and pentaerythritol, since abietic acid is available in abundant quantities and renewable resources in nature as shown in Figure 7.4. (Scharrer & Epstein, 1979). Zhuang et al. (Xu et al., 2019) conducted one of the very few studies about the effect of

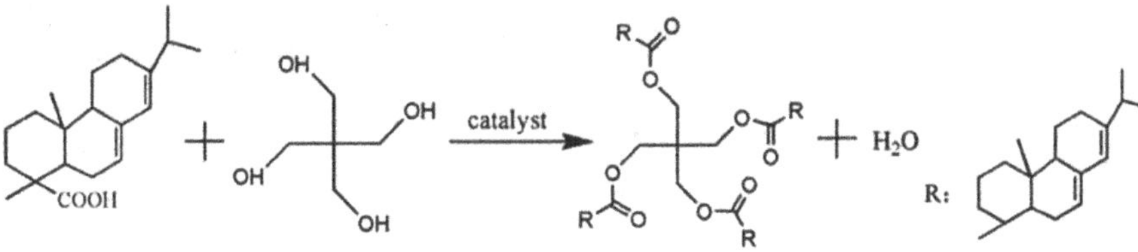

FIGURE 7.4 Synthesis of PRE (Xu et al., 2019).

PRE as an additive for vegetable oils. Their findings proved that PRE could step up the dynamic viscosity of the vegetable oils significantly. Therefore, PRE-based oils displayed a superior shear resistance. Furthermore, the addition of PRE can outstandingly boost the thermal stability of the vegetable oils (Rahman et al., 2017; Yao et al., 2011). In terms of oxidation properties, the vegetable oils containing PRE showed an enhanced oxidation resistance in comparison with the raw oils.

7.4 SUSTAINABILITY EVALUATION OF NANOFLUIDS

The health impact of cutting fluids has been studied extensively in the past few years. Primary or direct skin irritation is the consequence of direct exposure to cutting fluids. Most cutting fluids, on the other hand, create a breeding environment for germs, which poses a risk to the machine's operator. It is proved that some kinds of skin cancers, especially scrotal cancer is consistently related to long-term exposure to certain cutting fluids (Kipling, 1977; Mackerer, 1989). The environmental aspect of the fluids is a major problem. Many studies have shown that roughly 85% of the cutting fluids used worldwide are mineral oils that are categorized as hazardous. For instance, polycyclic aromatic hydrocarbons, chlorinated hydrocarbons, and nitrosamines, which have been recognized as carcinogens among the most hazardous chemicals. Thus, these kinds of fluids need further chemical treatment to remove the toxic constituents before disposal (Debnath, Reddy, & Yi, 2014; Howes, Toenshoff, Heuer, & Howes, 1991; Zhang, Rao, & Eckman, 2012). Surprisingly, the cost of the chemical treatment of the disposal oils could cost more than two times the purchasing costs (Shokrani, Dhokia, & Newman, 2012). However, the traditional cutting fluids could not meet the global requirements in energy-saving or machining outputs.

The nanofluids could decrease the energy consumption during the machining operations as shown in Figure 7.5. The use of palm oil as a base fluid with graphene nanoplatelets as an additive with percentages of 0.03, 0.1 0.2, 0.3, and 0.4 wt.% saved the cutting energies by 68.16%, 91.78%, 88.63%, 87.54%, and 86.14%, respectively, in the comparison with dry cutting (Ibrahim et al., 2020). The use of very small amounts of nanofluids of about 1 l/h lessens the costs of production of tens of liters of the conventional cutting fluids at the same operating time. However, the probability of direct exposure to the MWFs decreases the side effects on workers' health.

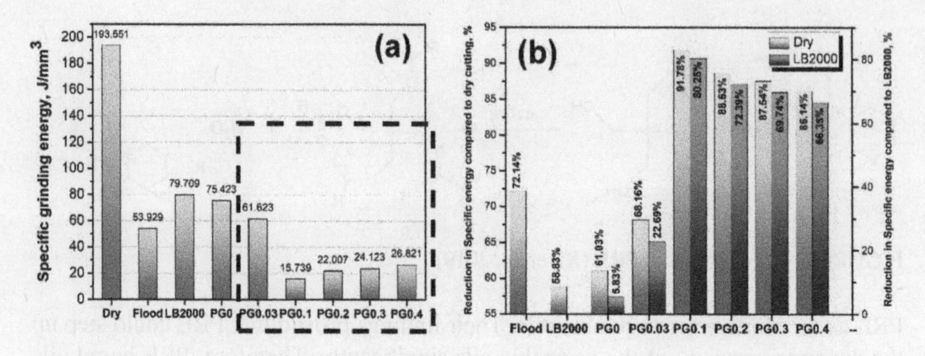

FIGURE 7.5 Energy consumption under different lubrication modes (Ibrahim et al., 2020).

7.5 APPLICATIONS OF NANOFLUIDS IN MODERN MACHINING OPERATIONS

The unique properties and benefits of functionalized nanofluids put them as potential alternatives to conventional fluids in the modern manufacturing processes. Table 7.1 shows the applications of different kinds of nanofluids in a wide range of machining processes, such as turning, grinding, drilling, milling, and the like.

7.6 CONCLUSION AND FUTURE TRENDS

The traditional fluids cannot meet the recent global standards in terms of environment, energy savings, and so on. Thus, there is an urgent demand for a new generation of fluids with unique operating performance and can fulfill the requirements of green manufacturing. Nanofluids with controlled properties attain much attention as an eco-friendly alternative. Different kinds of functionalized nano additives can accomplish the mission of cooling and lubrication due to the distinguished thermal and anti-friction characteristics. The nanofluids are being used with very small amounts (0.4–1 l//min) while the traditional fluids waste tens of liters per hour to dissipate the heat from the cutting zone. Thus, nanofluids decrease the costs significantly. Nonetheless, the nanofluids save the costs of the chemical treatments that are compulsory for conventional fluids. Regarding the hygienic sides, the nanofluids lessen the chances of causing severe health problems for workers. Nowadays, the global industries give great efforts to develop new generations of nanofluids with no or negligible environmental and health impact. This goal can be achieved by inventing and developing new techniques that can control the nanofluids during the operation and eliminate or prevent any leak of the nanoparticles to the surrounding.

ACKNOWLEDGMENTS

The work was supported and funded by the National Natural Science Foundation of China (52005174, 51875192), Natural Science Foundation of Hunan Province (2020JJ4193, S2021JJQNJJ0756), and Changsha Municipal Natural Science Foundation (kq2014048). The authors acknowledge the financial supports.

TABLE 7.1
The Applications of Nanofluids in Different Machining Operations

No.	Base oil	Material to be Cut	Nano Additive and Percentage	Machining Operation	Friction Coefficient	Machining Output		
						Ft (N)	Fn (N)	Ra (μm)
1	Canola oil (Singh, Sharma, & Dogra, 2020)	TI-6AL-4V-ELI	GNPs (1.5 wt.%)	Grinding	0.253	3.567	14.094	0.250
2	Palm oil (Wang et al., 2018)	Ductile cast iron 100-70-03	Al_2O_3 (2 wt.%)	Grinding	0.24	–	–	0.11
3	Canola oil (Singh, Sharma, Singh, & Dogra, 2019)	TI-6AL-4V-ELI	MoS_2 (2 wt.%)	Grinding	0.29	15.75	5.16	0.328
4	Distilled water (Sinha, Madarkar, Ghosh, & Rao, 2017)	Inconel 718 (IN718) superalloy	Ag 10% by volume of colloidal Ag NPs solution in DI water	Grinding	0.21	82	21	0.5
5	palm oil (Y. Zhang et al., 2015)	45 steel	MoS_2 (2 wt.%)	Grinding	0.3	–	–	0.7
6	Soybean oil (Kalita, Malshe, Kumar, Yoganath, & Gurumurthy, 2012)	Ductile Cast iron	MoS_2 (8 wt.%)	Grinding	0.31	–	–	–
7	Deionized water (Mao et al., 2013)	AISI 52100	AL_2O_3 (5 wt%)	Grinding	–	–	–	0.22
8	Distilled water (Setti et al., 2015)	Ti-6Al-4V	Al_2O_3 (1 wt.%) CuO (1 wt.%)	Grinding	0.26 0.28	45 37	20 18	0.75 0.62
9	Castor oil (Wang et al., 2016)	Ni-based alloy GH4169	NA	Grinding	0.30	–	–	0.36
10	Palm oil (Wang et al., 2017)	440C steel	Al_2O_3 (2 vol.%)	Grinding	0.34	–	–	–
11	vegetable-based oil (Şirin & Kıvak, 2019)	Inconel X-750 superalloy	hBN (0.5vol.%)	Milling	–	373.5 Resultant force	–	0.16
12	Emulsion of canola oil and distilled water (Shabgard, Seyedzavvar, & Mohammadpourfard, 2017)	AISI 1045 hardened steel	CuO 0.35 wt.%	Grinding	0.56	–	–	–
13	Paraffin oil (Nam, Lee, & Lee, 2011)	Aluminum 6061	Nano-diamond (1 vol.%)	Drilling	–	0.95 (Thrust force)	–	–

REFERENCES

Bai, X., Li, C., Dong, L., & Yin, Q. (2019). Experimental evaluation of the lubrication performances of different nanofluids for minimum quantity lubrication (MQL) in milling Ti-6Al-4V. *The International Journal of Advanced Manufacturing Technology*, *101*(9), 2621–2632.

Balandin, A. A. (2011). Thermal properties of graphene and nanostructured carbon materials. *Nature Materials*, *10*(8), 569–581.

Berman, D., Erdemir, A., & Sumant, A. V. (2013). Few layer graphene to reduce wear and friction on sliding steel surfaces. *Carbon*, *54*, 454–459.

Calvert, G. M., Ward, E., Schnorr, T. M., & Fine, L. J. (1998). Cancer risks among workers exposed to metalworking fluids: A systematic review. *American Journal of Industrial Medicine*, *33*(3), 282–292.

Chaudhari, S., Chakule, R., & Talmale, P. (2019). Experimental study of heat transfer characteristics of Al2O3 and CuO nanofluids for machining application. *Materials Today: Proceedings*, *18*, 788–797.

Chen, J., & Collier, C. P. (2005). Noncovalent functionalization of single-walled carbon nanotubes with water-soluble porphyrins. *The Journal of Physical Chemistry B*, *109*(16), 7605–7609.

Comyn, J. (1995). Surface characterization of pentaerythritol rosin ester. *International Journal of Adhesion and Adhesives*, *15*(1), 9–14.

Debnath, S., Reddy, M. M., & Yi, Q. S. (2014). Environmental friendly cutting fluids and cooling techniques in machining: A review. *Journal of Cleaner Production*, *83*, 33–47.

Howes, T., Toenshoff, H., Heuer, W., & Howes, T. (1991). Environmental aspects of grinding fluids. *CIRP Annals*, *40*(2), 623–630.

Ibrahim, A. M. M., Li, W., Xiao, H., Zeng, Z., Ren, Y., & Alsoufi, M. S. (2020). Energy conservation and environmental sustainability during grinding operation of Ti–6Al–4V alloys via eco-friendly oil/graphene nano additive and Minimum quantity lubrication. *Tribology International*, *150*, 106387.

Ibrahim, A. M. M., Shi, X., Zhai, W., & Yang, K. (2015). Improving the tribological properties of NiAl matrix composites via hybrid lubricants of silver and graphene nano platelets. *RSC Advances*, *5*(76), 61554–61561.

Ibrahim, A. M. M., Shi, X., Zhai, W., Yao, J., Xu, Z., Cheng, L., … Wang, Z. (2014). Tribological behavior of NiAl–1.5 wt% graphene composite under different velocities. *Tribology Transactions*, *57*(6), 1044–1050.

Ibrahim, A. M. M., Shi, X., Zhang, A., Yang, K., & Zhai, W. (2015). Tribological characteristics of NiAl matrix composites with 1.5 wt% graphene at elevated temperatures: An experimental and theoretical study. *Tribology Transactions*, *58*(6), 1076–1083.

Ilyas, S. U., Pendyala, R., & Marneni, N. (2014). Preparation, sedimentation, and agglomeration of nanofluids. *Chemical Engineering & Technology*, *37*(12), 2011–2021.

Jailani, S., Franks, G. V., & Healy, T. W. (2008). ζ potential of nanoparticle suspensions: Effect of electrolyte concentration, particle size, and volume fraction. *Journal of the American Ceramic Society*, *91*(4), 1141–1147.

Johnson, D. W., Dobson, B. P., & Coleman, K. S. (2015). A manufacturing perspective on graphene dispersions. *Current Opinion in Colloid & Interface Science*, *20*(5–6), 367–382.

Kalita, P., Malshe, A. P., Kumar, S. A., Yoganath, V., & Gurumurthy, T. (2012). Study of specific energy and friction coefficient in minimum quantity lubrication grinding using oil-based nanolubricants. *Journal of Manufacturing Processes*, *14*(2), 160–166.

Kipling, M. (1977). Health hazards from cutting fluids. *Tribology International*, *10*(1), 41–46.

Klocke, F., & Eisenblätter, G. (1997). Dry cutting. *Cirp Annals*, *46*(2), 519–526.

Kristiansen, K., Zeng, H., Wang, P., & Israelachvili, J. N. (2011). Microtribology of aqueous carbon nanotube dispersions. *Advanced Functional Materials*, *21*(23), 4555–4564.

Kumar Singh, R., Sharma, A. K., Mandal, V., Gaurav, K., Nag, A., Kumar, A., … Das, A. K. (2018). Influence of graphene-based nanofluid with minimum quantity lubrication on surface roughness and cutting temperature in turning operation. *Materials Today: Proceedings*, *5*(11), 24578–24586.

Lee, P.-H., Lee, S. W., Lim, S.-H., Lee, S.-H., Ko, H. S., & Shin, S.-W. (2015). A study on thermal characteristics of micro-scale grinding process using nanofluid minimum quantity lubrication (MQL). *International Journal of Precision Engineering and Manufacturing*, *16*(9), 1899–1909.

Li, Y., Tung, S., Schneider, E., & Xi, S. (2009). A review on development of nanofluid preparation and characterization. *Powder Technology*, *196*(2), 89–101.

Mackerer, C. R. (1989). Health effects of oil mists: A brief review. *Toxicology and Industrial Health*, *5*(3), 429–440.

Mao, C., Zhang, J., Huang, Y., Zou, H., Huang, X., & Zhou, Z. (2013). Investigation on the effect of nanofluid parameters on MQL grinding. *Materials and Manufacturing Processes*, *28*(4), 436–442.

Moyano, M. A., París, R., & Martín-Martínez, J. M. (2016). Changes in compatibility, tack and viscoelastic properties of ethylene n-butyl acrylate (EBA) copolymer–pentaerythritol rosin ester blend by adding microcrystalline wax, Fischer–Tropsch wax and mixture of waxes. *International Journal of Adhesion and Adhesives*, *65*, 47–53.

Nam, J. S., Lee, P.-H., & Lee, S. W. (2011). Experimental characterization of micro-drilling process using nanofluid minimum quantity lubrication. *International Journal of Machine Tools and Manufacture*, *51*(7–8), 649–652.

Pal, A., Chatha, S. S., & Sidhu, H. S. (2021). Performance evaluation of the minimum quantity lubrication with Al_2O_3-mixed vegetable-oil-based cutting fluid in drilling of AISI 321 stainless steel. *Journal of Manufacturing Processes*, *66*, 238–249.

Pathak, Y., & Dorle, A. (1987). Study of rosin and rosin derivatives as coating materials for controlled release of drug. *Journal of Controlled Release*, *5*(1), 63–68.

Pei, X., Hu, L., Liu, W., & Hao, J. (2008). Synthesis of water-soluble carbon nanotubes via surface initiated redox polymerization and their tribological properties as water-based lubricant additive. *European Polymer Journal*, *44*(8), 2458–2464.

Pop, E., Varshney, V., & Roy, A. K. (2012). Thermal properties of graphene: Fundamentals and applications. *MRS Bulletin*, *37*(12), 1273–1281.

Rahman, M. A., Lokupitiya, H. N., Ganewatta, M. S., Yuan, L., Stefik, M., & Tang, C. (2017). Designing block copolymer architectures toward tough bioplastics from natural rosin. *Macromolecules*, *50*(5), 2069–2077.

Sahu, N. K., Andhare, A. B., & Raju, R. A. (2018). Evaluation of performance of nanofluid using multiwalled carbon nanotubes for machining of Ti–6AL–4V. *Machining science and Technology*, *22*(3), 476–492.

Sarkar, S., & Datta, S. (2021). Machining performance of inconel 718 under dry, MQL, and nanofluid MQL conditions: Application of coconut oil (base fluid) and multi-walled carbon nanotubes as additives. *Arabian Journal for Science and Engineering*, *46*(3), 2371–2395.

Scharrer, R. P., & Epstein, M. (1979). Oxygen-stable rosin-primary polyhydric aliphatic alcohol esters and a method for preparing the same utilizing arylsulfonic acid catalysis: Google Patents.

Setti, D., Sinha, M. K., Ghosh, S., & Rao, P. V. (2015). Performance evaluation of Ti–6Al–4V grinding using chip formation and coefficient of friction under the influence of nanofluids. *International Journal of Machine Tools and Manufacture*, *88*, 237–248.

Shabgard, M., Seyedzavvar, M., & Mohammadpourfard, M. (2017). Experimental investigation into lubrication properties and mechanism of vegetable-based CuO nanofluid in MQL grinding. *The International Journal of Advanced Manufacturing Technology*, *92*(9–12), 3807–3823.

Sheorey, D., & Dorle, A. (1994). Preparation and study of release kinetics of rosin pentaerythritol ester microcapsules. *Journal of Microencapsulation*, *11*(1), 11–17.

Shokrani, A., Dhokia, V., & Newman, S. T. (2012). Environmentally conscious machining of difficult-to-machine materials with regard to cutting fluids. *International Journal of Machine Tools and Manufacture*, *57*, 83–101.

Singh, R., Dureja, J. S., Dogra, M., Gupta, M. K., & Mia, M. (2019). Influence of graphene-enriched nanofluids and textured tool on machining behavior of Ti-6Al-4V alloy. *The International Journal of Advanced Manufacturing Technology*, *105*(1), 1685–1697.

Singh, R., Dureja, J., Dogra, M., Gupta, M. K., Mia, M., & Song, Q. (2020). Wear behavior of textured tools under graphene-assisted minimum quantity lubrication system in machining Ti-6Al-4V alloy. *Tribology International*, *145*, 106183.

Singh, H., Sharma, V. S., & Dogra, M. (2020). Exploration of graphene assisted vegetables oil based minimum quantity lubrication for surface grinding of TI-6AL-4V-ELI. *Tribology International*, *144*, 106113.

Singh, H., Sharma, V. S., Singh, S., & Dogra, M. (2019). Nanofluids assisted environmental friendly lubricating strategies for the surface grinding of titanium alloy: Ti6Al4V-ELI. *Journal of Manufacturing Processes*, *39*, 241–249.

Sinha, M. K., Madarkar, R., Ghosh, S., & Rao, P. V. (2017). Application of eco-friendly nanofluids during grinding of Inconel 718 through small quantity lubrication. *Journal of Cleaner Production*, *141*, 1359–1375.

Şirin, Ş., & Kıvak, T. (2019). Performances of different eco-friendly nanofluid lubricants in the milling of Inconel X-750 superalloy. *Tribology International*, *137*, 180–192.

Uysal, A., Demiren, F., & Altan, E. (2015). Applying minimum quantity lubrication (MQL) method on milling of martensitic stainless steel by using nano MoS2 reinforced vegetable cutting fluid. *Procedia-Social and Behavioral Sciences*, *195*, 2742–2747.

Wang, Y., Li, C., Zhang, Y., Li, B., Yang, M., Zhang, X., … Zhai, M. (2017). Comparative evaluation of the lubricating properties of vegetable-oil-based nanofluids between frictional test and grinding experiment. *Journal of Manufacturing Processes*, *26*, 94–104.

Wang, Y., Li, C., Zhang, Y., Yang, M., Li, B., Dong, L., & Wang, J. (2018). Processing characteristics of vegetable oil-based nanofluid MQL for grinding different workpiece materials. *International Journal of Precision Engineering and Manufacturing-Green Technology*, *5*(2), 327–339.

Wang, Y., Li, C., Zhang, Y., Yang, M., Li, B., Jia, D., … Mao, C. (2016). Experimental evaluation of the lubrication properties of the wheel/workpiece interface in minimum quantity lubrication (MQL) grinding using different types of vegetable oils. *Journal of Cleaner Production*, *127*, 487–499.

Wang, Y., Li, C., Zhang, Y., Yang, M., Zhang, X., Zhang, N., & Dai, J. (2017). Experimental evaluation on tribological performance of the wheel/workpiece interface in minimum quantity lubrication grinding with different concentrations of Al_2O_3 nanofluids. *Journal of Cleaner Production*, *142*, 3571–3583.

Winer, W. O. (1967). Molybdenum disulfide as a lubricant: A review of the fundamental knowledge. *Wear*, *10*(6), 422–452.

Witharana, S., Palabiyik, I., Musina, Z., & Ding, Y. (2013). Stability of glycol nanofluids—the theory and experiment. *Powder Technology*, *239*, 72–77.

Wong, E. W., Sheehan, P. E., & Lieber, C. M. (1997). Nanobeam mechanics: Elasticity, strength, and toughness of nanorods and nanotubes. *Science*, *277*(5334), 1971–1975.

Xu, Z., Lou, W., Zhao, G., Zhang, M., Hao, J., & Wang, X. (2019). Pentaerythritol rosin ester as an environmentally friendly multifunctional additive in vegetable oil-based lubricant. *Tribology International*, *135*, 213–218.

Yao, K., Wang, J., Zhang, W., Lee, J. S., Wang, C., Chu, F., … Tang, C. (2011). Degradable rosin-ester–caprolactone graft copolymers. *Biomacromolecules*, *12*(6), 2171–2177.

Young, R. J., Kinloch, I. A., Gong, L., & Novoselov, K. S. (2012). The mechanics of graphene nanocomposites: A review. *Composites Science and Technology*, *72*(12), 1459–1476.

Yu, C., Chen, C., Gong, Q., & Zhang, F. A. (2012). Preparation of polymer microspheres with a rosin moiety from rosin ester, styrene and divinylbenzene. *Polymer International*, *61*(11), 1619–1626.

Yu, W., & Xie, H. (2012). A review on nanofluids: Preparation, stability mechanisms, and applications. *Journal of nanomaterials*, *2012*, 1–17.

Zhang, Y., Li, C., Jia, D., Zhang, D., & Zhang, X. (2015). Experimental evaluation of MoS2 nanoparticles in jet MQL grinding with different types of vegetable oil as base oil. *Journal of Cleaner Production*, *87*, 930–940.

Zhang, J. Z., Rao, P., & Eckman, M. (2012). Experimental evaluation of a bio-based cutting fluid using multiple machining characteristics. *Wear*, *12*, 13–14.

Zhu, H.-T., Lin, Y.-S., & Yin, Y.-S. (2004). A novel one-step chemical method for preparation of copper nanofluids. *Journal of Colloid and Interface Science*, *277*(1), 100–103.

8 Ionic Liquids as a Potential Sustainable Green Lubricant for Machining in the Era of Industry 4.0

Upendra Maurya and V. Vasu
National Institute of Technology, Warangal, India

CONTENTS

8.1 INTRODUCTION

Industry 4.0 is unfolding in front of us, and the pace of change is increasing rapidly [1]. Fields of the Internet of Things (IoT), artificial intelligence (AI), machine learning (ML), and big data analysis (BD) are being used in a cyber-physical system (CPS) to adopt and automate the machines [2]. These technologies not only can enable manufacturing systems to communicate with each other but also share real-time data that can be used for rapid and accurate decision-making. The maintenance crews can use real-time data to be better prepared for preventive maintenance

DOI: 10.1201/9781003284574-9

resulting in saving and persistent machine operation. The big data technology for machine operation can facilitate a predictive maintenance approach by accurately identifying the failure points and saving big on resources.

In short, Industry 4.0 aspires to develop smart robust machines with sophisticated CPS capable of self-sense and self-act. These technologies in cyber-physical production system (CPPS) will also revolutionize the manufacturing industry with the introduction of smart design (computer-aided design [CAD] and computer-aided machining [CAM], virtual reality, augmented reality, 3D printing), smart machining (smart tools, robots), smart monitoring, and control (various sensors and actuators) [2]. Industry 4.0 will enable the efficient flow of energy, and appropriate power can be given to machine components to minimize their resources and environmental impact, resulting in energy and operational savings. The benefits of CPPS, that is, energy efficiency, resource savings, automated machining, smart monitoring, and control, cannot be fully realized without future-ready metalworking fluids (MWFs). The performances of MWFs are application-specific and highly depend on machining parameters, and new formulations need to be developed to cater to particular applications for future machining centers.

8.2 FUNDAMENTAL OF MWFS

8.2.1 Definition, Purpose, and Types of MWFs

MWFs are liquid compounds that are supplied between tool and workpiece interface during machining with an overall aim to enhance productivity by colling and/or lubrication. As per the DIN-51385 standard, the processing fluids (MWFs) may be classified into two types as per their formulation, that is, oil-based and water-based. Generally, the base oil may be derived from mineral oil or synthetic oil or be bio-based (vegetable oil). The selection of any particular MWF is critical and depends on various factors, such as the nature of the machining process, the properties of the tool and workpiece material, and others. For instance, water-based MWFs are suitable when the prime objective is to dissipate the heat generated while oil-based MWFs are most appropriate when lubrication between tool and workpiece interface is the prime motive. The bulk of the MWFs properties depends on the base fluid and generally base fluids on their own are not sufficient to meet the dynamic demanding condition of machining. To enhance the machining performance, several chemical compounds (additives) are added to the base fluid, such as corrosion inhibitors, anti-weld agents, friction modifiers, extreme pressure emulsifiers, stabilizers, dyes, odorant, fragrances, and others. Figure 8.1 gives the classification of MWFs based on their composition as per DIN-51385 and common additive groups used in formulating these MWFs.

8.2.2 Supply Methods of MWFs

Considering economic and environmental aspects, various efficient MWFs supply methods for lubricating the tool–workpiece interface have been investigated over decades. Generally, for machining operation, a liquid coolant in large quantity (flood

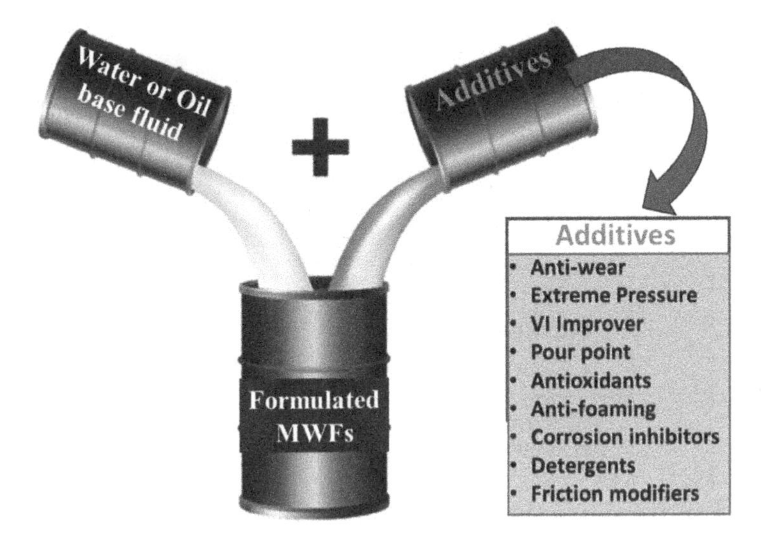

FIGURE 8.1 Metalworking fluid (MWF) formulation composed of a base fluid and common additives. Base fluids may be water-based or oil-based. Generally, commercial MWFs may consist of more than a dozen additives.

mode) is used to cool the heat generated due to plastic deformation and subsequent shearing of workpiece material. Other than cooling, these MWFs may also facilitate friction reduction, chip removal, and corrosion prevention of tool and workpiece materials. Several studies regarding flood coolant have been reported with an aim to enhance productivity [3, 4], develop efficient supply methods [5, 6] and develop effective MWFs formulations by altering its compositions [7, 8].

During conventional machining operation, the workpiece material is sheared by a tool and gets removed in the form of chips. The plastic deformation of the material in the primary shear zone releases energy in the form of heat due to the breakage of intermolecular bonds of the workpiece material as shown in Figure 8.2. Hence, the heat generated in the primary shear zone cannot be avoided, and the majority of the generated heat gets removed with the chip while some fraction is transferred to the tool and the workpiece. In the secondary shear zone, heat is generated primarily due to friction between tool and chip, which further heats the tool. This is the single biggest factor contributing to tool wear, and several MWFs compositions and supply methods have explored techniques to minimize friction and subsequent heat generation in the secondary shear zone. Due to their higher heat capacity, water-based MWFs have found wide application as flood coolants in various manufacturing processes. But they are not always effective in terms of longer tool life and efficiency. During the cutting operation, the tool edge becomes very hot, and with flood coolant, the tool gets quenched (rapidly cooled). These repetitive heating and cooling (thermal shock) of the tool cutting edge create microcracks, subsequently making the tool weaker.

To limit the profuse volume (30,000–60,000 mL/hr in flood cooling) or eliminate the MWFs and manage the thermal shock of the tool in machining, several different

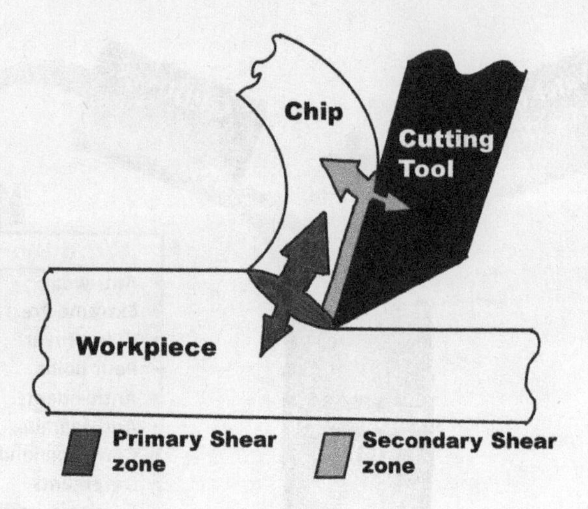

FIGURE 8.2 The primary and secondary shear zones during machining. Heat generated in primary shear zone cannot be avoided since it is released due to breakage of chemical bonds while effective lubrication can reduce heat generation in secondary shear zone by reducing friction between the chip and the tool.

approaches are being explored and have been implemented for over decades. The primary focus has been to develop harder tools by tailoring tool compositions and/or introducing new tool materials [9, 10]. In addition, various high-temperature-resistant multilayered coatings are being explored to further enhance the tool's hardness [11–13]. The other approaches revolve around either by softening the workpiece locally using thermally enhanced machining (e.g., laser-assisted machining) [14, 15] or reducing cutting temperature by using enhanced cooling techniques (e.g., cryogenic coolants) [16–18] or lubricating tool–workpiece interface (MQL, minimum quantity lubrication) with overall aim to reduce friction and enhance productivity [19–23]. The cryogenic cooling of the tool–workpiece interface is achieved by a cryogenic medium such as liquid nitrogen (boiling point—196 °C). cryogenic cooling is particularly applied for difficult-to-cut materials such as Inconel, titanium, and others [24–26], and it can further be used to enhance the machined surface quality [24, 27–29]. Few studies have reported the advantages of combining cryogenic methods with MQL and claimed improved performance [27, 30]. MQL is preferable in turning and forming operations and is rarely used for abrasive machining [31, 32]. In addition, MQL can easily replace existing MWFs setups with minor modifications [33].

As the name implies, MQL requires a very small quantity of lubricants to reduce the friction between tool and workpiece interface resulting in an overall decrease of cutting temperature, cutting forces, and enhancement of surface finish. The "minimum" quantity is subjective and may range from 20 mL/hr (German DIN specification) to 100 mL/hr (as per a few studies); nevertheless, the quantity is several 100 times lessor than flood cooling [33, 34]. Generally, in flood cooling, the primary objective is to remove the heat, for that large volume of fluid is transferred to the cutting zone. On the other hand, MQL aims to manage the overall heat, in particular elimination of heat generated due to friction between the tool and the workpiece [33]. MQL is suitable for almost all types of machining, but there are certain machining conditions where MQL shines such as open machines, intermittently used machine (less frequently used machines), micromachining, general machining, and high-speed machining [33].

The majority of existing MWFs are not sustainable due to performance limitations and negative impacts on the environment (soil and water pollution due to leakage and disposal) and operator's health (cancer, skin irritation, chronic respiratory problems). New effective and efficient additives need to be developed for future-ready MWFs. Room temperature ionic liquids (RTILs) are one of the promising candidates for effective lubrication of tool–workpiece interface and an alternative to harmful additives [35–38].

8.3 WHAT ARE IONIC LIQUIDS?

Ionic liquids (ILs) are innovative compounds with tunable physical, chemical, biological, and thermal properties. These inherent properties have promising potential to revolutionize many disciplines of scientific and engineering research such as solvents, catalysts, machining, electrolytes, cell biology, and many more at both laboratory and commercial scale [39, 40]. Ionic liquids are entirely made up of ions (cation and anion). Technically, molten table salt can be categorized as an ionic liquid, having a melting temperature of 801 °C, but at standard temperature and pressure, it remains in the solid crystalline phase. Nowadays, in scientific literature and general use, the term *IL* refers to the ionic compounds that are in liquid phase below 100 °C while the ionic compounds having a melting temperature above 100 °C are loosely termed *molten salt* [40, 41]. *"Room temperature ionic liquid" RTIL* is another term used for these ILs, generally consisting of organic cations and organic or inorganic anions.

ILs are also called "designer solvents," for the possibility of obtaining desired physicochemical properties by altering cations and anions (tunability). For instance, by altering cations and anions or both, viscosity, water miscibility, corrosive resistance, and polarity can be tuned for task-specific applications [42]. ILs inherit several desirable properties, such as negligible vapor pressure or nonvolatility at ambient conditions, noncorrosive, nonflammable, excellent thermal stability, tunable polarity, solubility, basicity, acidity, and others [40, 43]. Compared to hazardous volatile organic compounds (VOCs) ionic liquids are often considered as green solvents since it has extremely low vapor pressure.

Since the invention of the first IL over a century ago, up to 100,785 articles have been published through May 2021 (as per Mendeley reference manager searched results with keyword "ionic liquids") in various publications, patents, books, book chapters, conference proceedings, and other forms. Among these publications, more than 50% (66,299) came in the previous decade, indicating a great interest in IL research across scientific disciplines. Figure 8.3 gives the number of publications in the field of ionic liquids over the last 15 years. Most of these publications are from the field of chemistry (physical, analytical, organic, multidisciplinary, electrochemistry) material science, chemical engineering, nanotechnology, and energy fuels [43].

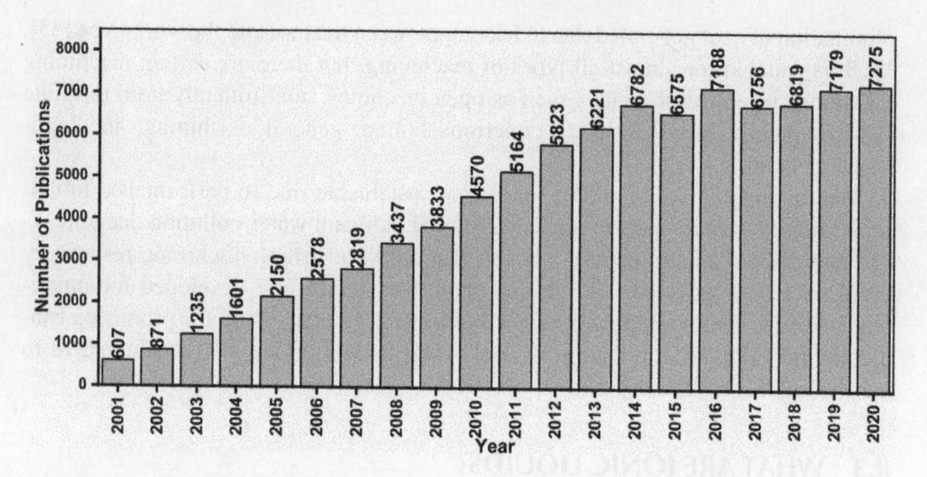

FIGURE 8.3 The publication over the last 20 years as per Mendeley reference manager searched results with keyword "ionic liquids" (ILs). IL research got momentum around 2000 when it became commercially available, meaning one need not be an expert in IL synthesis to use it for one's research.

Figure 8.4 shows the molecular structures of the most common cations and anions used in the IL research for friction reduction.

8.4 POTENTIAL OF ILS IN MACHINING

During machining, the temperature in the primary and secondary shear zone may reach very high and MWFs with lower thermal stability may disintegrate and evaporate at higher temperatures and may expose nascent surfaces resulting in poor machining. To enhance machinability and productivity new high-performance lubricants are being relentlessly explored.

ILs with properties such as lower volatility, nonflammability, desirable viscosity, low toxicity, physicochemical stability, excellent lubricity, polarity (very useful for adsorption film formation), and others make them suitable candidates for potential high-performance future-ready MWFs [49–52]. Due to excellent fluidity over a wide range of temperatures, that is, as high as 450 °C to as low as −100 °C, ILs can be suitable for almost all types of machining operations. Mist formation during machining generally results in oily premises, equipment, and repeated emission exposer is harmful to the operator's health.

A significant proportion of ILs are biodegradable and possess very low toxicity, which makes them benign not only on operator health but also on the environment [53, 54]. In addition, due to negligible vapor pressure, ILs do not pose any concern related to emission and evaporation during machining at a relatively higher temperature. On the other hand, ILs are nonflammable, meaning they can be easily stored and transported from the lubricant manufacturer to end user. Moreover, due to the possibility of altering both cation and anion and obtaining tunable desired properties of the ILs, depending on the particular application, ILs can be a very potent lubricants group [44, 51, 55].

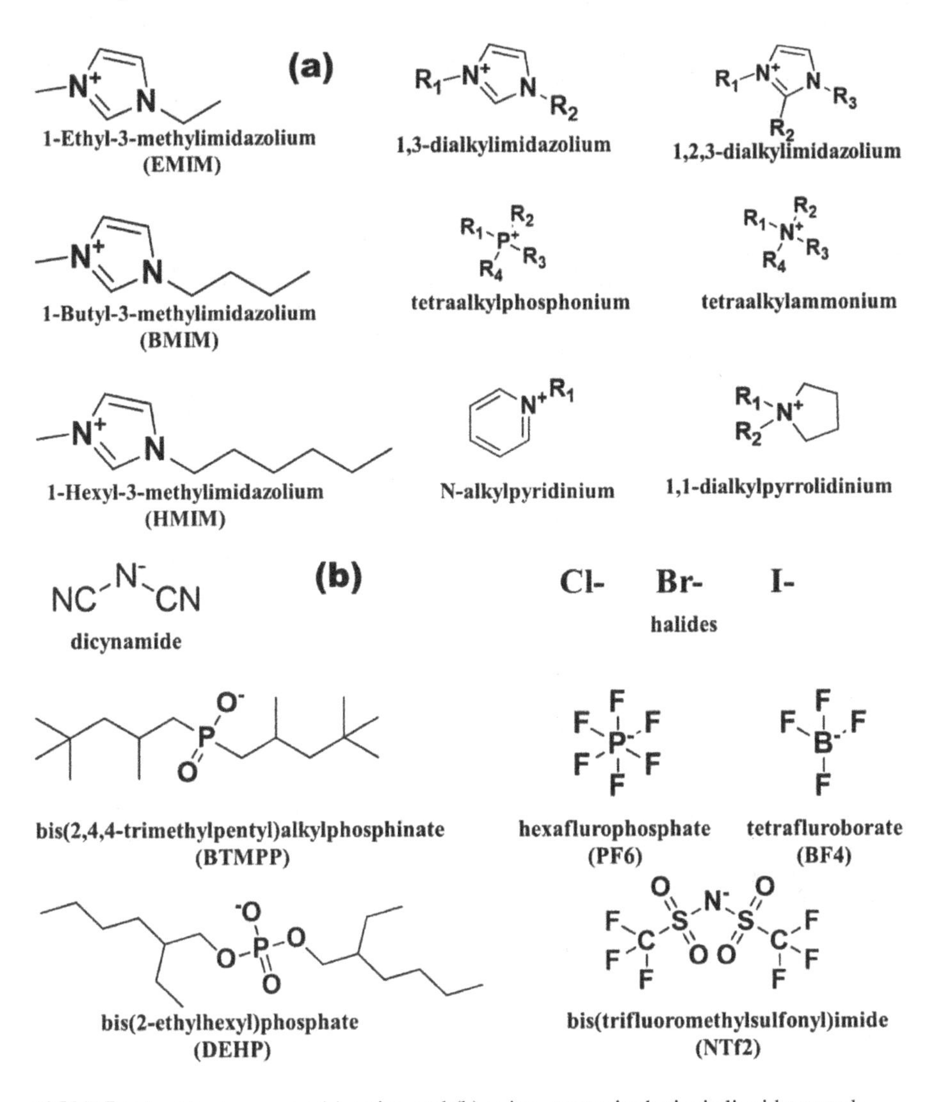

FIGURE 8.4 Most common (a) cation and (b) anion groups in the ionic liquid research.

The first study of ILs in machining was reported by Pham et al. in 2014 [37]. This research group used two ILs having different viscosities for machining an aluminum workpiece as neat oil and observed better performance over conventional oil. A couple of other groups attempted to use ILs as an additive in MWFs, but most of the initial ILs were immiscible in base oil, so their emulsion was used for machining purposes [35, 36, 44]. Just over half a dozen studies exploring ILs' potential for machining oil application have been reported to date as shown in Table 8.1. The table lists the pair of cation and anion, type of base oil, blend/emulsion concentration, tool and workpiece material, and machining methods, along with important findings. These reported studies show an addition of even small ILs concentration (0.5–10 wt.%) can enhance machining performance.

TABLE 8.1
Ionic Liquids, Miscibility, Base Oil, and Machining Performances

Sr. No	Cation	Anion	MWF	% of ILs	Workpiece and Tool Material	Machining Method	Important Findings	Ref.
1	EMIM BMIM	NTf_2 Iodide	-----	100%	Aluminum & Tungsten carbide	Three axis micro milling CNC machining	Performed better or equal to commercial conventional machining oils. Negligible evaporative loss (1%) compared to 75–85% of conventional oils.	Pham et al. [37]
2	BMIM	PF_6	Deionized Water (emulsion)	0.5%	Titanium & CBN	Turning operation Using machining center	Significant tool wear reduction. IL emulsion showed the smallest cutting forces, largest shear band spacing, best surface finish.	Devis et al. [44]
3	BMIM BMIM BMIM	PF_6 BF_4 NTf_2	Canola Oil (emulsion)	1 wt%	medium carbon steel & uncoated Tungsten carbide	CNC vertical milling	**PF_6** & BF_4 cation-based ILs gave the smoothest surface and lowest cutting force. In contrast, NTf_2 cation based IL showed performance similar to the vegetable oil and flood cooling system	Goindi et al. [35]
4	BMIM BMIM P_{4449}	PF_6 BF_4 DEHP	Conola Oil (emulsion) & PEG400 (miscible)	3 wt% 1 wt% 0.5 and 1 wt%	medium carbon steel & uncoated Tungsten carbide	CNC vertical milling	**PF_6** & BF_4 cation-based ILs showed improvement only at higher cutting speeds. Phosphonium IL showed improvement only at a lower cutting speed.	Goindi et al. [36]
5	BMIM	BF_4	Neem oil (emulsion) & Neem oil+water (emulsion)	1 and 2 wt%	carbon steel & Titanium Nitride Coated and Uncoated Tungsten Carbide	Turning operation using a geared lathe	Surface roughness was improved in the range of 40–62% compared to neat Neem oil	Panneer et al. [45]

6	N_{1888} P_{66614}	NTf_2 BTMPP	Modified Jatropha based ester (miscible)	1, 5, 10 wt%	medium carbon steel & uncoated Cermet	Turning operation using NC lathe machine	**O**ptimum concentration of 10wt% and 1wt% for ammonium phosphonium ILs respectively. **O**verall reduction of 4–5% cutting force, 7–10% cutting energy, 2–3% friction coefficient, 8–11% tool chip contact, 22–25% chip thickness, 1–2% friction angle over synthetic esters	Abdul Sani et al. [38]
6	N_{1888} P_{66614}	NTf_2 BTMPP	Modified Jatropha based ester (miscible)	1, 5, 10 wt%	medium carbon steel & uncoated Cermet	Turning operation using NC lathe machine	**10** wt% ammonium IL concentration reduced cutting force up to 12%, cutting temperature by 9%, surface roughness by 7%, improved tool life by up to 50%. **1** wt% phosphonium IL concentration reduced cutting force up to 11%, cutting temperature by 7%, surface roughness by 4%, improved tool life by up to 40%.	Abdul Sani et al. [46]
7	P_{66614}	Cl	Coconut oil (miscible)	1 wt%	Steel & multilayer-coated cermet	Turning operation using CNC lathe	**E**xcellent surface finish. **N**o significant benefit for flank wear was observed for 1 wt% IL concentration **R**ecommends use of higher ILs concentration (above wt%).	Pandey et al. [47]
8	BMIM	PF_6	Deionized Water & coconut oil	0.5 wt%	Inconel 825 Steel & carbide	Turning operation using CNC Lathe	**R**eduction of surface roughness by 88%, cutting temperature 74%, chip thickness 89%, minimized tool wear.	M. Naresh Babu [48]

8.5 EFFECT OF ILS ON MACHINING PARAMETERS

ILs, as neat MWFs, are economically restrictive; hence, other than Pham et al., researchers have explored it as an additive to MWFs. Studies have shown that even tiny concentrations of ILs may improve the machining performance such as surface roughness, tool wear, cutting zone temperature, cutting forces, and so on. The following sections discuss the effect of ILs on the machining of different tool–workpiece interfaces lubricated with various IL groups.

8.5.1 Effect of ILs on Machining Forces

Figure 8.5 illustrates the typical experimental setup for machining using ionic liquids [44]. During machining, shearing stresses between tool and workpiece interface fracture the material layers, resulting in the chip formation. Effective lubrication may

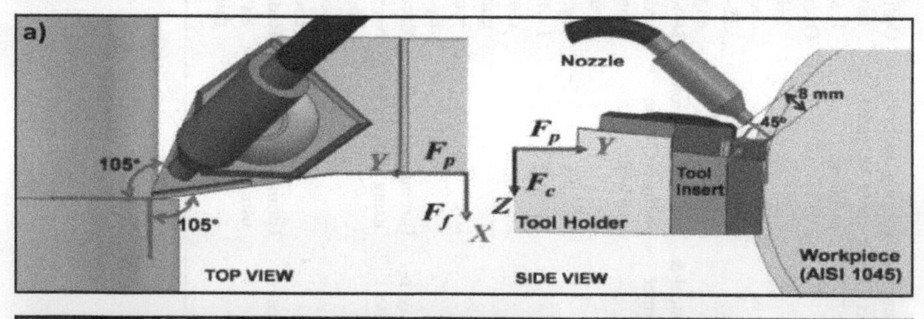

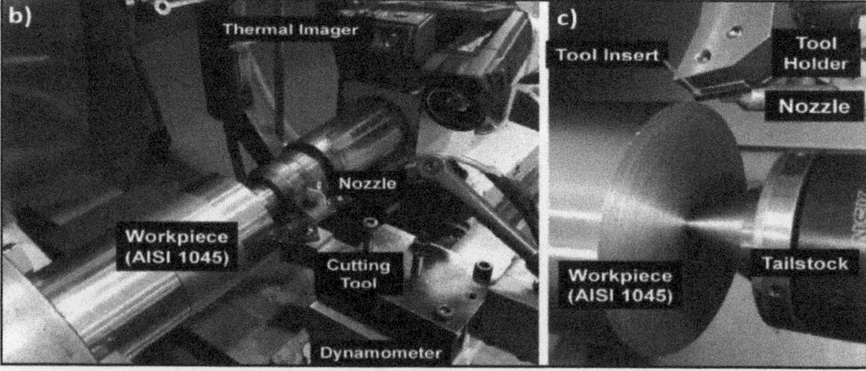

FIGURE 8.5 Experimental setup for machining, medium carbon steel workpiece with cermet tool (a) simplified diagram (top and side view) of tool and workpiece arrangement with MQL lubrication technique. The performance of ionic liquid (IL) blended metalworking fluids was tested in two phases: (b) In the first phase, the workpiece was machined using NC (numerical control) lathe machine with provision to record cutting force, surface roughness, and cutting temperature; (c) in the second phase, a computer numerical control (CNC) lathe machine was used for understanding tool wear mechanism and assessment of tool life.

(Reproduced with permission from copyright Tribology International 2019 reference number 51)

reduce friction and cutting zone temperature and enhance machinability. Pham et al. compared two imidazolium ILs (i.e., [BMIM][I] and [EMIM][TFSI]) with other lubrication conditions such as dry, distilled water, and two conventional commercial machine cutting fluid (i.e., ST501, TC#1) [37]. As expected, dry and distilled water machining conditions gave the highest cutting force while both the ILs gave a similar performance to that of TC#1. Viscosity influences the lubricant film formation, and a higher viscosity may form a thicker film on the surface interface, but despite higher viscosity of [BMIM][I] (viscosity of 401 cP at 25°C as compared to 32 of [EMIM][TFSI]), both ILs performed nearly similar. The similar performance of both the ILs was attributed to the reduction of viscosity of [BMIM][I] due to the rise in temperature while the increase in viscosity of [EMIM][TFSI] due to dispersion of nano/micro-sized chip particles during machining. Cation and anion both significantly affect the machining parameters, and altering any parameter can change performance. Goindi et al. synthesized three unique ILs with common cation (BMIM) and different fluorine-containing anions (PF_6, BF_4, NTf_2). An emulsion in canola oil of PF_6, BF_4 anion-based ILs gave the lowest cutting forces in both smooth and rough machining conditions [35]. While [BMIM][NTf_2] in canola oil performed in the same range of flood cooling and MQL with neat canola oil, they reported a similar trend of peak and mean cutting forces for different lubrication method, and their order of performance with respect to cutting forces were dry machining > flood cooling > MQL with neat vegetable oil > MQL with vegetable oil + IL emulsion. Cutting speed, depth of cut, and feed rate influence the performance of the lubricants. Some ILs can perform better in light operating conditions while others are more suited for heavy operating conditions. It was observed that PF_6 and BF_4 anion-based imidazolium ILs emulsion in canola oil reduced the cutting force at higher cutting speed at both (light and heavy) operating conditions. On the other hand, they observed decreased cutting force only at lower cutting speeds provided operating conditions were heavy.

While negligible performance improvement was observed at low cutting speed in light operating conditions. They proposed that these ILs require heavy operating conditions to generate high enough contact zone temperature for decomposition of ILs and subsequently release of tribo-active elements for protective tribofilm formation [36]. Moreover, at lower cutting speeds and lighter operating conditions, the temperature may not reach the decomposition temperature of ILs, resulting in weaker tribofilm. In contrast, the oil miscible phosphonium IL, when mixed in vegetable oil, showed excellent performance at lighter operating conditions due to lower decomposition temperature while negligible improvement in cutting force performance was observed at higher cutting speed. Generally, protic ionic liquids (PILs) forms relatively stronger hydrogen bonds with anion compared to aprotic ionic liquids (AILs). Which can significantly influence the lubricant film formation. Sani et al. reported a 12% improvement in cutting force performance when 10 wt% ammonium-based AIL mixed in modified Jatropha oil. While the optimum performance of 11% for PIL phosphonium-based IL was observed at 1 wt.% but not at 10 wt.% concentration. They concluded that there may be competition between base oil and IL for adsorption on the nascent surfaces. The adsorption depends on the polarity and concentration, and the base oil (modified jatropha oil), because of its excellent wettability, may have reduced the effectiveness of the additive, that is, PIL. They also observed a direct correlation between cutting forces and peak cutting temperature during machining. Both ILs decreased the peak cutting temperature in the range of 7–9% over the synthetic ester base oil.

8.5.2 Effect of ILs on Surface Roughness

Surface roughness values can give insight into the quality and performance of the machining process; manufacturing industries strive for lower surface roughness. In addition, it may also directly affects cutting forces, wear rate, cutting zone temperature, and obviously surface finish [47]. The surface tension of the MWFs may influence the machining specifically surface roughness. Pham et al. argued that MWFs with Higher surface tension remove the chip particles, minimizing the three-body abrasion and resulting in a smoother surface [37]. They reported that [BMIM][I] gave the lowest surface roughness compared to [EMIM][TFSI] and two conventional machining formulated oil due to their higher free-surface energy. Higher surface tensions of [BMIM][I] cleansed the chips due to the well-known "lotus effect," resulting in the lowest surface roughness. On the other hand, MWFs with a lower surface tension wet the workpiece surface but remove relatively fewer chip particles, resulting in increased three-body abrasion and subsequently poor surface roughness. If lubrication is inadequate or does not provide sufficient protective film the tool and workpiece may undergo severe adhesive wear. Goindi et al. observed vigorous scouring on the workpiece surfaces due to relatively higher adhesion rate on the surfaces lubricated under dry, flood cooling, and MQL with vegetable oil [35]. While the addition of ILs in the vegetable base oil showed very low adhesive scouring probably due to protective film formation on the tool flank and workpiece surfaces resulting in a smoother surface. Sani et al. reported a 4–7% improvement in surface roughness performance over synthetic easter for both PIL and AIL at optimum concentrations. They also attributed this improved surface roughness performance to nonflammable, thermally stable, and nonvolatile properties of the ILs, which enhanced the adsorption of protective layer on tool–workpiece interface subsequently reducing friction wear and peak temperature [46]. Pandey et al. observed excellent surface roughness improvements by blending imidazolium-based ILs in coconut oil; they also made similar conclusions about the protective film formation of the tool–workpiece interface. ILs performance depends on its miscibility, polarity, and other physicochemical properties. Devis et al. observed marginal improvement in surface roughness by the addition of 0.5% IL in the water–IL (emulsion) MQL lubrication system. Vegetable-based MWFs provide excellent lubricity generally better than mineral oil resulting in lower cutting force, cutting zone temperature, and most importantly improved surface roughness properties [56, 57]. Pandey et al. blended phosphonium-based ILs in coconut oil and claimed superiority of coconut oil-based MWFs over mineral oil and reported IL improved surface roughness by 3.1 and 2.7 times compared to dry and spray cooling conditions, respectively [47]. They also made observations using the TOPSIS (Technique for Order Preference by Similarity to the Ideal Solution) method that feed rate and depth-of-cut parameters influenced the surface roughness equally (each with 48.98%).

8.5.3 Mechanism of Tool Wear Lubricated with ILs

Devis et al. reported that 0.5 wt.% IL emulsion in water (water + IL) reduced tool wear by 60% and 15% over dry and water machining, respectively [44] (see Figure 8.6a–b). Furthermore, EDS (energy-dispersive X-ray spectroscopy) analysis

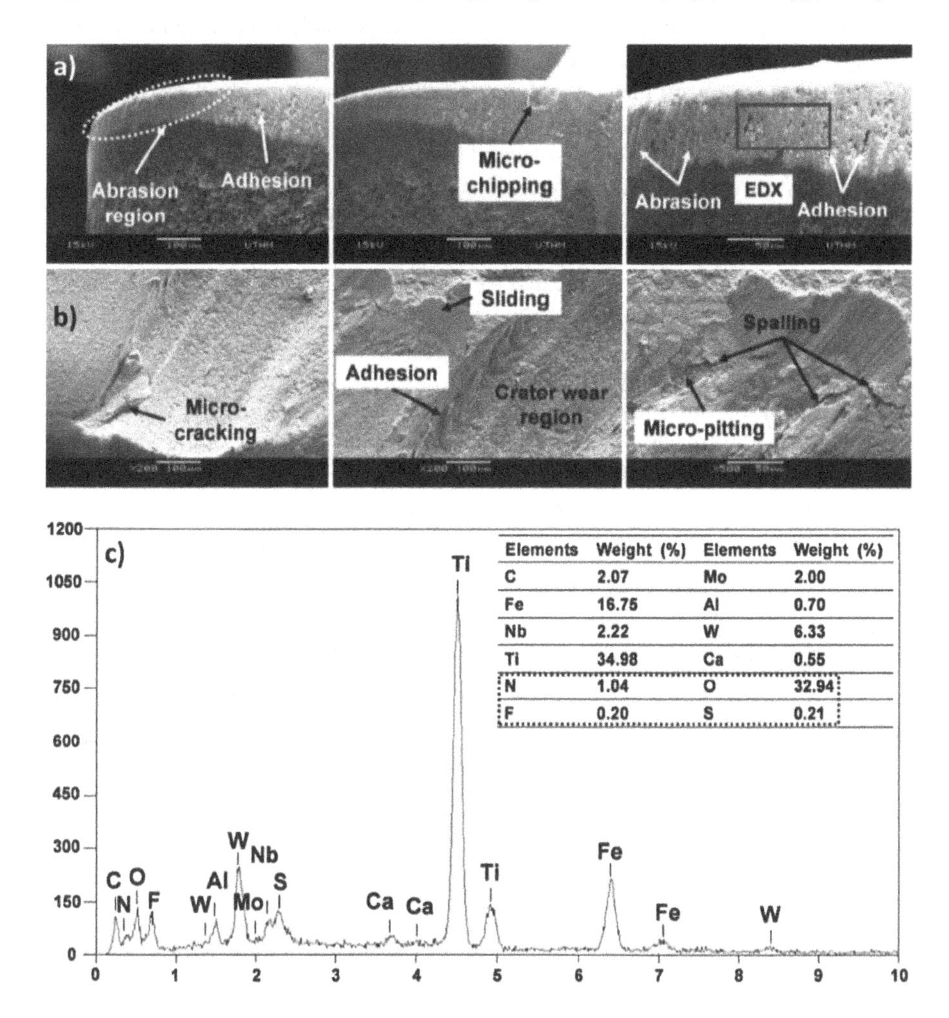

Elements	Weight (%)	Elements	Weight (%)
C	2.07	Mo	2.00
Fe	16.75	Al	0.70
Nb	2.22	W	6.33
Ti	34.98	Ca	0.55
N	1.04	O	32.94
F	0.20	S	0.21

FIGURE 8.6 Scanning electron microscopy and EDS analysis of the cutting tool lubricated with 10 wt% [$N_{1,8,8,8}$] [NTf2]: (a) combination of wear mechanisms (adhesion, abrasion, microchipping) were observed on the tool flank face, (b) while on tool rake face, primarily crater wear, along with traces of adhesion, pitting, spalling, attrition wear, was also observed. (c) EDS analysis shows a strong concentration of Fe and C, which originated from workpiece and adhered on the tool. Furthermore, O, N, F, and S presence on tool surface confirms the degradation of ILs at elevated temperature and the formation of a strong protective film rich in Fe_2O_3, FeO, sulfides, fluorides, and others.

(Reproduced with permission from copyright Tribology International 2019 reference number 51.)

of tool cutting edge gave 75% and 55% titanium (workpiece material) residue for the dry and water lubrication, respectively, indicating the dominant adhesive wear mechanism of the tool (see Figure 8.6c). On the other hand, the EDS analysis gave merely 15% titanium residues, indicating a dominant abrasive wear mechanism for tool wear. They proposed protective tribofilm formation due to tribo-chemical reaction between the IL and tool/workpiece surfaces responsible for reduced tool wear; see Figure 8.7. Longer tool life depends on the ability of the lubricant to be successfully directed toward the cutting zone, resulting in decreased friction. Goindi et al. observed the workpiece material adhered strongly to the cutting edge of the tool due to weak or no film in the case of flood cooling and dry condition, respectively. While relative average tool chip contact length was reduced due to the presence of ILs, resulting in decreased adhesion between the tool and the workpiece. The EDS analysis of the tool cutting edge shows the presence of fluorine; during machining, at tool chip interface, ILs disintegrate and tribo-active elements, that is, fluorine, react with the nascent surface and form stronger fluorides. the density of active elements deposition on the surface highly depends on the machining condition, and the more severe the condition, the higher the deposition.

Goindi et al. reported higher fluorine density in rough machining than in finish machining since the machining condition was severe during rough machining due to higher depth of cut. Sani et al. reported a 50% improvement of tool life with the addition of 10 wt.% AIL while 5 wt.% PIL addition enhanced tool life by 20% compared

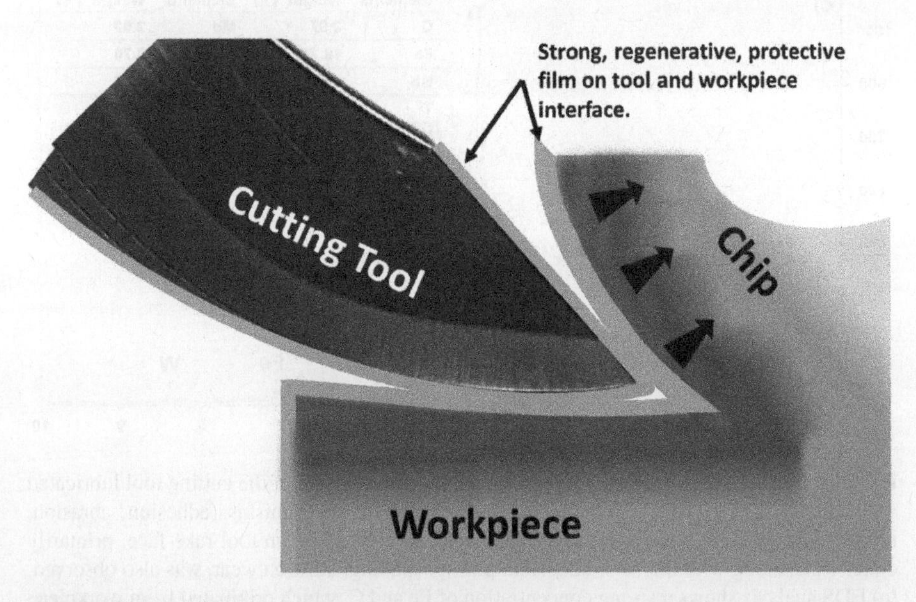

FIGURE 8.7 The protective film formation on the tool and workpiece surfaces. Due to polarity, ionic liquids (ILs) adhere to the tool and workpiece surfaces and form a weaker physisorption film, reducing friction. Furthermore, at elevated temperatures and high shear stress, ILs decompose, and active elements (P, O, N, F, S) react with the tool surface, forming a stronger regenerative metal oxide protective film. These weak physisorption and chemically reacted stronger oxide films improve the machining performance.

to neat synthetic ester oil. They also reported disintegration of ILs at high temperature and formation of scratch-resistant protective film on surfaces composed of iron oxides (FeO, Fe_2O_3), phosphides, sulphates [46]. The EDS spectra showed the presence of active elements P, O, N, F, and S on the tool flank surface lubricated with the respective formulated MWFs. They proposed two predominant mechanisms for film formation one is weaker physisorption (physical adsorption) due to the polar nature of ILs while the other is a stronger chemical reaction due to shear stress and higher cutting zone temperature. On the other hand, Pandey et al. did not observe any significant improvement in the tool flank wear with 1 wt.% of [$P_{6,6,6,14}$][Cl] in coconut oil and suggested the use of a higher concentration of ILs in the base oil for the positive impact the tool flank wear [47]. They reported with the use of TOPSIS analysis that cutting speed is the single most important parameter (90.1%) that influences tool wear.

8.5.4 Physicochemical Properties of ILs on Machining Conditions

Generally during machining, the contact temperature may reach very high and influence the surface morphology and machining forces. Additives can provide performance enhancement in certain criteria, but the bulk of the lubricant performance still depends on the base oil. That means that if the base oil is not thermally stable, then no additive concentration can enhance its thermal stability, and the formulated lubricant may perform poorly. For instance, Goindi et al. observed that even though phosphonium ILs showed maximum viscosity improvement in base oil, its overall performance was poor at high-speed operation due to the lower thermal stability of the PEG (polyethylene glycol) base oil [36]. They observed the direct correlation between the thermal decomposition temperature of ILs and machining forces. They reported that at lighter operating conditions, the temperature did not show any effect on machining forces since the cutting zone temperature may not reach degradation temperature. But at higher operating speeds, a strong correlation was observed; ILs with higher thermal decomposition temperatures performed better with respect to machining forces at higher cutting speeds. Most of the ILs have a higher specific-heat capacity than the base oil and may enhance the overall heat transfer rate and subsequently a better cooling rate. Furthermore, due to the inherent polarity of ILs, they adsorb on the cutting surfaces and forms an effective protective layer resulting in lower cutting zone temperature. Sani et al. reported 7–9% lesser temperature of the cutting zone lubricated by MWFs consisting of ILs and attributed this improvement to enhanced wettability, heat transfer capacity, and reduced temperature of the cutting zone due to the effective lubrication provided by ILs [46]. Furthermore, they also concluded that higher viscosity positively influenced the surface roughness and machining forces.

8.5.5 Effect of ILs Concentrations on Machining

Pandey et al. blended [P_{66614}] [Cl] in coconut oil and reported excellent improvement in surface roughness while no improvement of tool flank wear was observed. Furthermore, they recommended higher ILs concentration for possible improvement in tool flank wear performance [47]. Based on the published literature, one thing is certain: that more is not always better in the case of ILs concentration and that multiple factors may affect performance such as machining condition, ILs chemistry, and tool and workpiece materials, among others. For instance, Sani et al. blended phosphonium and ammonium-based ILs in modified Jatropha oil and reported different optimum concentrations for each [46, 58]. Phosphonium ILs showed improvement up to 1 wt.% concentration beyond that performance deteriorated while ammonium ILs showed improvement up to 10 wt.%. Few studies have reported certain ILs may not show any improvement in the performance of the base fluid. Goindi et al. observed that NTf_2 anion-based imidazolium ILs did not show any improvement over a canola base oil while two other imidazolium ILs improved machining performances [35]. Similarly in another study, they reported $[P_{4449}]$[DEHP] that ILs performed better only at lighter machining conditions due to lower degradation temperature while the other two imidazolium-based ILs showed excellent performance only at heavier machining conditions while no improvement in lighter machining conditions [36]. Similar results have also been reported in tribological studies of ILs in different base oils. For instance, Gusain et al. did not observe any significant change in wear reduction by increasing $[N_{8888}]$[BScB] concentration from 1–2.5 wt.% in PEG200 base oil [59]. X. Fu et al. reported no further improvement in friction and wear was observed by increasing $[N_{18910}]$[dioctyl phosphite], ZDDP, and $[N_{18910}]$[didodecyl phosphite] concentration from 0.5 wt.% to 3 wt.% [60].

8.6 CONCLUSION AND OUTLOOK

MWFs are one of the most important components of any manufacturing system. It can be called a complex compound and can significantly affect the productivity, resource efficiency, and energy of any manufacturing process. There cannot be a single universal MWF composition that suits all types of applications due to the wide range of performance parameters and dynamic demanding conditions, prompting a multidisciplinary approach for MWF development. As Industry 4.0 is unfolding, the demand for efficient, effective, and environmentally benign MWFs is ever increasing. Investigations have shown the promising potential of using ILs for enhancing the performance parameters of the MWFs. Most of the earlier concerns related to first and second generations of ILs, such as toxicity, thermal stability, cost, miscibility, corrosive nature, and others, are now minimized or eliminated with the development of the third generation of ILs. These ILs are composed of long alkyl chain length ammonium and phosphonium cations paired with long-branched phosphonate, sulphonate, and carboxylates anions, which diffuse the charges and give excellent miscibility even in nonpolar base fluids.

To date, fewer than a dozen studies have been published, which is negligible if compared to the vastness of feasible ILs. Moreover, over half of these studies involve non-miscible ILs in the base fluid. Nevertheless, almost all of these reported studies have shown excellent improvement in terms of cutting zone temperature, cutting forces, surface finish, and tool wear. The published research discussed in this chapter has repeatedly confirmed that even a very small concentration is sufficient to get the benefit of ILs in MWFs since neat ILs may be economically restrictive. Furthermore, the comparative study of ILs added MWFs with conventional machining oils has shown superior performance of ILs opening new direction in MWFs research. The positive results have generated great interest in ILs research aiming replacement or liming some of the harmful compounds of formulated MWFs. The working mechanism of ILs is still not clearly understood but deduced from surface characterization techniques such as scanning electron microscopy, EDS at post-machining tool, and workpiece surface analysis. Generally, it is believed that physisorption and chemical reactions are two dominant mechanisms associated with ILs operation, but the real process is still not known due to the complexities in machining because of chemical, physical, mechanical, and thermal interactions.

Since the synthesis of the first oil miscible ILs in 2012, several unique ILs with excellent miscibility in polar and nonpolar base fluids have been reported in the field of tribology; these ILs should be explored in the field of machining considering their friction performance and cost. All the published literature of ILs research in machining deals with mixing ILs in base fluids while no report is available considering the compatibility of ILs with other additives. Future studies should explore the synergy between ILs and other common MWFs additives, such as antioxidants, corrosion inhibitors, detergents, dispersants, and so on, first at laboratory scale and then dynamic conditions of the actual industrial scale. Furthermore, ILs' performance in fully formulated MWFs should also be explored to understand its compatibility with bulk additive packages for actual machining applications. Almost all the IL research in machining is focused on the MQL approach. ILs should also be incorporated with other lubricants supply methods, such as cryogenic cooling, laser-assisted machining, and others. Research on the long-term stability of ILs in formulated MWFs is also not available, particularly regarding its interaction with impurities and decomposition products, such as water, acids, soot, micro/nanoparticles, and others. Furthermore, ILs' compatibility with machining tools and workpiece materials should also be studied since ILs may cause accelerated wear due to their chemistry. It is generalized that ILs are green, biocompatible, and biodegradable, which is not always the case; future studies should explore these properties as well.

ABBREVIATIONS

[EMIM]	Ethyl-methyl-imidazolium
[BMIM]	Butyl-methyl-imidazolium
[NTf_2]	Bis(trifluoromethanesulfonyl)amide
[PF_6]	Hexafluorophosphate
[BF_4]	Tetrafluoroborate
[DEHP]	Bis(2-ethylhexyl)phosphate
[P_{4449}]	Tributyl(nonyl)phosphonium
[N_{1888}]	Trioctylmethylammonium
[P_{66614}]	Tetradecyltrihexylphosphonium
[BTMPP]	Bis(2,4,4-trimethylpentyl)phosphinate

REFERENCES

1. Ghosh, A.K., Sharif Ullah, A.M.M., Kubo, A., Akamatsu, T., D'Addona, D.M.: Machining phenomenon twin construction for industry 4.0: A case of surface roughness. *J. Manuf. Mater. Process.* 4, (2020). https://doi.org/10.3390/jmmp4010011
2. Zheng, P., Wang, H., Sang, Z., Zhong, R.Y., Liu, Y., Liu, C., Mubarok, K., Yu, S., Xu, X.: Smart manufacturing systems for Industry 4.0: Conceptual framework, scenarios, and future perspectives. *Front. Mech. Eng.* 13, 137–150 (2018). https://doi.org/10.1007/s11465-018-0499-5
3. Aoyama, T., Kakinuma, Y., Yamashita, M., Aoki, M.: Development of a new lean lubrication system for near dry machining process. *CIRP Ann. - Manuf. Technol.* 57, 125–128 (2008). https://doi.org/10.1016/j.cirp.2008.03.094
4. Denkena, B., Helmecke, P., Hülsemeyer, L.: New production technologies in aerospace industry - 5th machining innovations conference (MIC 2014) energy efficient machining with optimized coolant lubrication flow rates. *Procedia CIRP*. 24, 25–31 (2014). https://doi.org/10.1016/j.procir.2014.07.140
5. Claudin, C., Mondelin, A., Rech, J., Fromentin, G.: Effects of a straight oil on friction at the toolworkmaterial interface in machining. *Int. J. Mach. Tools Manuf.* 50, 681–688 (2010). https://doi.org/10.1016/j.ijmachtools.2010.04.013
6. Mondelin, A., Claudin, C., Rech, J., Dumont, F.: Effects of lubrication mode on friction and heat partition coefficients at the tool-work material interface in machining. *Tribol. Trans.* 54, 247–255 (2011). https://doi.org/10.1080/10402004.2010.538489
7. Skerlos, S.J., Hayes, K.F., Clarens, A.F., Zhao, F.: Current advances in sustainable Metalworking Fluids research. *Int. J. Sustain. Manuf.* 1, 180–202 (2008). https://doi.org/10.1504/IJSM.2008.019233
8. Tawakoli, T., Hadad, M.J., Sadeghi, M.H.: Investigation on minimum quantity lubricant-MQL grinding of 100Cr6 hardened steel using different abrasive and coolant-lubricant types. *Int. J. Mach. Tools Manuf.* 50, 698–708 (2010). https://doi.org/10.1016/j.ijmachtools.2010.04.009
9. Peng, Y., Miao, H., Peng, Z.: Development of TiCN-based cermets: Mechanical properties and wear mechanism. *Int. J. Refract. Met. Hard Mater.* 39, 78–89 (2013). https://doi.org/10.1016/j.ijrmhm.2012.07.001
10. Xu, C.H., Feng, Y.M., Zhang, R.B., Zhao, S.K., Xiao, X., Yu, G.T.: Wear behavior of Al2O3/Ti(C,N)/SiC new ceramic tool material when machining tool steel and cast iron. *J. Mater. Process. Technol.* 209, 4633–4637 (2009). https://doi.org/10.1016/j.jmatprotec.2008.10.017
11. Biksa, A., Yamamoto, K., Dosbaeva, G., Veldhuis, S.C., Fox-Rabinovich, G.S., Elfizy, A., Wagg, T., Shuster, L.S.: Wear behavior of adaptive nano-multilayered AlTiN/MexN PVD coatings during machining of aerospace alloys. *Tribol. Int.* 43, 1491–1499 (2010). https://doi.org/10.1016/j.triboint.2010.02.008
12. Zhang, S., Guo, Y.B.: An experimental and analytical analysis on chip morphology, phase transformation, oxidation, and their relationships in finish hard milling. *Int. J. Mach. Tools Manuf.* 49, 805–813 (2009). https://doi.org/10.1016/j.ijmachtools.2009.06.006
13. Fox-Rabinovich, G.S., Kovalev, A.I., Aguirre, M.H., Beake, B.D., Yamamoto, K., Veldhuis, S.C., Endrino, J.L., Wainstein, D.L., Rashkovskiy, A.Y.: Design and performance of AlTiN and TiAlCrN PVD coatings for machining of hard to cut materials. *Surf. Coatings Technol.* 204, 489–496 (2009). https://doi.org/10.1016/j.surfcoat.2009.08.021
14. Rahman Rashid, R.A., Sun, S., Wang, G., Dargusch, M.S.: An investigation of cutting forces and cutting temperatures during laser-assisted machining of the Ti-6Cr-5Mo-5V-4Al beta titanium alloy. *Int. J. Mach. Tools Manuf.* 63, 58–69 (2012). https://doi.org/10.1016/j.ijmachtools.2012.06.004

15. Schubert, A., Nestler, A., Pinternagel, S., Zeidler, H.: Influence of ultrasonic vibration assistance on the surface integrity in turning of the aluminium alloy AA2017. *Materwiss. Werksttech.* 42, 658–665 (2011). https://doi.org/10.1002/mawe.201100834
16. Sun, S., Brandt, M., Dargusch, M.S.: Machining Ti-6Al-4V alloy with cryogenic compressed air cooling. *Int. J. Mach. Tools Manuf.* 50, 933–942 (2010). https://doi.org/10.1016/j.ijmachtools.2010.08.003
17. Umbrello, D., Micari, F., Jawahir, I.S.: The effects of cryogenic cooling on surface integrity in hard machining: A comparison with dry machining. *CIRP Ann. - Manuf. Technol.* 61, 103–106 (2012). https://doi.org/10.1016/j.cirp.2012.03.052
18. Dhar, N.R., Kamruzzaman, M.: Cutting temperature, tool wear, surface roughness and dimensional deviation in turning AISI-4037 steel under cryogenic condition. *Int. J. Mach. Tools Manuf.* 47, 754–759 (2007). https://doi.org/10.1016/j.ijmachtools.2006.09.018
19. Kuzu, A.T., Berenji, K.R., Ekim, B.C., Bakkal, M.: The thermal modeling of deep-hole drilling process under MQL condition. *J. Manuf. Process.* 29, 194–203 (2017). https://doi.org/10.1016/j.jmapro.2017.07.020
20. Sultan, A.A., Okafor, A.C.: Effects of geometric parameters of wavy-edge bull-nose helical end-mill on cutting force prediction in end-milling of Inconel 718 under MQL cooling strategy. *J. Manuf. Process.* 23, 102–114 (2016). https://doi.org/10.1016/j.jmapro.2016.05.015
21. Weinert, K., Inasaki, I., Sutherland, J.W., Wakabayashi, T.: Dry machining and minimum quantity lubrication. *CIRP Ann. Manuf. Technol.* 53, 511–537 (2004). https://doi.org/10.1016/S0007-8506(07)60027-4
22. Emami, M., Sadeghi, M.H., Sarhan, A.A.D.: Investigating the effects of liquid atomization and delivery parameters of minimum quantity lubrication on the grinding process of Al2O 3 engineering ceramics. *J. Manuf. Process.* 15, 374–388 (2013). https://doi.org/10.1016/j.jmapro.2013.02.004
23. Liew, W.Y.H.: Low-speed milling of stainless steel with TiAlN single-layer and TiAlN/AlCrN nano-multilayer coated carbide tools under different lubrication conditions. *Wear.* 269, 617–631 (2010). https://doi.org/10.1016/j.wear.2010.06.012
24. Wang, Z.Y., Rajurkar, K.P.: Cryogenic machining of hard-to-cut materials. *Wear.* 239, 168–175 (2000). https://doi.org/10.1016/S0043-1648(99)00361-0
25. Aramcharoen, A., Chuan, S.K.: An experimental investigation on cryogenic milling of inconel 718 and its sustainability assessment. *Procedia CIRP.* 14, 529–534 (2014). https://doi.org/10.1016/j.procir.2014.03.076
26. Bermingham, M.J., Palanisamy, S., Kent, D., Dargusch, M.S.: A comparison of cryogenic and high pressure emulsion cooling technologies on tool life and chip morphology in Ti-6Al-4V cutting. *J. Mater. Process. Technol.* 212, 752–765 (2012). https://doi.org/10.1016/j.jmatprotec.2011.10.027
27. Kenda, J., Pusavec, F., Kopac, J.: Analysis of residual stresses in sustainable cryogenic machining of nickel based alloy - Inconel 718. *J. Manuf. Sci. Eng. Trans. ASME.* 133, 1–7 (2011). https://doi.org/10.1115/1.4004610
28. Shokrani, A., Dhokia, V., Muñoz-Escalona, P., Newman, S.T.: State-of-the-art cryogenic machining and processing. *Int. J. Comput. Integr. Manuf.* 26, 616–648 (2013). https://doi.org/10.1080/0951192X.2012.749531
29. Zisman, W.A.: Relation of the Equilibrium Contact Angle to Liquid and Solid Constitution. *Adv. Chem.* 43, 1–51 (1964). https://doi.org/10.1021/ba-1964-0043.ch001
30. Madanchi, N., Winter, M., Thiede, S., Herrmann, C.: Energy Efficient Cutting Fluid Supply: The Impact of Nozzle Design. *Procedia CIRP.* 61, 564–569 (2017). https://doi.org/10.1016/j.procir.2016.11.192
31. Bay, N., Azushima, A., Groche, P., Ishibashi, I., Merklein, M., Morishita, M., Nakamura, T., Schmid, S., Yoshida, M.: Environmentally benign tribo-systems for metal

forming. *CIRP Ann. - Manuf. Technol.* 59, 760–780 (2010). https://doi.org/10.1016/j.cirp.2010.05.007
32. Attanasio, A., Gelfi, M., Giardini, C., Remino, C.: Minimal quantity lubrication in turning: Effect on tool wear. *Wear.* 260, 333–338 (2006). https://doi.org/10.1016/j.wear.2005.04.024
33. Walker, T.: The MQL handbook-A guide to machiing with Minimum Quantity Lubrication. 1–41 (2011)
34. da Silva, L.R., Bianchi, E.C., Fusse, R.Y., Catai, R.E., França, T.V., Aguiar, P.R.: Analysis of surface integrity for minimum quantity lubricant-MQL in grinding. *Int. J. Mach. Tools Manuf.* 47, 412–418 (2007). https://doi.org/10.1016/j.ijmachtools.2006.03.015
35. Goindi, G.S., Sarkar, P., Jayal, A.D., Chavan, S.N., Mandal, D.: Investigation of ionic liquids as additives to canola oil in minimum quantity lubrication milling of plain medium carbon steel. *Int. J. Adv. Manuf. Technol.* 94, 881–896 (2018). https://doi.org/10.1007/s00170-017-0970-1
36. Goindi, G.S., Jayal, A.D., Sarkar, P.: Application of ionic liquids in interrupted minimum quantity lubrication machining of plain medium carbon steel: Effects of ionic liquid properties and cutting conditions. *J. Manuf. Process.* 32, 357–371 (2018). https://doi.org/10.1016/j.jmapro.2018.03.007
37. Pham, M.Q., Yoon, H.S., Khare, V., Ahn, S.H.: Evaluation of ionic liquids as lubricants in micro milling - Process capability and sustainability. *J. Clean. Prod.* 76, 167–173 (2014). https://doi.org/10.1016/j.jclepro.2014.04.055
38. Abdul Sani, A.S., Rahim, E.A., Sharif, S., Sasahara, H.: Machining performance of vegetable oil with phosphonium- and ammonium-based ionic liquids via MQL technique. *J. Clean. Prod.* 209, 947–964 (2019). https://doi.org/10.1016/j.jclepro.2018.10.317
39. Hallett, J.P., Welton, T.: Room-temperature ionic liquids: Solvents for synthesis and catalysis. 2. *Chem. Rev.* 111, 3508–3576 (2011). https://doi.org/10.1021/cr1003248
40. Welton, T.: Room-Temperature Ionic Liquids. Solvents for Synthesis and Catalysis. *Chem. Rev.* 99, 2071–2084 (1999). https://doi.org/10.1021/cr980032t
41. Welton, T.: Ionic liquids: a brief history. *Biophys. Rev.* 10, 691–706 (2018). https://doi.org/10.1007/s12551-018-0419-2
42. Rogers, R.D., Seddon, K.R.: Ionic Liquids - Solvents of the Future? *Science (80-).* 302, 792–793 (2003). https://doi.org/10.1126/science.1090313
43. Singh, S.K., Savoy, A.W.: Ionic liquids synthesis and applications: An overview. *J. Mol. Liq.* 297, 112038 (2020). https://doi.org/10.1016/j.molliq.2019.112038
44. Davis, B., Schueller, J.K., Huang, Y.: Study of ionic liquid as effective additive for minimum quantity lubrication during titanium machining. *Manuf. Lett.* 5, 1–6 (2015). https://doi.org/10.1016/j.mfglet.2015.04.001
45. Karthikey, V.V.: Investigations on the effect of Ionic Liquid (BMIMBF4) based Metal Working Fluids applied using Minimum Quantity Lubrication on Surface Roughness. *Int. J. Pure Appl. Math.* 118, 2283–2293 (2018)
46. Abdul Sani, A.S., Rahim, E.A., Sharif, S., Sasahara, H.: The influence of modified vegetable oils on tool failure mode and wear mechanisms when turning AISI 1045. *Tribol. Int.* 129, 347–362 (2019). https://doi.org/10.1016/j.triboint.2018.08.038
47. Pandey, A., Kumar, R., Sahoo, A.K., Paul, A., Panda, A.: Performance analysis of trihexyltetradecylphosphonium chloride ionic fluid under MQL condition in hard turning. Int. *J. Automot. Mech. Eng.* 17, 7629–7647 (2020). https://doi.org/10.15282/IJAME.17.1.2020.12.0567
48. Naresh Babu, M., Anandan, V., Dinesh Babu, M.: Performance of ionic liquid as a lubricant in turning inconel 825 via minimum quantity lubrication method. *J. Manuf. Process.* 64, 793–804 (2021). https://doi.org/10.1016/j.jmapro.2021.02.011

49. Saurín, N., Minami, I., Sanes, J., Bermúdez, M.D.: Study of the effect of tribo-materials and surface finish on the lubricant performance of new halogen-free room temperature ionic liquids. *Appl. Surf. Sci.* 366, 464–474 (2016). https://doi.org/10.1016/j.apsusc.2016.01.127
50. Cai, M., Yu, Q., Liu, W., Zhou, F.: Ionic liquid lubricants: When chemistry meets tribology. *Chem. Soc. Rev.* 49, 7753–7818 (2020). https://doi.org/10.1039/d0cs00126k
51. Gusain, R., Mungse, H.P., Kumar, N., Ravindran, T.R., Pandian, R., Sugimura, H., Khatri, O.P.: Covalently attached graphene-ionic liquid hybrid nanomaterials: Synthesis, characterization and tribological application. *J. Mater. Chem.* A. 4, 926–937 (2016). https://doi.org/10.1039/c5ta08640j
52. Upendra, M., Vasu, V.: Synergistic Effect Between Phosphonium-Based Ionic Liquid and Three Oxide Nanoparticles as Hybrid Lubricant Additives. *J. Tribol.* (2020). https://doi.org/10.1115/1.4045769
53. Nagendramma, P., Khatri, P.K., Thakre, G.D., Jain, S.L.: Lubrication capabilities of amino acid based ionic liquids as green bio-lubricant additives. *J. Mol. Liq.* 244, 219–225 (2017). https://doi.org/10.1016/j.molliq.2017.08.115
54. Rohlmann, P., Munavirov, B., Furó, I., Antzutkin, O., Rutland, M.W., Glavatskih, S.: Non-halogenated ionic liquid dramatically enhances tribological performance of biodegradable oils. *Front. Chem.* 7, 1–8 (2019). https://doi.org/10.3389/fchem.2019.00098
55. Saurín, N., Sanes, J., Bermúdez, M.D.: New Graphene/Ionic Liquid Nanolubricants. *Mater. Today Proc.* 3, S227–S232 (2016). https://doi.org/10.1016/j.matpr.2016.02.038
56. Fernando, W.L.R., Sarmilan, N., Wickramasinghe, K.C., Herath, H.M.C.M., Perera, G.I.P.: Experimental investigation of Minimum Quantity Lubrication (MQL) of coconut oil based Metal Working Fluid. *Mater. Today Proc.* 23, 23–26 (2019). https://doi.org/10.1016/j.matpr.2019.06.079
57. Belluco, W., De Chiffre, L.: Performance evaluation of vegetable-based oils in drilling austenitic stainless steel. *J. Mater. Process. Technol.* 148, 171–176 (2004). https://doi.org/10.1016/S0924-0136(03)00679-4
58. Abdul Sani, A.S., Rahim, E.A., Sharif, S., Sasahara, H.: Machining performance of vegetable oil with phosphonium- and ammonium-based ionic liquids via MQL technique. *J. Clean. Prod.* 209, 947–964 (2019). https://doi.org/10.1016/j.jclepro.2018.10.317
59. Gusain, R., Singh, R., Sivakumar, K.L.N., Khatri, O.P.: Halogen-free imidazolium/ammonium-bis(salicylato)borate ionic liquids as high performance lubricant additives. *RSC Adv.* 4, 1293–1301 (2014). https://doi.org/10.1039/c3ra43052a
60. Fu, X., Sun, L., Zhou, X., Li, Z., Ren, T.: Tribological Study of Oil-Miscible Quaternary Ammonium Phosphites Ionic Liquids as Lubricant Additives in PAO. *Tribol. Lett.* 60, 1–12 (2015). https://doi.org/10.1007/s11249-015-0596-0

9 Sustainable Electrical Discharge Machining Process

A Pathway

Satish Mullya

Annasaheb Dange College of Engineering and Technology, Ashta, India

Tribeni Roy

Birla Institute of Technology and Science Pilani (BITS Pilani), Rajasthan, India

London South Bank University, London, UK

CONTENTS

DOI: 10.1201/9781003284574-10

9.1 INTRODUCTION

The surging population and the ever-increasing demand for consumer products with new features have boosted the manufacturing sector. The reduction in product life-cycle time and a wide variety have led to technological advancements. New machining methods with high process capabilities are being developed for shaping advanced materials with a high strength-to-weight ratio. This is essential for fabricating compact and flexible products to meet customer requirements. Advanced machining processes (AMPs) are suitable for processing hard and tough materials using energy in its direct form. In the present competitive market, higher production, fast supply, and reduced cost have led to the manufacturing sector ignoring the environmental effects of machining processes. With strict guidelines from governments, industries are bound to reduce harmful emissions and solid wastes and avoid their direct disposal to the environment. Recycling plants and waste management are undertaken to reduce the environmental impact.

Machining processes, including conventional and nonconventional, must be sustainable in terms of power consumption, economic, environmental, social, and health aspects. Sustainable machining processes will have certain gain to the society. The electrical-discharge machining (EDM) process was invented in the late 1940s and has an almost 80-year history [1]. It is one of the most extensively used processes for fabricating complicated features in electrically conductive high-strength materials with a high aspect ratio. The thermal erosion process in EDM can cut practically any known material with a high level of accuracy and precision. Hence, it is used in die-making, automotive, aircraft, electronics, and biomedical applications. EDM has become indispensable in the industry because of its ability to machine intricate shapes, achieve a high surface finish, and its being economical and environmentally friendly with a compact system. The dielectric breakdown, ionization, and generation of spark melt and fuse the tool and workpiece material. This emits harmful gases, toxic wastes, and aerosols that pollute air and water. Inhaling such gases and working with conventionally used dielectric may lead to severe health issues and skin problems. With the growing demand for EDM machines and their variants, which is estimated to be about USD 10 billion by 2027, the problem of environmental sustainability will be exacerbated. Therefore, it is very important to find alternative ways of machining using EDM that will reduce the pollutants and ultimately lead to green and sustainable manufacturing ecosystems.

9.2 DESCRIPTION OF EDM PROCESS

EDM is a thermoelectric process that uses a pulsating direct current supply with an on/off cycle. The tool is connected to the negative terminal and the workpiece to the positive terminal while submerged under the dielectric fluid as shown in Figure 9.1. During the pulse-on time, voltage drops between the two electrodes and the current increases. The flow of ions and its avalanche form a plasma channel that is restricted to grow by the surrounding dielectric fluid increasing the energy density at a small point. Temperature rises to about 10000 K and pressure of 3 Kbar in a small inter-electrode gap (IEG) of 50 μm or less [2]. The tool and workpiece surface partly melt

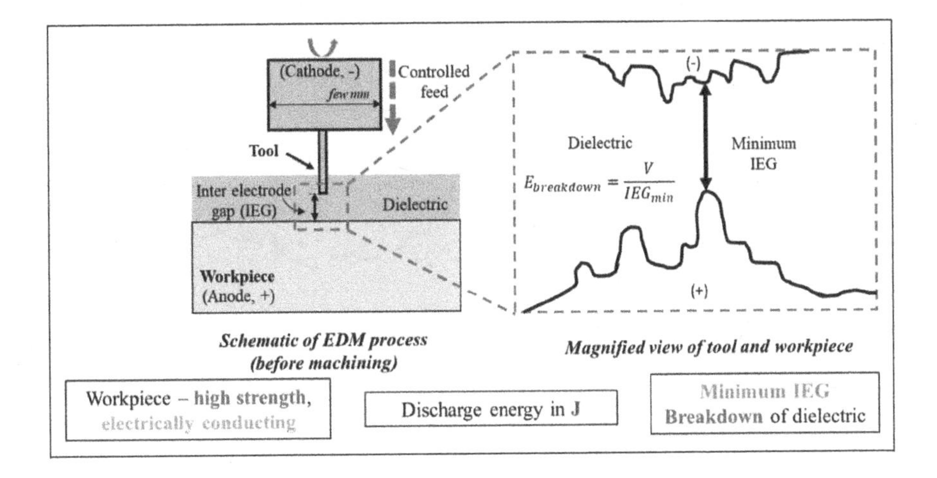

FIGURE 9.1 Schematic of the electrical-discharge machining process.

and vaporize. During the pulse-off time, the pressure drops, and the bubble enclosing the plasma channel collapses and removes the molten material in the form of debris particles as shown in Figure 9.2. A small crater is formed on both the electrodes, the fresh dielectric flushes the debris particles, and the subsequent cycle continues for the next sparking. Material is removed intermittently in the form of thousands of craters formed at the smallest gap of the electrode. A larger volume of material is removed from the anode, that is, the workpiece, due to the higher mobility of electrons. Sparking occurs where the tool faces the workpiece and the negative replica of tool is formed on the workpiece. Various electrically conductive materials that can be machined by EDM are steel, titanium, nitinol, Inconel, hindalium, cobalt, chromium, nickel, and high-strength alloys [3].

9.2.1 Conventional EDM and Its Variants

The conventional EDM process is popularly known as die-sink EDM process, in which the tools sink in the downward direction to cut the desired cavity in the workpiece. Two or more processes either conventional or nonconventional are combined to form a hybrid process in order to use the advantages and overcome the limitations of each process. EDM is combined with conventional grinding to form electrical-discharge grinding (EDG) and combined with milling forms electrical-discharge milling (ED milling). Wire EDM (WEDM) and reverse EDM (REDM) are the variants being developed to be applicable for various machining needs. The schematic diagram of all the processes is shown in Figure 9.3.

EDG uses the phenomenon of EDM in which the material is removed due to sparking and grinding. Sparking occurs between the workpiece and the rotating grinding wheel, and the material is removed in the form of debris particles. The high rotational surface speed of 180 m/min flushes the dielectric in the IEG and removes the debris from the gap. The conductive wheel is usually made up of graphite as it is cheap and machinable. As the material removal is not due to physical shearing action

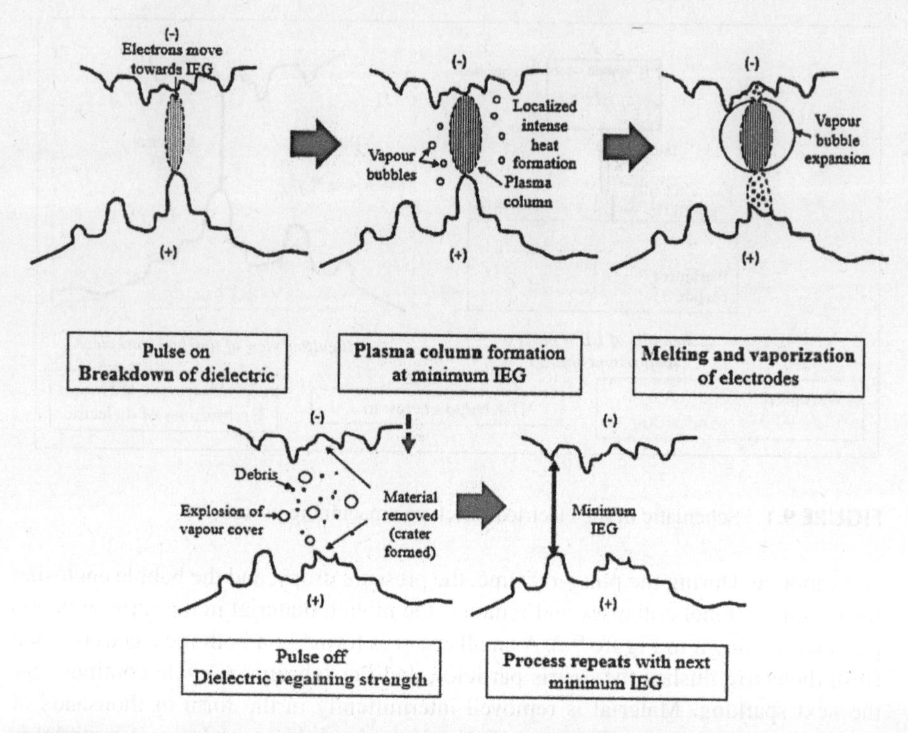

FIGURE 9.2 Breakdown phenomenon in electrical-discharge machining.

as in conventional grinding, faster material removal with high accuracy and surface finish is achieved.

ED milling uses a simple cylindrical tool to cut complex 3D geometry with a high aspect ratio. The CNC control regulates the motion of the tool along the trajectory of the cut in all directions. The servo control system maintains the gap between the tool and the workpiece, and continuous spark generated, along the circumference of the tool, removes the material due to melting and vaporization. The tool motion is similar to conventional milling, and hence, it is called ED milling. The tool rotation predominantly flushes the dielectric and removes the debris from the IEG. The dimensional accuracy of the machined channel greatly depends on the tool wear, which is the underlying limitation of the process, and it is a relatively slow process.

In industry, the most popular variant of EDM is WEDM. Its operation is similar to ED milling; however, it uses a continuous wire, and the tool (wire) does not rotate. Commonly used material of wire is copper, brass, and molybdenum steel of diameter 150–300 μm. During operation, fresh wire is fed continuously with high tension along the surface of workpiece, and the material is removed due to sparking. The intense heat partly melts the wire, and hence, it is not reused and is collected by the take-up reel. The material removal rate (MRR) is enhanced by using stratified wire that consists of copper core and zinc coating. The wire can be fed in an angular direction to cut complex shapes with high accuracy and finish. The only drawback of the process is that machining starts from the edge of the workpiece and in case of cavity

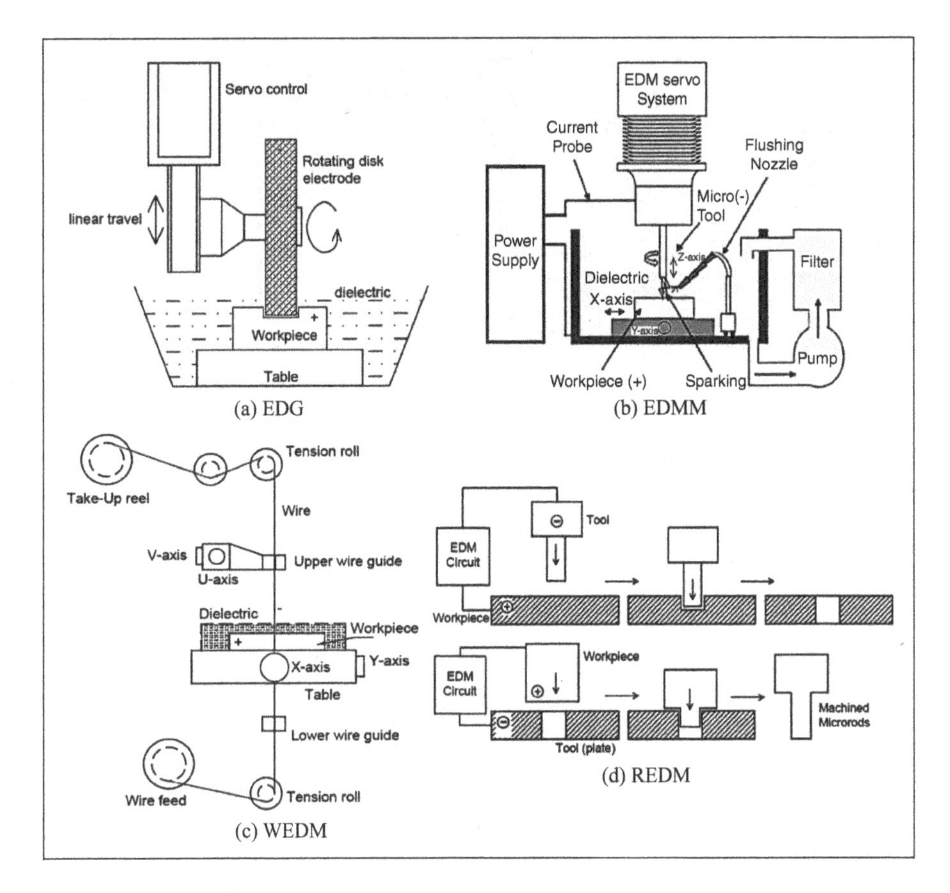

FIGURE 9.3 Variants of EDM.

at the center, a predrilled hole is required. Another limitation is that blind hole or cavities cannot be formed by WEDM.

REDM is a variant of EDM process in which arrayed patterns are fabricated on the workpiece by replicating the cavity pattern of the tool. In the conventional die-sink EDM, the tool forms the cavity on the workpiece whereas in REDM, the cavity in the tool is used to form projection on the workpiece. This is useful in making micro tools, micro-arrayed patterns that are used as tools in micro EDM, micro ECM, and biomedical applications. The existing microfabrication processes have constraints in machining high-aspect-ratio-arrayed patterns of various sections on the metallic surface, which can be easily produced using REDM.

9.2.2 Micro-EDM Process

Nowadays miniaturized products are in great demand, increasing the usability of the product. Miniaturization reduces the entire system's size, saves energy, and improves system response. Due to the growth of micro-electromechanical system (MEMS) and microsystems, such as micro pumps, micro actuators, and micro valves, the necessity

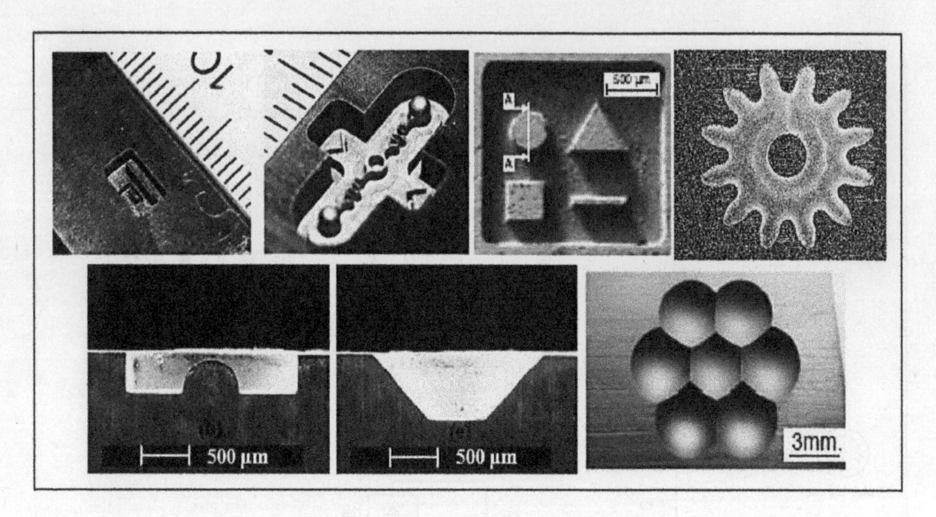

FIGURE 9.4 Various complex cavities formed by ED milling [5–9].

Reproduced with permission.

for microfabrication has expanded in the industrial sector throughout time, and there has been continuous research in the area of newer machining techniques. Because of their unique material removal mechanism, nonconventional methods such as laser beam machining (LBM), electron beam machining (EBM), and plasma arc machining (PAM) can be regarded as ideal for micro machining. However, to transfer technology from macro to micro domain, specific changes in input conditions, such as energy, source/diameter, motion accuracy, and system precision, are required. LBM, EBM, PAM, and EDM are examples of micromachining processes [4].

Micromachining is nothing but fabricating at dimensions less than 1 mm; it does not imply that the part manufactured is micro size or so. Scaling down the machining domain greatly affects the ease of machining, tooling requirements, and metrology. The present technology of micro EDM makes it easy to machine intricate 3D geometry with a defined level of accuracy and precision as shown in Figure 9.4. The process is well developed to fabricate micro holes, micro slots, micro pillars, and micro cavities in hard and tough materials. This requires micro tools, lower input energy (μJ), ultra-precise machine slides, and sophisticated machine control.

9.3 CLASSIFICATION OF DIELECTRIC

Dielectric plays a vital role in the EDM process. It acts as an insulator between the two electrodes and is responsible for sparking. It also has other functions, such as cooling the electrodes, modulating the shape of the plasma channel, and removing debris. The tool and workpiece are immersed in a dielectric tank that is constantly refilled with new dielectric. To increase material removal, the dielectric is injected at a high velocity near the sparking zone through the nozzle.

Oil-based dielectrics, water-based dielectrics, and gaseous dielectrics are the three types of dielectrics. Mineral oil, kerosene, mineral seals, transformer oil, distilled

TABLE 9.1
Physical Properties of Kerosene, Deionized Water, and Air

Physical Property	Kerosene	Deionized water	Air
Dielectric strength (MV/cm)	1	0.7	0.03
Dielectric constant	1.8	80	1.02
Electric conductivity (μs/cm)	0.015	1.33	5×10^{-11}
Kinematic viscosity (cm^2/s)	1.16×10^{-2}	0.852×10^{-2}	0.148
Specific heat (J/kg°C)	2100	4200	1.005
Thermal conductivity (W/mk)	0.14	0.62	0.016
Density (kg/m^3)	860	100	1.29
Boiling point (°C)	200	100	–

water, distilled water with additives, and gaseous dielectric are some of the commonly used dielectrics used in EDM [10]. Physical parameters, such as dielectric strength, dielectric constant, electric conductivity, viscosity, specific heat, thermal conductivity, density, and boiling point, are some of the parameters used for selecting the desired type of dielectric depending on the application. Table 9.1 lists the physical parameters of various dielectric fluids. It shows that kerosene has a higher dielectric strength than water and air, making it a good option for many EDM operations. But kerosene is no longer commercially used because of environmental issues.

Moisture and impurities must be kept out of the dielectric. The chemical stability of a liquid dielectric is the most important factor to be considered when while selecting one the dielectric. Because the liquids are easily polluted, their breakdown strength is poor. Dust, moisture, dissolved gases, and ionic impurities are the most common contaminants in liquid dielectric. As the liquid dielectric becomes charged due to the electric field, the presence of dust particles diminishes its breakdown strength. The local field is increased when a single conducting particle is present in the electrode gap. When this field surpasses the liquid's breakdown strength, local breakdown occurs near the particle, resulting in the creation of gas bubbles and eventually the liquid's breakdown. The breakdown strength decreases as particle size increases. Due to the difference in permittivity of two materials and the applied field, particles in the dielectric experience a force. When a high number of these particles are present, these forces cause them to align and create a stable chain, which bridges the gap and causes a dielectric breakdown. Impurities in the dielectric oil, such as small droplets of water suspended in it, make it more susceptible to impurity. The dielectric strength of the oil is significantly reduced when a small amount of water is present. The dielectric strength of liquids is also affected by dissolved gases such as oxygen and carbon dioxide. Ionic impurities, such as water vapor, are easily dissociated, resulting in extremely high conductivity and, depending on the electric field, heating of the liquid.

Different metallic additive powders, such as copper, graphite, titanium, aluminum, silicon carbide, carbon nanotubes, chromium, nickel, silicon, and others, are added into the dielectric fluid to enhance its thermal and electrical properties [3]. It increases the MRR, reduces the tool wear rate (TWR), improves surface integrity while operating at a low value of current. Nonmetallic powders are ineffective in improving the thermal and electrical properties of dielectric, but they reduce the dielectric constant causing frequent breakdown. Particle size and its concentration affect the sparking phenomenon and response. The large particle size results in poor ionization and delays discharge time. Smaller particles (70–80 nm) enhances the MRR and surface finish. Higher concentration of particles creates unstable arcing and deposition on the machined surface. Furthermore, hydrocarbon-based dielectrics release some unsaturated carbon particles, fumes, toxic reagents, and the like, which is unhygienic, causing breathing and skin problems and can affect the operator's health [11]. The use of vegetable oils can eliminate these problems and offer certain advantages. The hydrocarbon oils often cause fire hazards due to a thickening of unsaturated hydrocarbons at the IEG. This is eliminated while using biofuels; also, comparatively, it releases fewer unsaturated particles due to the presence of short hydrocarbon chains [12]. It increases the MRR and reduces the TWR and power consumption because of high density, viscosity, and convective heat transfer coefficient. It ultimately leads to green manufacturing due to easy availability and biodegradability [13]. The water–oil (W/O) emulsion, waste vegetable oils (WVOs), and glycols are also tested for EDM application and showed better response. However, due to poor heat dissipation, it results in higher heat-affected zones.

9.3.1 Hydrocarbon-Based Dielectric

Hydrocarbon oils outperform distilled or deionized water in die-sink applications [14]. These types of dielectrics are utilized in precision machining because of their greater insulating qualities. It is a hydrogen-and-carbon-based chemical compound that decomposes at high temperatures during sparking, forming a carbide coating on the work surface as well as carbon (tar) particles. Deionized water, meanwhile, forms an oxide coating on the work surface. Because carbide has a higher melting point than oxides, and hydrocarbon oils require a higher discharge energy than deionized water to remove material. Similarly, the impulsive force of discharge in hydrocarbon oil is unstable, resulting in a decrease in the MRR. A strong carbide coating grows over the tool surface, providing a layer of protection against tool wear. As a result, tool wear is lower in this case as compared to deionized water. The main issue with hydrocarbon oil is that it has a low flash point and is significant volatility, which can cause fires [15]. Mineral oils, mineral seal, and transformer oil are some of the hydrocarbon oils used.

9.3.2 Water-Based Dielectric

Tap water, deionized water, and water-in-oil emulsion are the most often utilized water-based dielectrics. Deionized water outperforms kerosene and urea solution in terms of processing. It increases the MRR, decreases the TWR, and enhances surface finish but at the expense of accuracy [16–18]. Deionized water debris particles have a higher average diameter than kerosene debris particles. Although deionized water is environmentally good, its use can damage machine parts due to corrosion.

TABLE 9.2
Machining Characteristics of Different Dielectric Fluid [10, 14, 19, 20]

Hydrocarbon Based	Water Based	Gaseous Based
• Efficient in die sink applications • Higher surface roughness • Due to the higher carbon concentration, the white layer has a higher microhardness • Prone to fire hazards. Decomposes to form carbon black, benzene (C_6H_6), benzopyrene ($C_{20}H_{12}$), acetylene, ethylene, hydrogen, carbon dioxide, carbon monoxide	• Higher performance and environmentally suitable than hydrocarbon oil • Lower surface roughness, higher MRR, poor machining accuracy, lower tool–electrode wear ratio • Thicker workpiece heat affected zone with a higher concentration of microcracks • Substances generated are water vapor, carbon monoxide, nitrogen oxide, ozone, and chloride	• Higher MRR than hydrocarbon oil • Lower tool–electrode wear ratio • Substances generated are fluorite, chlorite, nitrite, bromide, nitrate, phosphate, sulfate, carbon dioxide, ozone, and carbon monoxide

9.3.3 Gaseous-Based Dielectric

Oxygen, nitrogen, air, helium, and argon are some of the gaseous dielectrics utilized in EDM. The physical state of the dielectric has a significant impact on the discharge phenomenon in the IEG. Regardless of the type of dielectric utilized, the recast diameter of the crater rises as the pulse duration increases. This is due to the materials' heat conduction and the discharge plasma's expansion. Because of the oxidation reaction between the molten metal and oxygen, the recast width and depth of crater are larger in oxygen than in air. The pressure of the discharge-generated bubble has an impact on the material removal process. At low external pressure, craters with a flat molten surface are formed with no central depression; however, at high external pressure, molten metal flows out of the crater, forming multiple central depressions. In liquid dielectric, this results in better material removal efficiencies than in gaseous dielectric. The tool and workpiece are subjected to an impulsive force as the discharged-generated bubble expands and contracts. The impulsive force in a liquid dielectric is greater than in a gaseous dielectric due to the liquid dielectric's higher density and viscosity [19]. Table 9.2 lists the characteristics of various dielectrics.

9.4 ENVIRONMENTALLY FRIENDLY DIELECTRICS FOR ACHIEVING SUSTAINABILITY IN EDM

9.4.1 Bio-Friendly Alternatives

The dependency of EDM on conventional dielectric, such as hydrocarbon oil, in spite of a high removal rate and low power consumption, has reduced the sustainability index. Besides these, fusion in the machining zone partially burns the tool/

workpiece material and dielectric to emit harmful oxides and monoxides and evolve gases that create unhygienic breathing atmosphere that is very harmful to the operator. In addition, ineffective flushing causes unstable sparking due to carbon particles, releasing carbon monoxide. The sustainability criteria include different aspects such as economic, environmental, social, and personal health of the operator. This section discusses alternative dielectrics such as emulsions, organic solutions, vegetable oils as a potential alternative to the conventional dielectric and its impact on removal rate, surface properties, and sustainability.

To decrease harmful emissions, vegetable oils such as jatropha, canola, castor, and neem are blended with regular diesel. Furthermore, these oils are employed in transformers and high-voltage capacitors as a dielectric [21]. As a result, the prospect of using these transesterified oils as an EDM dielectric has grown. The process of extracting biofuels from tree seeds and converting them to methyl ester is known as transesterification. Experiments with vegetable oils such as jatropha, canola, neem, sunflower, palm, coconut, and WVO as a dielectric in EDM have yielded promising results [11]. These oils are denser and have higher oxygen content, which results in a higher MRR, a lower TWR, and a lower surface roughness (SR)and they are biodegradable and less flammable compared to hydrocarbon oils. Among these oils, for the palm oil, the TWR is more while all other responses are improved for all the oils [22]. The aqueous solution of urea and various types of emulsions have improved the machining performance [23]. In the case of W/O emulsion, the machined surface showed a harder recast surface and fewer micro cracks, however, it resulted more micro voids [24]. Table 9.3 shows various nonconventional dielectric fluid used in EDM.

Various factors considered for the sustainability of EDM are represented in Figure 9.5. The social, economic, and environmental issues are addressed, taking into account various factors such as time consumption, harmful gas emissions, dependency on conventional dielectric such as kerosene, and solid wastes.

Figure 9.6 shows the number of dielectrics used in EDM for the last 10 years. It is seen that initially only five dielectrics—kerosene, EDM oil, water, air, and air–oil mixture—were used as a dielectric for EDM until between 2010 and 2012. Later,

TABLE 9.3
Nonconventional Dielectrics

Category	Different Oils	Machining Performance
Vegetable Oils	Jatropha, Canola, Neem, Sunflower, Palm, Coconut, and Waste vegetable oil (WVO) [5]	Better machining performance so useful for EDM.
Aqueous solution	Urea $[Co(NH_2)_2]$ [5]	Good surface integrity is achieved.
Emulsions	Emulsion, Water-oil emulsion (W/O), Oxygen mixed emulsion, Nano emulsion [5]	Better machining performance with harder recast surface and fewer micro-cracks, however, results in more micro voids.

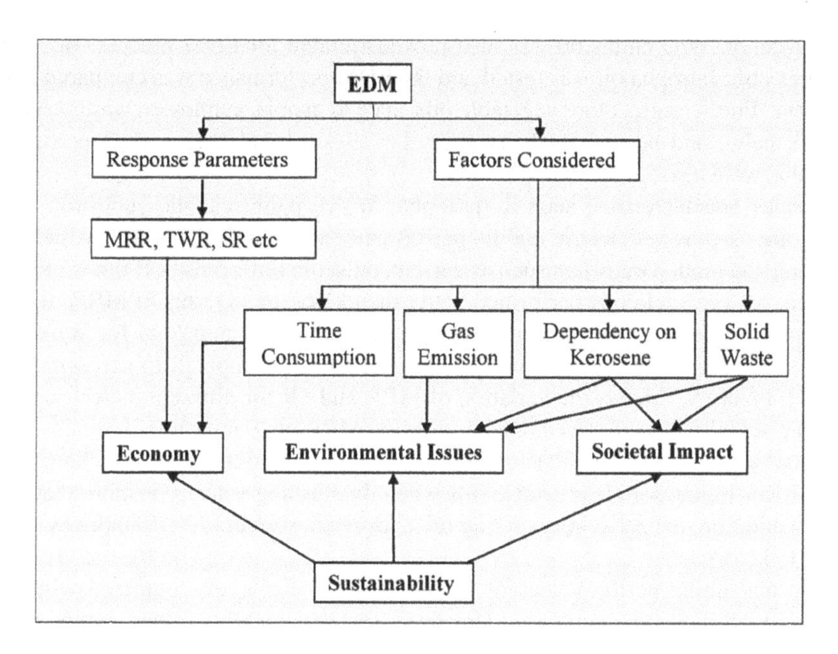

FIGURE 9.5 Sustainability in electrical-discharge machining.

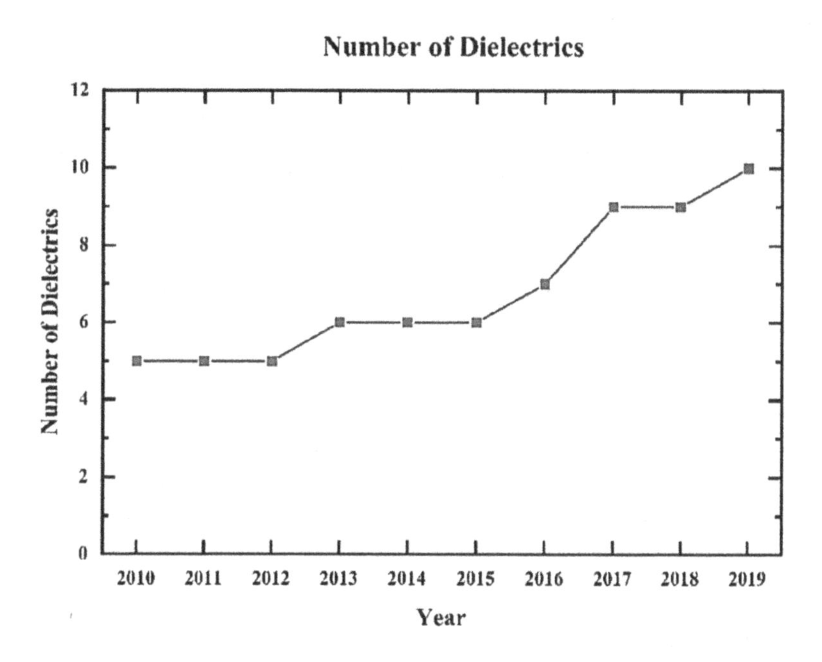

FIGURE 9.6 Number of dielectrics used in electrical-discharge machining.

urea, alcohol, W/O emulsion was also proved suitable for EDM process. In 2016, bio-dielectric Jatropha oil was tested and the EDM performance was compared with kerosene. Furthermore, other vegetable oils, such as canola, sunflower, waste vegetable oil, palm, and neem, which have similar physical and chemical properties to Jatropha, were also tested.

Various vegetable oils, such as jatropha, WVO, palm, canola, sunflower, and neem, are used as a dielectric and its performance was compared with kerosene by changing the input parameters such as current, pulse-on time, pulse-off time, and gap voltage. All vegetable oils performed better than kerosene in terms of MRR, with a maximum value of 53.33 mm^3/min for jatropha and 29.21 mm^3/min for kerosene. The lower SR is also achieved using vegetable oils as compared to kerosene [22, 25–27]. Figure 9.7 shows the variation of MRR and SR for different dielectrics and flushing velocity. It is observed that a maximum MRR of 19 mm^3/min is obtained for jatropha, whereas a lower SR is obtained for canola. In comparison with kerosene, the MRR is higher for all vegetable oils when the flushing velocity is more than 0.3 m/s. In addition, the SR is lower for all oils; however, overcut is 2–5% more as compared to kerosene.

The machining performances of various vegetable oils are compared with the widely used dielectric kerosene. However, for sustainability other commercial, social, and environmental factors, such as cost-effectiveness, harmful emissions, solid waste products, noise, and operator health are studied and compared with conventional dielectrics. Table 9.4 shows probable future dielectrics and their suitability considering various factors, such as availability, cost-effectiveness, machining performance, and its effect on the environment and society. Among all, vegetable oils and W/O emulsion prove to be better in terms of machining performance with very little or no adverse effect on the operator's health, society, and environment. Vegetable oils are expensive due to costly extraction and transesterification processes whereas a W/O emulsion is relatively cheap and is a suitable dielectric for EDM, leading to green manufacturing.

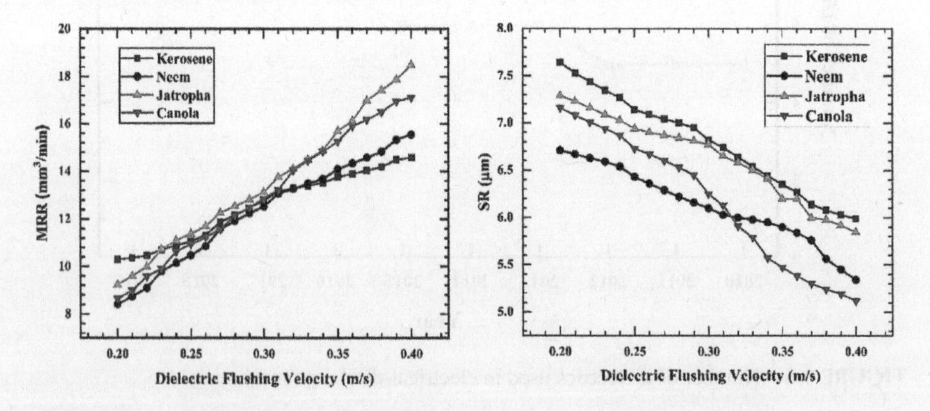

FIGURE 9.7 Comparison of material removal rate and surface roughness for various dielectrics.

TABLE 9.4
Different Criteria for Sustainability

Dielectrics	Availability	Cost	Performance	Environmental impact	Societal impact
Kerosene	Easy	Less	Good	Yes	Yes
EDM oil	Easy	More	Average	No	No
Urea solution	Easy	Less	Average	No	No
Water	Easy	Less	Average	No	No
$CaCl_2$ solution	Not easy	Less	Average	No	No
Air	Easy	More	Average	No	No
Air with glycol	Not easy	More	Good	No	No
Vegetable oils	Available	More	Best	No	No
W/O emulsion	Available	Less	Best	No	No

9.5 DEVELOPMENT OF DRY TO NEAR-DRY EDM FOR SUSTAINABILITY

9.5.1 Dry EDM

Dry EDM (DEDM) is an alteration of the oil EDM process in which instead of liquid dielectric, a gaseous medium is used. In DEDM, various gases used as dielectric media are air, oxygen, argon, nitrogen, and others. High-velocity gas is passed through a hose to hollow an electrode into an IEG. This high-velocity gas assists in removing burr and debris particles from the IEG as well as providing a cooling effect to the tool and the workpiece, thereby avoiding extreme heating of the electrodes. Figure 9.8 illustrates the DEDM process. The tubular tool is used to provide passage for the high-velocity gas to flow into the IEG. The gas behaves as a dielectric medium during electrical discharges, and debris particles formed during and after discharges are flushed away from the IEG by the high velocity of the gas. This process makes the IEG clean and revamps the gap for further sparking. During DEDM machining process, toxic

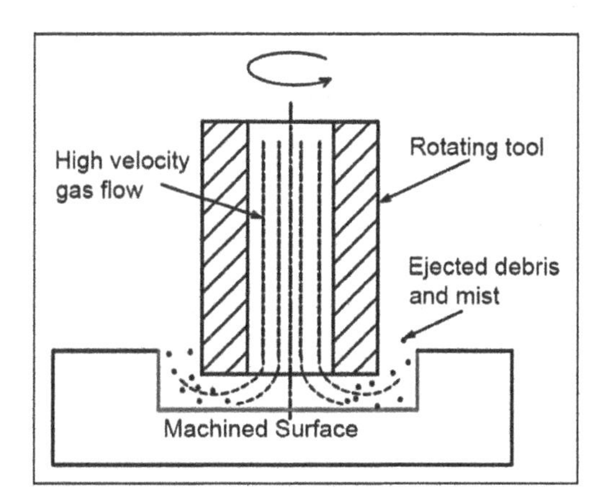

FIGURE 9.8 Dry electrical-discharged machining process.

fumes will not be generated due to the use of gas as a dielectric medium; hence, it can be considered to be an environmentally friendly EDM process. Furthermore, it avoids fire hazards due to the absence of flammable material in the IEG.

9.5.2 Near-Dry EDM

The near-dry EDM (NDEDM) process is the next generation of the EDM process to overcome a few drawbacks of the DEDM process. Out of the various gases, only air and oxygen show better machining efficiency in terms of MRR and TWR. In addition, DEDM has a severe debris accumulation problem, which subsequently results in poor surface finish and dimensional accuracy [28]. Due to debris accumulation, arcing occurs, and it affects directly the MRR, severely damaging the workpiece surface and the tool. These drawbacks are eliminated by using the NDEDM process.

Figure 9.9 illustrates the NDEDM setup. In the NDEDM process, a mist flow, that is, a combination of gas and liquid dielectric, is used as a dielectric medium. The potential advantages of NDEDM are the extensive choices of gas and liquid dielectric available and flexibility to adjust the proportion of liquid in gas. The mist flow formation takes place by using the minimum quantity lubrication system. The dielectric (mist flow) properties in NDEDM process support achieving various machining needs, such as a high MRR, a lower TWR, and a high surface finish as compared to the DEDM process. The NDEDM process is also environmentally friendly as it eliminates harmful gases and results in favorable working conditions.

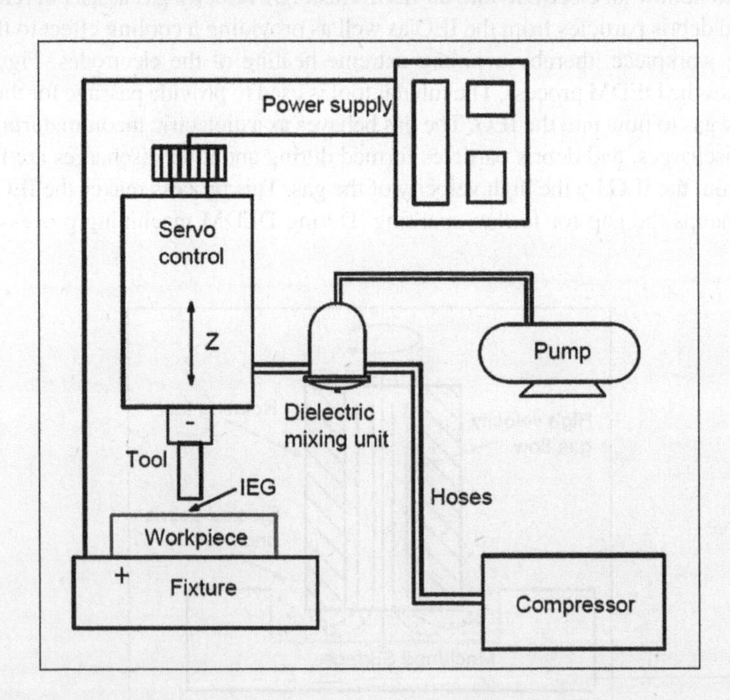

FIGURE 9.9 Schematic set up of near-dry electrical-discharge machining process.

9.6 CONCLUSION

This chapter has highlighted different conventional dielectrics and the possible bio-dielectrics in terms of machining performance in EDM process. Besides these, various factors leading to a sustainable EDM process is discussed to identify a compromised solution that will enhance EDM performance and will not result in pollution. Conventional hydrocarbon-based dielectrics are efficient but result in pollution and an unhealthy working atmosphere. Hence, to avoid this, oils from natural seeds are tested and found suitable for EDM applications. Vegetable oils and W/O emulsion were found to be the better alternatives, among all others. These are biodegradable and showed better performance in comparison to kerosene. As compared to wet EDM, DEDM and NDEDM are promising and environmentally friendly as they are cleaner and avoid pollution in all aspects, such as air, water, and land, and are safer for the operator. Extensive research is required to study different aspects of various bio-dielectrics and variation of their properties over the machining time and at elevated temperature. To reduce the usage of conventional dielectric, it can be blended with vegetable oils to achieve a balance between output performance and sustainability, which can be a future scope for research. At the outset, sustainable EDM can be achieved by using dry and near-dry dielectric media with vegetable oils and W/O emulsion, resulting in green manufacturing.

REFERENCES

1. Mohd N., Solomon D.G., Bahari M.F. (2007) "A review on current research trends in electrical discharge machining (EDM)," *Int J Mach Tools Manuf* 47: 1214–1228.
2. DiBitonto D.D. (1969) Theoretical models of the electrical discharge machining process. I. A simple cathode erosion model. *J Appl Phys* 66(9): 4095–4103.
3. Das S., Paul S., Doloi B. (2020) Feasibility assessment of some alternative dielectric mediums for sustainable electrical discharge machining: a review work. *J Braz Soc Mech Sci Eng* 42(148): 1–21.
4. Jain V.K., Sidpara A., Balasubramaniam R., Lodha G.S., Dhamgaye V.P. and Shukla R. (2014) "Micromanufacturing: A review - Part I," *Proc IMechE Part B: J Eng Manuf* 228(9): 973–994.
5. Richard J., Demellayer R. (2013) Micro-EDM-milling development of new machining technology for micro-machining. *Procedia CIRP* 6: 292–296.
6. Uhlmann E., Perfilov I. (2018) Machine tool and technology for manufacturing of micro-structures by micro dry electrical discharge milling. *Procedia CIRP* 68: 825–830.
7. Gotoh H., Tani T., Okada M., Goto A., Masuzawa T., Mohri N. (2013) Wire electrical discharge milling using a wire guide with reciprocating rotation. *Procedia CIRP* 6: 199–202.
8. Wang J., Qian J., Ferraris E., Reynaerts D. (2016) Precision Micro-EDM milling of 3D cavities by incorporating *in-situ* pulse monitoring. *Procedia CIRP* 42: 656–661.
9. Schulze V., Ruhs C. (2010) On-Machine Measurement for the Micro-EDM-Milling Process Using a Confocal White Light Sensor. In *Proc. 10th Int. Conf. Eur. Soc. Precis. Eng. Nanotechnology*, EUSPEN 2: 37–40.
10. Chakraborty S., Dey V., Ghosh S.K. (2015) A review on the use of dielectric fluids and their effects in EDM characteristics. *Precis Eng* 40: 1–6.

11. Valaki J.B., Rathod P.P., Sidpara A.M. (2019) Sustainability issues in electric discharge machining. In: Gupta K (ed) Innovations in Manufacturing for Sustainability. Materials forming, machining and tribology. Springer, Cham.
12. Das S., Paul S., Doloi B. (2019) Investigation of the machining performance of neem oil as a dielectric medium of EDM: a sustainable approach. In: *IOP conference series: materials science and engineering*, vol 653, no 1. IOP Publishing, p 012017.
13. Valaki J.B., Rathod P.P., Sankhavara C.D. (2016) Investigations on technical feasibility of Jatropha curcas oil based bio dielectric fluid for sustainable electric discharge machining (EDM). *J Manuf Process* 22: 151–160.
14. Leão F.N., Pashby I.R. (2004) A review on the use of environmentally-friendly dielectric fluids in EDM. *J Mater Process Technol* 149: 341–346.
15. Zhu G., Bai J., Guo Y., Cao Y., Huang Y. (2014) A study of the effects of working fluids on micro-hole arrays in micro-electrical discharge machining. *Proc IMechE, Part B: J Eng Manuf* 228(11): 1381–1392.
16. Chen S.L., Yan B.H., Huang F.Y. (1999) Influence of kerosene and distilled water as dielectrics on the electric discharge machining characteristics of Ti–6A1–4V. *J Mater Process Technol* 87: 107–111.
17. Jeswani M.L. (1981) Effect of the addition of graphite powder to kerosene used as the dielectric fluid in electrical discharge machining. *Wear* 70: 133–139.
18. Jahan M.P., Rahman M., Wong Y.S. (2010) Modelling and experimental investigation on the effect of nano powder mixed dielectric in micro-electro discharge machining of tungsten carbide. *Proc IMechE, Part B: J Eng Manuf* 224: 1725–1739.
19. Ndaliman M.B., Khan A.A., Ali M.Y. (2013) Influence of dielectric fluids on surface properties of electrical discharge machined titanium alloy. *Proc IMechE, Part B: J Eng Manuf* 227(9): 1310–1316.
20. Zhang Y., Liu Y., Shen Y., Ji R., Li Z., Zheng C. (2014) Investigation on the influence of the dielectrics on the material removal characteristics of EDM. *J Mater Process Technol* 214: 1052–1061.
21. Mariprasath T., Kirubakaran V. (2016) A critical review on the characteristics of alternating liquid dielectrics and feasibility study on pongamia pinnata oil as liquid dielectrics. *Renew Sustain Energy Rev* 65: 784–799.
22. Valaki J.B., Rathod P.P. (2016) Investigating feasibility through performance analysis of green dielectrics for sustainable electric discharge machining. *Mater Manuf Process* 31(4): 541–549.
23. Yan B.H., Tsai H.C., Huang F.Y. (2005) The effect in EDM of a dielectric of a urea solution in water on modifying the surface of titanium. *Int J Mach Tools Manuf* 45(2): 194–200.
24. Zhang Y., Liu Y., Ji R., Cai B. (2011) Study of the recast layer of a surface machined by sinking electrical discharge machining using water-in-oil emulsion as dielectric. *Appl Surf Sci* 257(14): 5989–5997.
25. Valaki J.B., Rathod P.P. (2016) Assessment of operational feasibility of waste vegetable oil based bio-dielectric fluid for sustainable electric discharge machining (EDM). *Int J Adv Manuf Technol* 87(5–8): 1509–1518.
26. Das S., Paul S., Doloi B. (2020) Feasibility investigation of neem oil as a dielectric for electrical discharge machining. *Int J Adv Manuf Technol* 106: 1179–1189.
27. Valaki J.B., Rathod P.P., Khatri B.C., Vaghela J.R. (2016) Investigations on palm oil based biodielectric fluid for sustainable electric discharge machining. In: *Proceedings of the international conference on advances in materials and manufacturing*, India.
28. Kao C.C., Tao J., Shih A. (2007) Near-dry electrical discharge machining. *Int J Mach Tools Manuf* 47: 2273–2281.

10 Sustainable Abrasive Jet Machining

Şenol Bayraktar and Cem Alparslan
Recep Tayyip Erdoğan University, Rize, Turkey

CONTENTS

10.1 INTRODUCTION

Depending on the development of technology and science, it is demanded to increase the machining quality and increase the machining efficiency in many fields such as engineering devices, precision machines, electronic instruments, and optical components. The demand for rapid prototyping in the manufacturing industry is increasing day by day. These demands allow the research and use of new technologies to quickly transform raw materials into usable products. In addition, the increase in material diversity has revealed the need for new machining processes. Nontraditional machining methods have emerged as a result of the inadequacy of traditional processing methods in certain areas. The abrasive jet machining (AJM) cutting technique is among the current techniques developed in recent years. The AJM cutting process is a system in which water with abrasive particles is used under high pressure as a cutter. Before the cutting process using AJM started, water jet cutting processes began to be described as fully industrial cutting devices in the 1930s. Carl Johnson developed the water jet for cutting materials with plastic geometry in 1956. Wood and plastic material cutting operations were carried out using AJM in the early 1970s [1]. The water jet was developed to perform the cutting of laminated paper in 1971. Then, the AJM method was found as a result of adding an abrasive to the water jet. The first AJM operation was carried out in 1979 by Mohamed Hashish. He added abrasive sand into the water jet to cut materials such as glass, steel, and concrete. It was first used in the cutting process of automobile glass in 1983 [2, 3]. A better cutting edge and machined surface quality [4–8] were obtained in cutting operations using this method. In addition, the range of material that can be cut is greatly

DOI: 10.1201/9781003284574-11

increased. The use of AJM in the industry has increased and it has also been used for cutting plates made of different materials since 1993. Today, the AJM process is preferred for cutting many materials, such as copper (Cu), aluminum (Al), lead (Pb), molybdenum (Mo), titanium (Ti), tungsten (Tu), ceramics, marble, granite, plastic, composite, stainless steel, super superalloy, and glass. AJM technology has started to be used in machining processes such as milling, turning, drilling, and deburring, as well as two-dimensional (2D) cutting [9]. Figures 10.1 and 10.2 illustrate the AJM cutting theory and different cutting applications in AJM, respectively.

The machined surface of the material is examined in three parts in the AJM machining process (Figure 10.3). The upper part of the cut surface represents the initial damage region (IDR). It occurs on top of Kerf. IDR arises due to the radial distance difference between the expansion of the jet and the energy of the jet before the abrasive particles striking the material surface. The abrasive impact angle in this region is significantly greater than the angle required for the remaining depth of cut. The smooth cutting region (SCR) and rough cutting region (RCR) are below the IDR. These two regions are distinguished by their roughness marks pattern. The surface properties formed in RCR consist of cutting parameters that affect the kinetic energy of the jet [10].

It is observed that environmental problems are increasing due to the effect of many factors. Cooling liquids are used to reduce heat in the machining of difficult to cut materials in industries where conventional cutting techniques are preferred. The

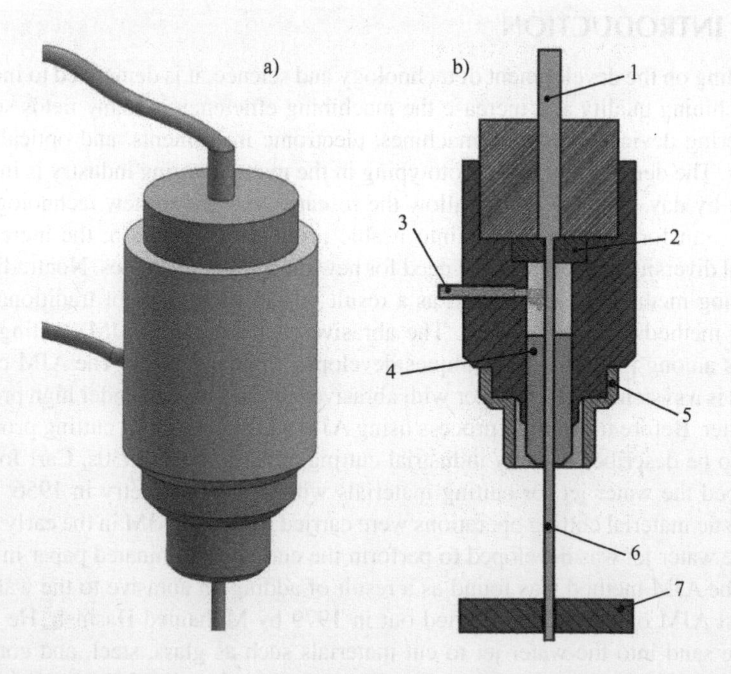

FIGURE 10.1 Abrasive jet machining (AJM) cutting theory, (a) three-dimensional image of AJM and (b) cross-sectional view of AJM: (1) high-pressure water inlet, (2) jewel (ruby or diamond), (3) abrasive (garnet), (4) mixing tube, (5) guard, (6) cutting water jet, (7) cutting material.

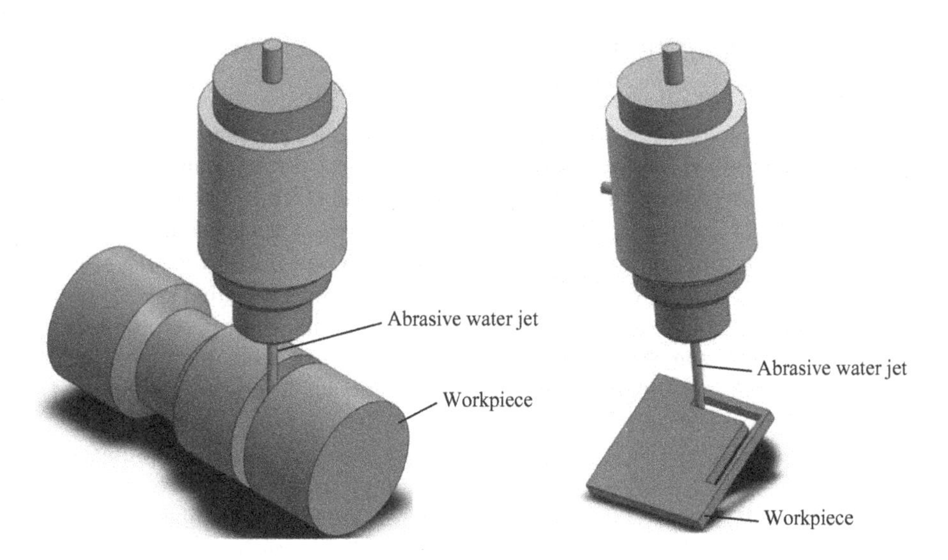

FIGURE 10.2 Different cutting applications in abrasive jet machining: (a) turning and (b) milling.

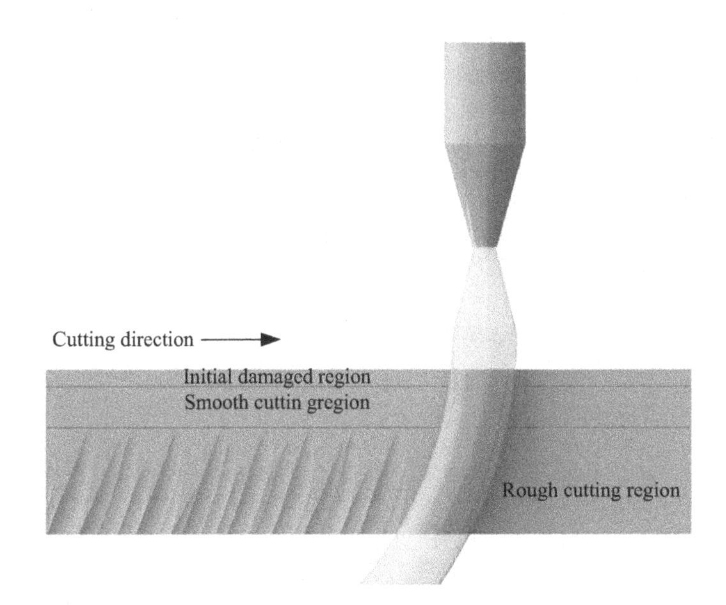

FIGURE 10.3 Cutting edge regions of cut material using abrasive jet machining.

disposal of these coolants is an environmental problem. On the other hand, it is not possible to use conventional methods in cutting materials such as rubber, textile, and glass. The AJM technique is preferred for eliminating these problems as an alternative and environmentally friendly cutting method. Therefore, studies on current technologies in the literature related to the AJM technique have been investigated in detail and presented in a comparative way in this chapter.

10.2 WHY IS AJM PREFERRED?

The importance of cost, time, and quality in the machining of materials increases depending on the development of the industry. It is aimed to obtain a part as close to the desired form in the world industry. This situation causes machining processes to be preferred because of their short time and cheap cost. The AJM process is used more than other machining methods among these methods. The most important reason for preferring the machining process using AJM is that problems such as burning, melting, and deformation are not encountered on the cutting surface of the material, since it is a cold-cutting technique. In other words, there is no heat generation in the cutting zone or less heat is generated when machining hard materials. It can be preferred for rough operations before conventional machining (Figure 10.4). Thus, cutting tool cost and machining time can be reduced.

Many benefits of the AJM method compared to other methods are as follow:

- This method can be integrated into more than one production line.
- Because AJM is a cold-cutting process, conditions such as burning and melting do not occur on the cutting surface compared to laser and plasma cutting.
- There is no chemical pollution due to the effect of burning and melting on the cutting edge during the machining.
- Chip volume to be removed from the material is less due to the small nozzle outlet diameter. Therefore, material losses are minimized. It also contributes to the easy cutting of very narrow and sharp corners.
- There is no operational and cutting tool sharpening cost as in conventional machining methods.
- It provides hygienic cutting standards in the areas where it operates in the food sector.

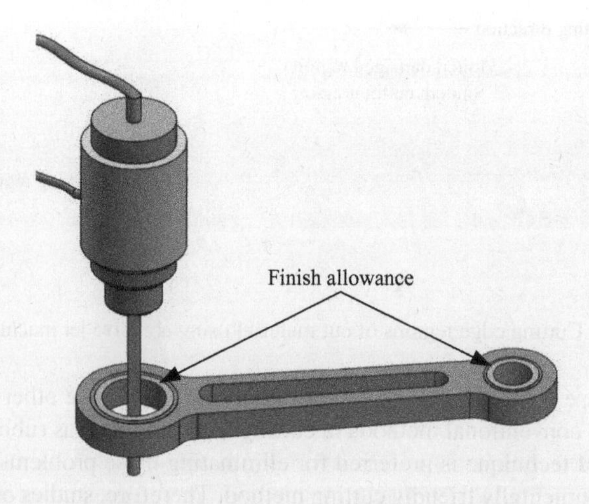

FIGURE 10.4 Application of abrasive jet machining before conventional machining

- Because there is no need to prepare mold and cutting tool equipment as in conventional machining methods, time losses are minimized.
- This cutting method does not require the cooling fluids used in conventional machining techniques. In other words, because no harmful liquids, gases, or oils are required, it protects the environment.
- There is no need for secondary machining or manufacturing operations after cutting.
- It allows precise cutting of soft or hard materials, such as tool steel, titanium, aluminum, brass, stone, marble, glass, stainless steel, granite, composite, carbon fiber, ceramic, and plastic.

10.3 EVALUATION OF AJM IN TERMS OF ENVIRONMENTALLY FRIENDLY CUTTING

AJM has emerged as an environmentally friendly method in terms of its general structure. Unlike conventional material cutting methods, the AJM process does not require coolants or lubricants. Therefore, dustlike chips formed during cutting are not chemically contaminated. Dust particles mixed with the air are formed less than other machining methods. Thus, no harmful fumes are observed due to a thermal effect in the cutting zone, such as laser or plasma. In addition, the particles used for abrasive purposes in water do not harm the environment. The abrasive waste is usually inert garnet with a small fraction of particles from the cut material in this method. As long as the cut materials are not hazardous materials, such as lead, the abrasive wastes are not harmful to the environment. In addition, abrasive wastes can be preserved in a secure storage area. The cutting width of AJM allows the close placement of parts for maximum material benefit. Therefore, the amount of junk remaining from the cutting process is at a minimum level. In addition, these junks do not contain chemical pollution and can be recycled. This provides raw material, energy, and cost savings for recyclable materials such as aluminum, steel, and titanium. The main cost expenses in cutting using AJM are abrasive particles, water, and electrical energy. Approximately 1.9–3.8 liters of water are used per minute in the AJM technique. The water can be recycled for repeated use with the closed-loop system. The wastewater used in the system is generally clean enough to be filtered or used in other systems [11]. According to all these criteria, the AJM cutting technique stands out as being environmentally friendly and economically beneficial compared to other conventional and unconventional machining techniques. Businesses operating in industrial areas need to plan the production system at the optimum level for manufacturing efficiency, taking into account the product, cost, and environmental factor relations.

10.4 THEORY OF AJM

The AJM technique is based on the principle of water erosion. High-pressured water is sprayed onto the surface of the material to be cut. Waste material is removed from this area by spraying. Pure water is used for cutting soft materials such as rubber, plastic, and paper with low hardness. Some abrasive particles are added to the water in order to cut harder materials. The cutting process using the abrasive is called AJM. The material is removed by the high-pressure liquid flow of the abrasive particles and the erosion effect in the cutting zone in the AJM cutting method. Abrasive particles (garnet, sand, Al_2O_3, etc.) with a grain size of 10–180 μm are generally used in these systems [12]. Erosion in AJM is the type of wear that occurs as a result of the continuous impact of abrasive particles on the material surface of the high-pressured liquid. Therefore, abrasive particles have a material removal function in the AJM process. Abrasive particles accelerate with the effect of high liquid pressure. Thus, the chip removal process is realized by increasing the impact so as to create kinetic energy on the material surface [13]. Abrasive particles contribute to the formation of wear with the effect of erosion on the cutting surface of the material due to cutting, fatigue, melting, and fracture sub-mechanisms [14].

The structure of the AJM cutting system generally consists of a hydraulic pump, hydraulic concentrator, accumulator, mixing chamber or tube, control valve, flow regulator or valve, nozzle, drain, and accumulator system (Figure 10.5). *Hydraulic Pump:* It is used to pump the water in the storage tank during cutting. This pump transfers water at a pressure of about 5 bars to the hydraulic intensifier. The water in the system can be sent to the intensifier by using an accelerator that increases the water pressure up to 11 bars [15]. *Hydraulic Intensifier:* Approximately 4000 bar pressure is required for cutting in AJM. Therefore, pressure accelerators are used to increase the water pressure to a higher pressure level. The hydraulic intensifier can increase the pressure of the water coming from the pump up to 3000–4000 bar at a pressure of 4 bar. *Accumulator:* High-pressured water in the system is temporarily stored. Fluctuations of the liquid that may occur with the effect of pressure during the cutting process can be eliminated due to the need for excessive pressure energy.

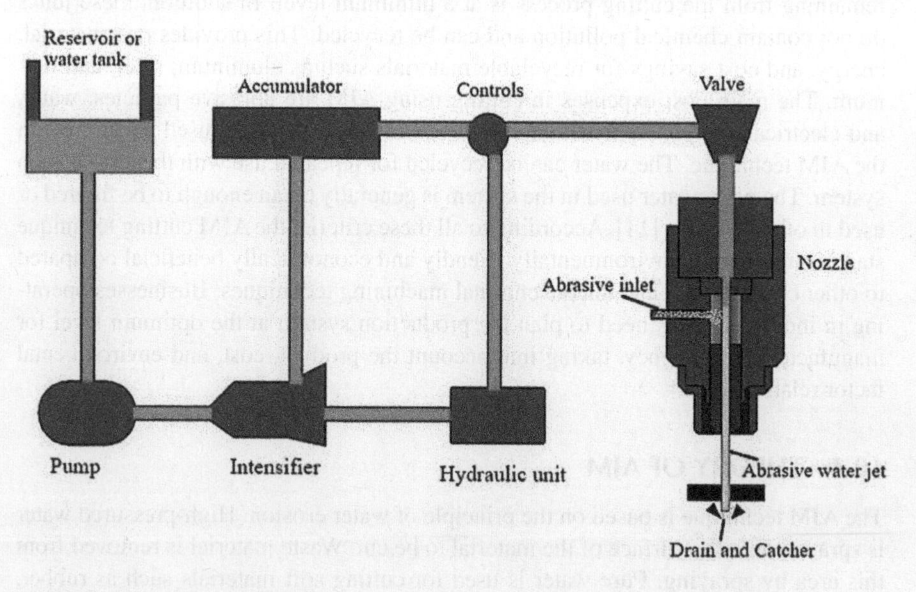

FIGURE 10.5 Schematic view of the cutting process using abrasive jet machining.

Mixing chamber or tube: It is the vacuum chamber used to mix abrasive particles with water [16]. *Control Valve:* It is used to control the pressure and direction of the water jet. *Nozzle:* It is used to convert the pressure energy of water into kinetic energy. In other words, it converts the pressure of the water jet into a high-velocity water jet. The size of the nozzle tip is about 0.2–0.4 mm. Ruby or diamond material is used to prevent wear of the nozzle tip [17]. *Drain and Accumulator System:* It is used to separate used particles and debris in water after machining with the help of the drain and holding system. Unwanted particles are removed from the water. It is returned to the reservoir for later use [18].

10.5 MACHINING QUALITY AND PERFORMANCE IN AJM

Because water is used during cutting with AJM, there are no thermal factors on the material cutting surface. Conditions that adversely affect the material structure such as burning and melting occur on the cutting surfaces due to thermal effects in cutting methods such as laser, plasma, and wire erosion. For example, magnetic silicon sheets used in electrical machines are affected by laser and wire erosion cutting methods. The efficiency of electrical machines decreases as a result of the reduction of magnetic properties of silicon sheets due to the effect of burning and melting at the cutting edge. Therefore, the AJM method emerges as an alternative solution. Because no heat is generated during machining, it is also referred to as a cold cutting process. On the other hand, this situation provides a positive contribution to the machining performance, surface quality, and cost during cutting compared to other cutting methods [13, 19]. There is a burr on the cutting edge with the effect of melting due to thermal effects. Burr formation causes a reduction in workpiece service life or quality. In addition, a second operation is required to remove the burr. This increases the cost. When the ideal parameters are determined in AJM, cutting can be performed without the need for a second manufacturing operation. AJM method is used in many fields, such as the paper, textile, rubber and plastic, cleaning, shoes and leather, glass, marble, granite, ceramics, electrical-electronics, automotive, space, and aviation industries [20]. While a channel with a width in the range of 0.1–0.4 mm can be cut using water jet, it increases by 1.5 mm with the addition of an abrasive into the water. Nonmetallic materials can be cut with pure water at pressure levels of up to 400 MPa using this method. Application areas include automobile carpets and fenders, foam materials, and cardboard. All hard materials, such as metal and ceramic, can be cut using AJM. Also, 300-mm thick steel can be cut using this method depending on the hydraulic power of the jet used. The only function of the water jet in AJM is to accelerate the abrasive particles and to remove the powder material from the surface after cutting. The average surface roughness measured under optimum cutting conditions can be less than Ra = 10 μm [21]. There are many parameters that affect cutting performance in AJM. Generally, the efficiency of the AJM process is controlled in two stages. These are the input and output parameters. Input parameters consist of independent variables, and output parameters are dependent variables. Independent variables include water jet pressure, abrasive material type, hardness, shape, flow rate and size, nozzle material type, nozzle diameter, jet impact angle, transverse rate and cutting parameters, and standoff distance (SOD). Dependent variables are generally material removal rate, surface roughness, machined surface microhardness, kerf angle, kerf width, and burr height. Dependent variables can be controlled by independent variable parameters that determine the machining quality that occurs as a result of cutting. Surface roughness is the irregular scratches that occur on the cutting surface of the material during machining. The average surface roughness is defined by Ra (μm). Kerf width is the gap created in the material during cutting. The minimum kerf width is important to achieve maximum workpiece quality. Kerf taper is the angle between the vertical and the cutting edge formed during machining in AJM [22–24]. Cutting parameters vary according to material type or machining conditions in AJM. The material machining process varies depending on the pressure level of the water. The water pressure is directly proportional to the material removal rate (MRR) and penetration depth. It is known that water jet pressure plays an important role on the Ra in the machining of many materials, such as cast iron, brass, and aluminum. Abrasive particles that become smaller with the effect of increasing water jet pressure improve surface quality [25, 26]. If the pressure is lower than the threshold pressure range, the material cannot be removed in the cutting. Therefore, jet pressure has an effect on water and abrasive dispersion [27, 28]. Water jet pressure and abrasive mass flow rate are the parameters that affect the MRR. As the abrasive mass flow rate increases, the surface roughness of the cutting edge decreases. In addition, the high abrasive mass flow rate contributes to the reduction of their kinetic energy by mixing large amounts of particles in the jet. Thus, the surface quality is increased [25]. The abrasive particle flow rate can vary depending on the diameter of the focusing nozzle. Better surface finish and cutting performance can be achieved in AJM by using the optimum combination of abrasive particles and parameters [29]. Garnet, SiC, and Al_2O_3 are generally used as abrasive particle types in AJM. Abrasive particle size, shape, and hardness have a significant impact on machining performance in cutting using AJM [30]. It has been observed that many studies have been carried out that the cutting performance changes depending on the shape of the abrasive particle. Fine-grained abrasive particles have a positive effect on penetration depth and kerf width [31]. One of the parameters that affect the cutting rate and cutting edge surface quality of the materials is the transverse rate. A low transverse rate increases the number of abrasive particles striking the material surface. Thus, experimental outputs that affect the cutting surface quality of the material such as surface roughness and kerf profile can be improved [27, 28]. In addition, the SOD parameter is effective on the kerf profile formed as a result of cutting using AJM. SOD is the distance between the nozzle and the material to be cut. The longer SOD increases the surface quality of the material in AISI H13 die steels [32]. This distance is usually chosen between 1–5 mm for optimum cutting quality [33]. Another parameter that affects the surface quality is the jet impact angle. It is defined as the angle between the material surface to be cut and the flow direction. The decrease in the jet impact angle generally increases the penetration depth and surface roughness. Oblique jet impact angle produces a higher MRR and better surface quality than the normal impact angle (90°) in terms of machining performance [34, 35]. Lower kerf, surface roughness and machining marks can be achieved by optimizing this parameter.

It has been determined that many researchers have examined Ra, MRR, kerf width, and microhardness as output parameters in the literature. In general, it has been shown that parameters such as water pressure, stand-off distance, nozzle diameter, abrasive type and abrasive flow rate have an effect on the outputs in AJM process. In these studies; Kumar et al. optimized the AJM parameters for cutting tungsten carbide (WC) reinforced (2, 4, 6, 8, and 10%) Al6082 alloy. The effects of SOD (2, 3, 4, 5, and 6 mm) and transverse speed (TS, 170, 190, 210, 230, and 250 mm/min) on MRR and Ra were investigated. It was determined that MRR was affected by TS, WC% and SOD, respectively. It was observed that the Ra was affected by %WC ratio, TS, and SOD, respectively. It has been revealed that cracks, ploughing, protrusion of the WC, and voids are observed on the machined surfaces (Figure 10.6). According to multi-response optimization based on desirability, the MRR is maximized and Ra is minimized. Optimum results were determined as SOD: 4.22 mm, TS: 223.28 mm/min, and WC rate: 2.10% [36].

Srivastava et al. comparatively investigated the machinability properties of A359/Al_2O_3/B4C hybrid metal matrix composite material in AJM and WED (wire electrical discharge)-turning processes using different cutting parameters. While a dull appearance was detected on the WED-turned surface due to the decomposition of the resolidified layer due to thermal effect, no cutting marks were observed. However, uneven cutting marks were observed in the AJM (Figure 10.7). While tensile stresses occurred in WED due to thermal effect, residual compressive stresses occurred in

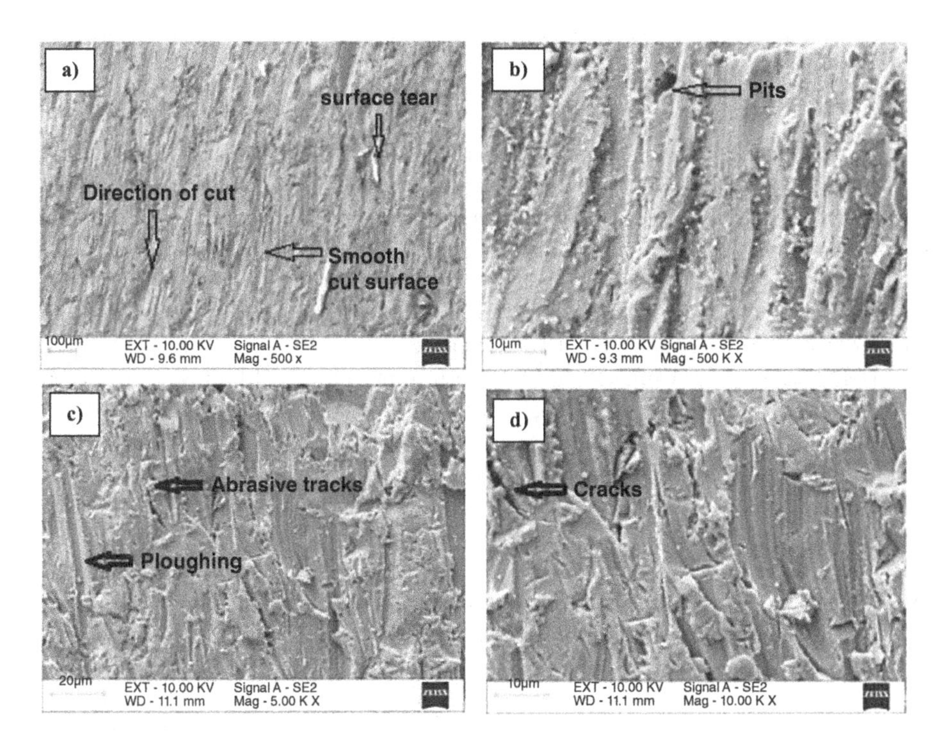

FIGURE 10.6 Machined surface structure: (a–b) standoff distance (SOD): 4 mm, transverse speed (TS): 210 mm/min, tungsten carbine (WC): 6%, material removal rate (MRR): 1.005 mm^3/min; surface roughness (SR): 4.21 µm and (c–d) SOD: 5 mm, TS: 190 mm/min, WC: 10%, MRR: 0.912 mm^3/min, SR: 4.29 µm [36].

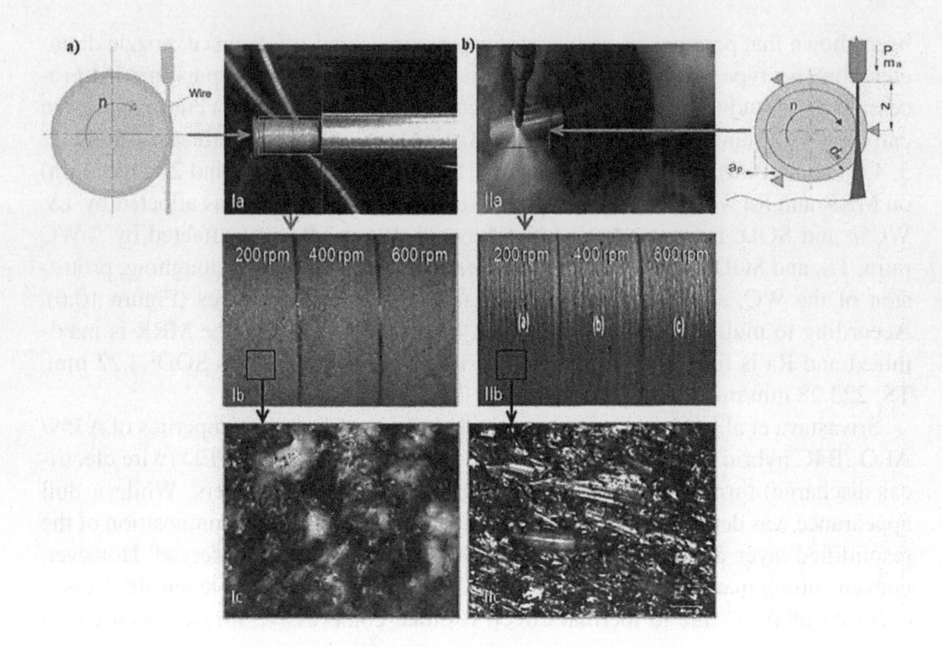

FIGURE 10.7 Comparison of machined surfaces: (a) wire electrical discharge and (b) abrasive jet machining [37].

AJM. It was determined that the Ra was higher in AJM than in WED at the same rotational speed values. It has been revealed that the WED-turning process requires a long machining time due to its low movement speed in the range of 0.1–0.3 mm/min, and accordingly, it is a slow cutting process. Melting and resolidification in WED turning occurred due to thermal effect as well as surface irregularities, while plastic deformation occurred due to erosion in AJM. More micro holes, micro cracks, and crater formations were detected in the cut surfaces using AJM than WED [37].

Holmberg et al. investigated surface finish in machining 718 alloy using EDM, AJM, LBM (laser beam machining), and conventional milling techniques. It has been stated that abrasive abrasives cause compression stresses due to erosion on surfaces machined using AJM, and the best surface quality in terms of cutting methods is obtained in AJM. It has been observed that shallow residual stresses and surface structure with high surface roughness occur on surfaces machined using EDM. Although AJM and EDM are an alternative technique to milling, additional machining may be required for finishing. This results from the removal of abrasives from cut surfaces using AJM. The reshaped edges need to be cleaned in EDM. It has been stated that the LBM technique cannot be considered as an alternative method due to its deep detrimental effect on the cutting edge (Figure 10.8) [30].

Wang et al. stated that nozzle movement speed and material thickness have a significant effect on the contour of the cutting front in cutting of AA 6061 alloy using AJM and that nozzle movement and jet lag vary in direct proportion. It has been revealed that the middle part of the cutting front becomes convex as the thickness of the cut material increases. It has been found that this situation causes the injection

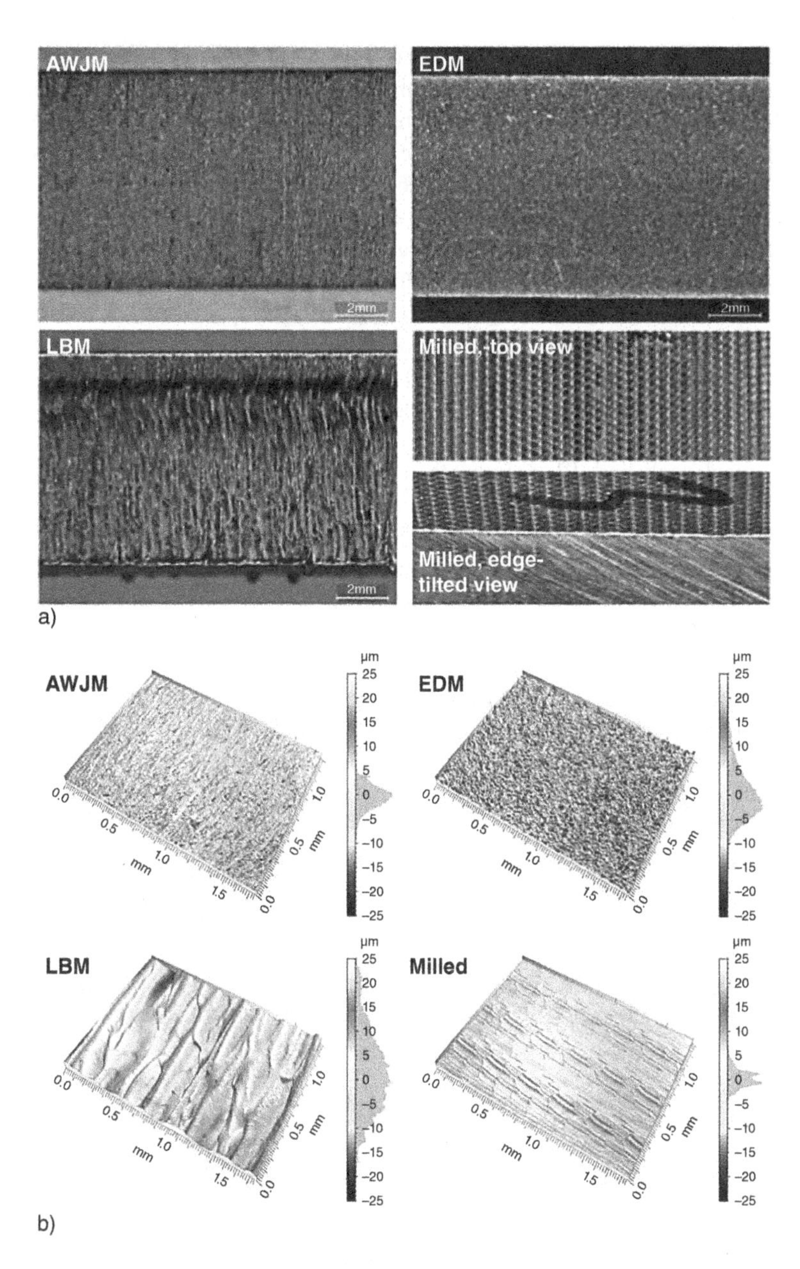

FIGURE 10.8 Machined surface structures: (a) cross section of machined surfaces and (b) surface roughness profile images for each cutting technique [30].

level to be exceeded in the penetration of the jet into the material. It was stated that AJM parameters also affect the cutting front, but these parameters are not considered separately as they will change the movement speed of the nozzle. They used the parabolic fitting technique for the prediction of the outputs obtained. It was stated that this technique could be a reference to improve the accuracy of the outputs [38].

Pradhan et al. investigated the cutting performance in terms of surface roughness, taper angle (TA), and MRR when machining AISI D2 workpiece using hot AJM (HAJM). Fifteen experiments using the response surface methodology (RSM) and Box–Behnken's experimental design were performed using different air pressure (3, 5, and 7 kgf/cm_2), abrasive temperature (60, 80, and 100 °C), and SOD (4, 6, and 8 mm) values. Predictive modeling and an analysis of variance (ANOVA) were performed using the obtained outputs. It was observed that the abrasive temperature effect was the most important factor on the surface quality and the roughness (Rz) values were measured in the range of 0.941–1.545 µm, typical plastic deformation and brittle fracture occurred in the surface structure that was abraded at high temperature. While low erosion amount and surface roughness occur at low temperatures, it has been stated that the most effective parameter on the improvement of MRR and TA is air pressure. The effectiveness of the study results has been revealed by statistical analysis [39]. Sabarinathan et al. used garnet and recycled alumina grinding wheel as abrasive in the AJM process. Kerf width, MMR, depth of cut, and cutting time were taken into account in determining the machining performance. Recycled alumina increased the cutting efficiency of the abrasive by 43% for marble and 63% for aluminum parts. It was determined that the kerf width formed in the marble and aluminum parts was measured more than the cut parts compared to the garnet abrasive and that the recycled alumina abrasives caused more surface roughness (Figure 10.9). It has been observed that recycled alumina abrasives provide deeper cutting than garnet abrasive under the same cutting conditions. While less cutting time is required to cut aluminum workpieces using recycled alumina abrasives than garnet abrasive, it has been shown that the cutting time is similar for marble workpieces [40].

Saravanan et al. optimized the SiC abrasive parameters for cutting Ti-6Al-4V alloy using AJM. They used Taguchi-Grey relational analysis for this. SiC volume, size, and abrasive flow rate were used as independent variables, while MRR and surface roughness were used as dependent variables. SiC containing garnet was found to improve MRR. According to the analysis, 48.81 mm^3/sec of MRR and 3.72 µm of surface roughness were obtained for SiC addition of 0.5% wt, SiC mesh size

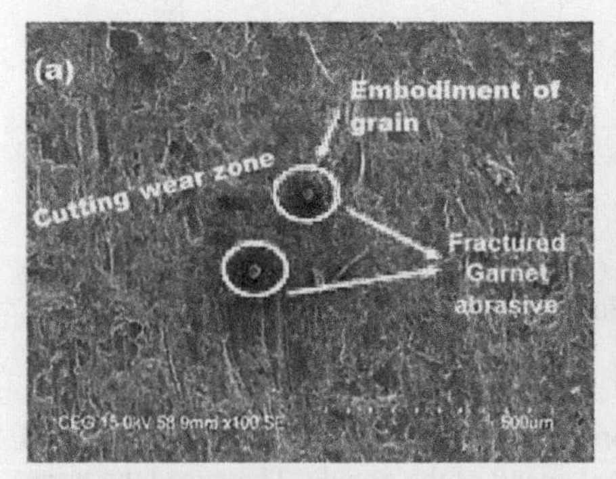

FIGURE 10.9 Surface conditions in the cutting of aluminum using different abrasives: (a) garnet and (b) recycled alumina [40].

of 80, and abrasive flow rate of 2 g/s parameters. It has been determined that the most effective parameter for low surface roughness and maximum MRR is particle size. It has been revealed that 1.91% improvement is achieved for MRR and Ra compared to the estimated values after validation experiments [41]. Paul and Mandal cut 5-mm-thick ceramic tiles with AJM using 60- and 120-μm-sized abrasives. Nozzle diameter (ND; 2, 2.5, and 3 mm), SOD (5, 7, and 9 mm), and pressure (4.2, 4.6, and 4.9 kgf/cm^2) were optimized for taper and MRR using RSM. According to the ANOVA results, it was determined that a maximum MRR of 0.0119 gm/s and a taper value of 0.02 mm were obtained by using 120-μm abrasive size. In addition, it has been revealed that the most important parameter on experimental outputs is SOD [42]. Chaitanya et al. conducted an experimental study on surface roughness using AJM. It was stated that high pressure and low SOD values should be used for higher sensitivity and penetration rate, and higher SOD should be preferred when MRR is important. It has been observed that Al_2O_3 abrasives perform better than SiC abrasive particles in terms of machined surface quality [43]. Kang et al. investigated the micro-dimples formed on alumina-based ceramics using micro AJM. It was stated that the processing time and air pressure were the most important factors among all the independent parameters for changing the dimple size. In addition, it was revealed that nozzle SOD and air pressure play a role in the dimple geometry. It has been observed that Zirconia-toughened alumina, also known as ZTA ceramic materials exhibit good machinability properties in micro AJM in terms of surface quality and dimple form accuracy when compared to high purity alumina. Also, SOD was found to be the most important parameter on TA and edge sharpness. It has been revealed that increasing the flow rate contributes to obtaining finer dimples [44]. Tomy and Hiremath cut hybrid silica glass fiber–reinforced composite (HFRC) materials using AJM. MRR, top hole circularity (TC), bottom hole circularity (BC), depth-averaged radial overcut (DAROC), TA and Ra properties were investigated under air pressure (5, 7 and 9 bar), abrasive grit size (#220, #320 and mixture [#220+#320]), SOD (1, 2 and 3 mm), and constant ND (760 μm) cutting conditions. It has been stated that the delamination that forms in the machining of composites using conventional methods does not form with AJM. Multi-objective GRA (Grey relational analysis) was used to optimize the experimental outputs. It has been determined that the optimum conditions are pressure of 7 bars, SOD of 1 mm, and abrasive type of #320. Experimental outputs under optimum conditions were MRR: 58.7 μg/s, TC and BC: 0.99, DAROC: 182.93 μm, TA: 1° and Ra: 4.95 μm. The most influential parameters on outputs using ANOVA were measured as 51%, 36%, and 9% for SOD, pressure, and abrasive grit size, respectively [45].

10.6 CONCLUDING REMARKS

AJM is a modular unconventional machining method. It can be used for applications such as cutting, drilling, milling, and turning. Almost all materials can be processed using this method in a wide range of materials. It can compete with other technologies in the industry due to minimum force, no heat-induced damage, and an environmentally friendly process. In addition, the flexibility, speed, and difficulty of machining materials significantly increase the use of AJM. Erosion forms with the effect of abrasive particles during cutting. In addition, plastic deformation and compressive stresses form with the effect of erosion on the cutting surface. However, the AJM method has been frequently preferred in the machining of materials, since it does not cause a heat unaffected zone and uses a lower cutting force and there is no negative difference in material properties in many studies in the literature. Cutting performance in AJM is generally determined by controlling the abrasive volume, shape, size, hardness, flow rate, traverse speed, jet impact angle, liquid pressure, SOD, and ND parameters. These parameters affect the experimental outputs that determine the cutting performance. These outputs are generally MRR, Ra, surface morphology, penetration ratio, kerf width, taper angle, and dimensional stability. It has been observed in many studies that the effects of SOD, air pressure, ND, and abrasive type on MRR, surface roughness, and TA were examined. It has been determined that the fatigue life of the workpiece is higher in machining using AJM when compared to other cutting methods. It has been revealed that this is due to the fact that the surface defects are less. Al_2O_3 abrasives generally showed better cutting performance than SiC abrasives. Taguchi, RSM and GRA, and multi-objective hybrid techniques were used to optimize the independent variables. ANOVA was used to reveal the effect of independent variables on dependent variables. Thus, in addition to optimization, data sets can be obtained at the independent variable points that are not tested.

REFERENCES

1. Kovacevic, R., Hashish, M., Mohan, R., Ramulu, M., Kim, J.T. and Geskin, S.E. (1997). State of the art of research and development in abrasive waterjet machining. *Journal of Manufacturing Science and Engineering*, 119, 776–785.
2. Zhu, T.H., Huang, Z.C., Wang, J., Li, L.Q. and Che, L.C. (2009). Experimental study on abrasive waterjet polishing for hard-brittle materials. *International Journal of Machine Tools and Manufacture*, 49, 569–578.
3. Davim, J.P., Rubio, J.C. and Abrao, A.M. (2007). A novel approach based on digital image analysis to evaluate the delamination factor after drilling composite laminates. *Composites Science and Technology*, 67(9), 1939–1945.
4. Bayraktar, Ş. and Hekimoğlu, A.P. (2020). Effect of zinc content and cutting tool coating on the machinability of the Al-(5–35) Zn alloys. *Metals and Materials International*, 26(4), 477–490.
5. Bayraktar, Ş. and Afyon, F. (2020). Machinability properties of Al–7Si, Al–7Si–4Zn and Al–7Si–4Zn–3Cu alloys. *Journal of the Brazilian Society of Mechanical Sciences and Engineering*, 42(4), 1–12.
6. Bayraktar, Ş. and Demir, O. (2020). Processing of T6 heat-treated Al-12Si-0.6 Mg alloy. *Materials and Manufacturing Processes*, *35*(3), 354–362.
7. Bayraktar, S. (2021). *Dry Cutting: A Sustainable Machining Technology*. In Sustainable Manufacturing, London, Elsevier.
8. Bayraktar, S. (2020). *Cryogenic Cooling-Based Sustainable Machining*. In High Speed Machining, London, Academic Press.
9. Ramulu, M. and Arora, D. (1993). Water jet and abrasive water jet cutting of undirectional graphite/epoxy composite. *Composites*, 24(4), 299–308.
10. Hoogstrate, A.M., and Van Luttervelt, C.A. (1997). Opportunities in abrasive water-jet machining. *Annals CIRP*, 46(2), 697–714.

11. Liu, H.T., Cutler, V., Raghavan, C., Miles, P., Schubert, E. and Webers, N. (2018). *Advanced Abrasive Waterjet for Multimode Machining, Abrasive Technology-Characteristics and Applications*, London, IntechOpen.
12. Folkes, J. (2009). Waterjet-an innovative tool for manufacturing. *Journal of Materials Processing Technology*, 209(20), 6181–6189.
13. Momber, A.W. and Kovacevic, R. (2012). *Principles of Abrasive Water Jet Machining*. USA, Springer.
14. Meng, H.C. and Ludema, K.C. (1995). Wear models and predictive equations: their form and content. *Wear*, 181, 443–457.
15. Miller, D.S. (2004). Micromachining with abrasive waterjets. *Journal of Materials Processing Technology*, 149, 37–42.
16. Youssef, H.A. and El-Hofy, H. (2008). *Nontraditional Machine Tools and Operations*, In: Machining Technology. Machine Tools and Operations, USA, CRC Press.
17. Hemanth, P., Naqash, P.V. and Babu, N.R. (2005). A simplified model for predicting the penetration of an abrasive water jet in a piercing operation. *International Journal of Manufacturing Technology and Management*, 72(4), 366–380.
18. Jain, V. (2008). *Advanced Non-traditional Machining Processes*. In: Machining. London, Springer.
19. Hlaváč, L.M., Hlaváčová, I.M., Gembalová, L., Kaličinský, J., Fabian, S., Měšťánek, J. and Mádr, V. (2009). Experimental method for the investigation of the abrasive water jet cutting quality. *Journal of Materials Processing Technology*, 209(20), 6190–6195.
20. Liu, H.T. (2017). "7M" advantage of abrasive waterjet for machining advanced materials. *Jornal of Manufacturing and Material Processing*, 1(1), 11.
21. Zaring, K. (1991). Procedure Optimization and Hardware Improvements in Abrasive Waterjet Cutting Systems. *6th American Water Jet Conference*. Houston, TX, USA.
22. Selvan, M.C. and Raju, N.M. (2012). Analysis of surface roughness in abrasive waterjet cutting of cast iron. *International Journal of Science Environment and Technology*, 1(3), 174–182.
23. Kolahan, F. and Khajavi, H. (2009). A statistical approach for predicting and optimizing depth of cut in AWJ machining for 6063-T6 al alloy. *International Journal of Mechanical and Mechatronics Engineering*, *3*(11), 1395–1398.
24. Todkar, M. and Patkure, J. (2014). Fuzzy modelling and ga optimization for optimal selection of process parameters to maximize MRR in abrasive water jet machining. *International Journal on Theoretical and Applied Research in Mechanical Engineering*, 3(1), 9–16.
25. Selvan, M.C.P., Raju, N.M.S. and Sachidananda, H.K. (2012). Effects of process parameters on surface roughness in abrasive water jet cutting of aluminium. *Frontiers of Mechanical Engineering*, 7(4), 439–444.
26. Babu, M.N. and Muthukrishnan, N. (2014). Investigation on surface roughness in abrasive water-jet machining by the response surface method. *Materials and Manufacturing Processes*, 29 (11–12), 1422–1428.
27. Uthayakumar, M., Khan, M.A., Kumaran, S.T., Slota, A. and Zajac, J. (2016). Machinability of nickel-based Superalloy by abrasive water jet machining. *Materials and Manufacturing Processes*, 31, 1733–1739.
28. Ay, M., Çaydaş, U. and Hasçalik, A. (2010). Effect of traverse speed on abrasive waterjet machining of age hardened Inconel 718 nickel based Superalloy. *Materials and Manufacturing Processes*, 25, 1160–1165.
29. Goutham, U., Hasu, B.S., Chakraverti, G. and Kanthababu, M. (2016). Experimental investigation of pocket milling on Inconel 825 using abrasive water jet machining. *International Journal of Current Engineering and Technology*, 6(1), 295–302.

30. Holmberg, J., Berglund, J., Wretland, A. and Beno, T. (2019). Evaluation of surface integrity after high energy machining with EDM, laser beam machining and abrasive water jet machining of alloy 718. *International Journal of Advanced Manufacturing Technology*, 1005(8), 1575–1591.
31. Karakurt, I., Aydin, G. and Aydiner, K. (2010). Effect of the abrasive grain size on the cutting performance of concrete in awj technology. *Technology*, 133, 145–150.
32. Sharma, A. and Lalwani, D.I. (2013). Experimental Investigation of Process Parameters Influence on Surface Roughness in Abrasive Water Jet Machining of AISI H13 Die Steel. *Proceeding of Conference on Intelligent Robotic Automation and Manufacturing* (pp. 333–341).
33. Deepak, D., Anjaiah, D. and Shetty, S. (2017). Optimization of Process Parameters in Abrasive Water Jet Drilling of D2 Steel to Produce Minimum Surface Roughness Using Taguchi Approach. *6th International Conference on Electronics, Computer and Manufacturing Engineering* (pp. 236–239).
34. Wang, J., Kuriyagawa, T. and Huang, C.Z. (2003). An experimental study to enhance the cutting performance in abrasive waterjet machining. *Machining Science and Technology*, 72, 191–207.
35. Shipway, P.H., Fowler, G. and Pashby, I.R. (2005). Characteristics of the surface of a titanium alloy following milling with abrasive waterjets. *Wear*, 258(1–4), 123–132.
36. Kumar, K.R., Sreebalaji, V.S. and Pridhar, T. (2018). Characterization and optimization of Abrasive Water Jet Machining parameters of aluminium/tungsten carbide composites. *Measurement*, 117, 57–66.
37. Srivastava, A.K., Nag, A., Dixit, A.R., Scucka, J., Hloch, S., Klichová, D., Hlaváček, P. and Tiwari, S. (2019). Hardness measurement of surfaces on hybrid metal matrix composite created by turning using an abrasive water jet and WED. *Measurement*, 131, 628–639.
38. Wang, S., Hu, D., Yang, F., Tang, C. and Lin, P. (2021). Exploring cutting front profile in abrasive water jet machining of aluminum alloys. *International Journal of Advanced Manufacturing Technology*, 112, 845–851.
39. Pradhan, S., Das, S.R., Nanda, B.K., Jena, P.C. and Dhupal, D. (2020). Experimental investigation on machining of hardstone quartz with modified AJM using hot silicon carbide abrasives. *Journal of the Brazilian Society of Mechanical Sciences and Engineering*, 42(11), 1–22.
40. Sabarinathan, P., Annamalai, V.E. and Rajkumar, K. (2020). Sustainable application of grinding wheel waste as abrasive for abrasive water jet machining process. *Journal of Cleaner Production*, 261, 121225.
41. Saravanan, K., Sudeshkumar, M.P., Maridurai, T., Suyamburajan, V. and Jayaseelan, V. (2021). Optimization of SiC Abrasive Parameters on Machining of Ti-6Al-4V Alloy in AJM Using Taguchi-Grey Relational Method. *Silicon*, 1–8.
42. Paul, T. and Mandal, I. (2020). Development and Parametric Optimization of Abrasive Jet Machining Setup. *Materials Today: Proceedings*, 22, 2306–2315.
43. Chaitanya, A.K., Babu, D.K. and Kumar, K.G. (2020). Experimental study on surface roughness by using abrasive jet machine. *Materials Today: Proceedings*, 23, 453–457.
44. Kang, C., Liang, F., Shen, G., Wu, D. and Fang, F. (2021). Study of micro-dimples fabricated on alumina-based ceramics using micro-abrasive jet machining. *Journal of Materials Processing Technology*, 297, 117181.
45. Tomy, A. and Hiremath, S.S. (2021). Machining and characterization of multidirectional hybrid silica glass fiber reinforced composite laminates using abrasive jet machining. *Silicon, 13*, 1151–1164.

11 Artificial Neural Networks for Machining

Şenol Bayraktar and Cem Alparslan
Recep Tayyip Erdoğan University, Rize, Turkey

CONTENTS

11.1 INTRODUCTION TO ARTIFICIAL NEURAL NETWORKS

Problems that people try to find solutions have increased depending on the developing technology. It has gained momentum for high reliability and quality in some studies done by humans using machines. It was predicted that the ability to create and solve algorithms could be given to machines by using the Turing machine, which was discovered by the scientist Alan M. Turing in 1936 [1]. Therefore, the way for computer technology has been paved. The computer has been included in human life and has started to be preferred in every field in the next technological developments. Scientists have developed the concept of "artificial intelligence," which enables machines to gain the ability to think, decide, learn, and interpret in ongoing studies. Artificial intelligence is a sub-branch of computer science that investigates and develops the functions of the human brain, how it works, and its information processing and storage capabilities. Studies in the field of artificial intelligence have focused on ANN, which imitates the ability of the human brain to make decisions and learn, in recent years. ANNs bring learning, remembering, and generalization features to systems as a result of being inspired by nerve cells in the human brain. It can be defined as a system designed to model the way the brain can perform any function. Learning in the human brain occurs as a result of the interaction of synaptic connections between neurons. In other words, people enter the process of learning by living from the day they are born. Everything felt using the five senses causes the synaptic connections to be adjusted. It even creates new connections between them. Thus, the learning process occurs. The same is also true for artificial neural networks (ANNs). Learning occurs by processing input/output data. In other words, data are used with algorithms. Then, the learning is formed by repeatedly adjusting the weights of the

DOI: 10.1201/9781003284574-12

synapse until convergence is achieved. It is formed as a result of the connection of artificial neural cells with each other. Usually, it consists of layers. It can be realized using electronic circuits and as software on computers. ANNs collect the necessary information after the learning process that was inspired by the brain's information-processing method. This information is stored thanks to the connections between the cells. Then, it is run as a processor capable of generalization. Basically, ANNs are mathematical systems consisting of many processing elements (neurons) that are communicated with each other. Process elements are structures that work using neuron logic and form a network within each other. For example, a processing element known as the transfer function receives signals from other elements (neurons) and combines them. It transforms the signals and produces a numerical result in the next step. Basically, a processing element generates an output value by a nonlinear transformation of an input (Figure 11.1).

ANNs are divided into feedforward and feedback networks. One-way signal passing is created in feedforward networks. Cells are arranged layer by layer. Input nodes consist of a hidden layer and an output layer. The outputs of the cells in a layer are transmitted as inputs to the next layer. The current network output is determined by processing the information in the intermediate and output layers. The output of at least one cell is sent as input to itself or to other cells in feedback networks. Information processing can be performed between cells in a layer or in cells between layers. Because of this structure, feedback networks are preferred to obtain a nonlinear dynamic behavior [2]. The first research on artificial intelligence in history was carried out by McCulloch and Pitts using Turing's computational model. Mathematical functions have been tried to be modeled using only the logical operators "and" and "or" [3]. It has been determined that artificial neural cells can gain learning abilities using neuron logic. Later, scientists such as Hebb, Minsky, Edmonds, McCarthy,

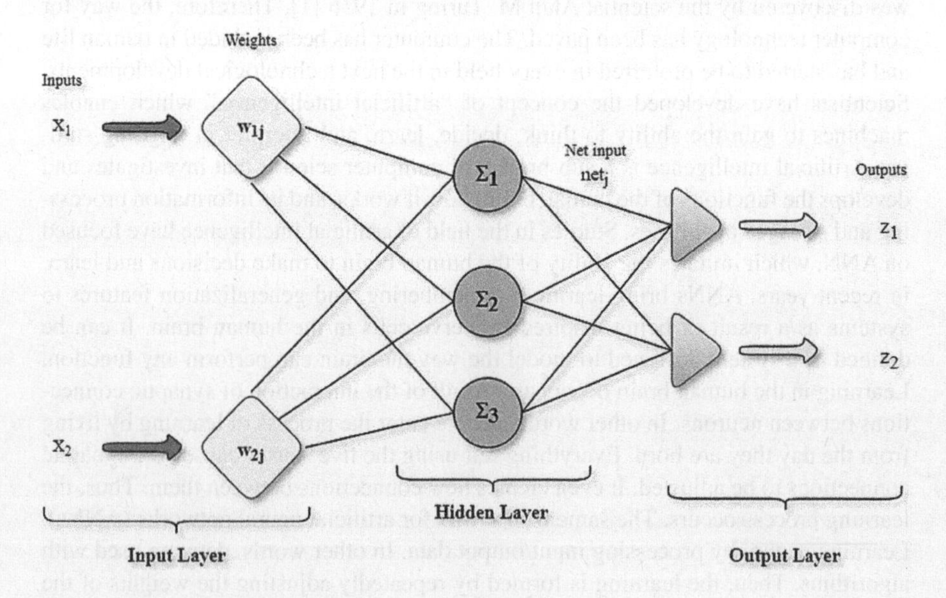

FIGURE 11.1 Multilayer neural network model.

Shannon, and Rochester conducted studies on ANN. The nonlinearly separable XOR (Exclusive OR) problem was solved with "back-propagation" in 1986. Thus, studies on ANN have been accelerated [4]. It has been observed that there are different studies from theoretical and laboratory studies on ANN since the 1990s. It has been used frequently for the solution of problems encountered in daily life due to the development of various and fast learning algorithms (Figure 11.2). There is no limit to the application areas of ANN that can be used in many industries today. It is used extensively in some areas such as prediction, modeling, classification. It is used by many disciplines due to its nonlinear data processing, error tolerance, learning ability, and ease of integration with existing systems [5]. It is widely used in manufacturing systems which are among the disciplines, due to its capabilities such as function modeling, clustering, optimization [6] classification, prediction, control, and monitoring [7]. It is generally preferred for problems such as engineering design, storage design, quality control, stock control, supply chain and management, monitoring, diagnosis, and demand prediction for different methods in the field of manufacturing.

The machinability properties of materials in the manufacturing sector are determined according to outputs such as surface roughness, tool wear, cutting force, torque, temperature, material removal rate (MRR), and kerf width. These outputs vary with the independent variable parameters such as cutting speed, feed, coolant, and depth of cut in conventional machining techniques. The independent variables are generally used as standoff distance, nozzle diameter, transverse speed, cutting power, abrasive type and size, gas pressure, pulse width, and pulse frequency in nonconventional techniques. Factors affecting surface roughness are given in Figure 11.3. These are generally cutting tool properties, machining parameters, workpiece properties, and cutting phenomena. The experimental outputs (surface roughness, cutting force, tool wear) by controlling the sub-parameters under these main independent variable parameters using ANNs can be predicted with different algorithms.

Experimental outputs can be predicted based on independent variables or control factors in machining theory using ANN. In addition, it contributes to the creation of data sets for non-experimented points. Thus, studies on the machinability properties of different materials depending on the independent variables in the literature are investigated in this chapter. A detailed analysis of the performance of ANNs has been performed and presented in a comparative way by using the results obtained.

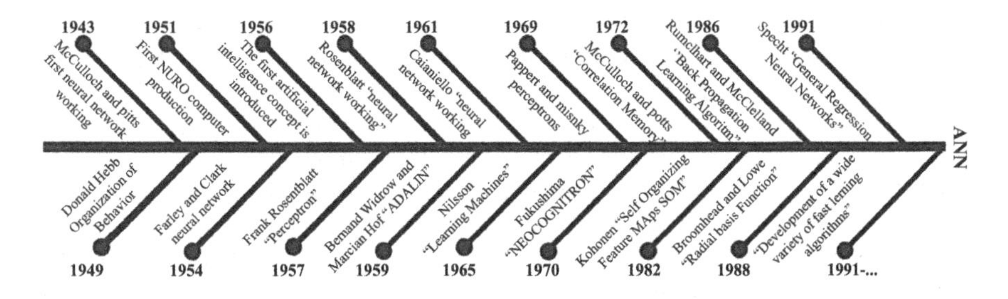

FIGURE 11.2 Chronological history of artificial neural networks.

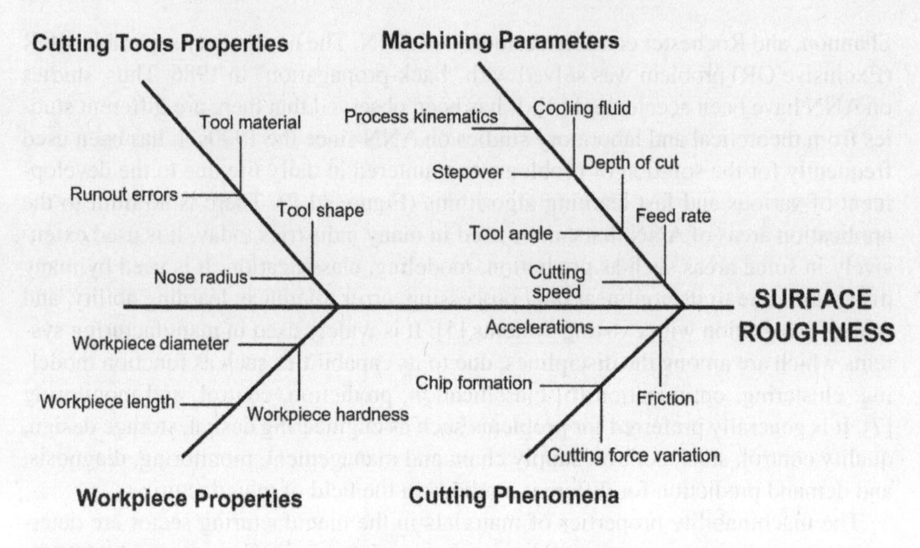

FIGURE 11.3 Control factors affecting surface roughness [8].

11.2 WHY ARE ANNS USED?

ANNs can produce solutions more easily than statistical techniques, thanks to their continuity in learning, generalization, and prediction. It offers a different solution method than conventional methods for the analysis of data and revealing the information in the data. It is a superior feature in that it has the ability to learn and can use different learning algorithms. It can work using adaptive and incomplete information. They also have the ability to make decisions under uncertainties and to be tolerant of errors. In addition, they have become one of the important tools in solving complex nonlinear problems. It can converge real-life nonlinear situations to any continuous function or its derivatives. Therefore, they are defined as the universal function convergent method [9]. While serial and sequential operations are generally used in information-processing methods, the processing method in ANNs consists of parallelism. The slowness of one unit does not affect the whole system thanks to this parallelism. Thus, it turns out that ANNs are faster and more reliable. It is necessary to develop an appropriate algorithm for the problem in order to solve any problem using conventional computational methods. Then these algorithms must be calculated by programming. Therefore, the solution ability of algorithms is limited to the ability of the expert who writes the code. Different algorithms are needed for each problem type. Conventional methods are not sufficient for solving complex problems due to the difficulty of algorithms. Also, it is necessary to display images of the object at all possible angles and distances in order to teach any object to a computer using conventional methods using all possible combinations of them. Learning in ANNs reaches the ability to generalize about the problem by using the examples whose characteristics are given. Thanks to this feature, solutions to problems that are complex for conventional methods can be found. ANNs can also be preferred for simple operations that people can easily do but may be impossible for conventional methods. Another point where it differs from conventional systems is continuous learning. It is possible for ANNs to learn new events continuously because they can learn new examples shown to them and adapt to new situations [10]. Information in conventional computational methods is stored in databases or in program codes. Information in ANNs is stored in weighted connections between neurons. In other words, the information is distributed in the network. The entire network represents the entire event it learned. For this reason, it is stated that ANNs can store information in distributed memory. ANNs can make generalizations about previously unexplored situations by using the examples given with self-learning ability. In other words, it can produce solutions for lost or erroneous data. Therefore, ANNs are good pattern recognition engines and robust classifiers [11]. ANNs can adjust their weights according to the changes in the problem for which a solution is sought. In other words, an ANN trained to solve a specific problem can be retrained according to the changing data in the problem. Even if the changes are continuous, training can be continued in real time. Due to this feature, ANNs can be used effectively in areas, such as adaptive sample recognition, signal processing, system diagnosis, and control [12]. A single main processor element in conventional processors performs each action in turn. Serial information processing is carried out in the operating system. Malfunction of any unit in a conventional system causes the entire system to malfunction. There are many simple processor elements, each of which deals with a part of a big problem in ANN models. Therefore, the connection weights can be adjusted according to the problems. The decrease in efficiency in a part of the network due to the effect of its flexible structures causes only a decrease in performance in the model. It does not pose a big problem in solving the problem, and it is not possible for the model to completely lose its function. Therefore, their ability to tolerate error is extremely high compared to conventional methods. In addition, the dense connection structure between the processor elements that share the total workload is the main power source of neural computation. The ANN method can be easily applied to the most complex problems and solutions can be provided due to this local operation structure. Contrary to conventional systems, they can work with incomplete information after training and generate results, even though there is incomplete information in new samples. The fact that an ANN works with incomplete information does not mean that their performance will decrease. The decrease in performance occurs due to the importance of incomplete information. Important information is determined during the training of the network itself. It is concluded that if the performance of the network is low, the incomplete information is important, while if the performance is high, the incomplete information is not important [13]. Unlike statistical methods, ANNs can work without the need for any extra transformation by using an unlimited number of variables and parameters. Thus, solutions can be found with excellent prediction accuracy [14]. The fact that it consists of cells that perform simple operations and that the connections are smooth provides great convenience in terms of realizing the networks. The structures of ANNs in different application fields also consist of these standard cells. Thus, ANNs used in different application fields can share similar learning algorithms and theories. Significant convenience is provided in solving problems using ANNs. Some of the benefits provided by ANN are given as follows:

- They do not require the use of a rule base.
- They have the ability to learn.
- They can learn using different learning algorithms.

- They can generate information for different independent variable points.
- They can be used in detection events.
- They can make pattern recognition and classification. They can complete the incomplete patterns.
- They have the ability to self-organize and learn by adapting to their environment.
- They have error tolerance. They may work using incomplete or unclear information. They show graceful degradation in error cases.
- They can work in parallel and process real-time information.
- They have distributed memory.

ANNs that are used to obtain information from complex or erroneous data can be used to solve problems that are difficult to be determined by people. It allows to obtain data sets at points not tested usually in the independent variable value ranges in the processing field.

11.3 ANNs IN MACHINING?

Today, different computer programs, decision support or -making systems, and optimization methods are used to make manufacturing processes more efficient and flexible [15]. Processes, customer demand, product quality, machining method, in production disruptions, and disruptions in raw material supply can be resolved flexibly by using these systems. It is very difficult to observe the changes that affect the system and to provide the necessary optimization in all nonlinear processes. Different methods are needed to follow these processes. ANNs are among these methods that will contribute to the manufacturing processes. The current status of manufacturing processes can be followed using ANNs. According to the experimental data obtained, it contributes to the working mechanism of the manufacturing processes. Therefore, the use of ANNs is very beneficial in terms of quality improvement in manufacturing processes [16]. Predicting the experimental output values and determining the optimum conditions for machining performance are extremely important in manufacturing. Modeling and optimization of manufacturing processes with less experiment are extremely important due to the high cost of testing. Developed mathematical models depend on the relationships between input and output parameters. Basically, models are examined in three categories. These are analytical, experimental, and artificial intelligence–based models. While analytical and experimental models can be developed with conventional approaches, such as Taguchi and response surface methodology (RSM), artificial intelligence–based models are developed with nonconventional approaches. The use of ANNs is very advantageous instead of nonconventional models due to the complexity of machining performance. Conventional methods can give good results in modeling machining processes. However, these techniques may not accurately identify the underlying nonlinear complex relationship between decision variables and responses [17]. The differences between conventional algorithms and ANNs are given in Table 11.1.

The importance of modeling and optimization studies has increased in various fields, such as machining operations, in recent years. In general, while performance values at the optimal point of machining conditions are predicted by optimization

TABLE 11.1
Differences between Conventional Algorithms and Artificial Neural Networks (ANNs)

Conventional algorithms	ANNs
• Outputs are obtained by applying the inputs to the set rules	• Rules are set by giving input and output information during learning
• Information and algorithms are precise	• Experience is utilized
• Calculation; central, simultaneous, and sequential	• Calculation; collective, asynchronous, and parallel after learning
• Memory packed and literal stored	• Memory allocated and spread over the network
• There is not error tolerance	• There is error tolerance
• It is relatively fast	• It is slow and hardware-dependent

processes, potential minimum values of machining performance are predicted by modeling. Experimental outputs or dependent variables proving machining performance in industry are cutting force, torque, tool wear, surface roughness, tool life, and machined surface hardness [18]. Experimental inputs or independent variables are used as cutting speed, feed rate, depth of cut, coolant, material hardness, cutting tool material, geometry, and coating type. Data sets can be obtained at points where no experiments are performed by using ideal learning, training, and testing algorithms suitable for the process and models obtained with different neuron and layer numbers. The use of ANNs is important in terms of cutting tool or experimental set costs. On the other hand, ANN machining conditions should be determined at an optimal level, and experimental outputs (surface roughness, cutting force, tool wear, etc.) should be monitored online in manufacturing. For this, the development of adaptive neurocontrollers supported by ANN contributes to the manufacturing sector. The main objective is to obtain quality and inexpensive parts in manufacturing processes. Surface roughness, fatigue resistance, lubrication, friction, and wear properties play an important role in determining the quality of the machined parts. Many studies have been conducted to predict experimental outputs with feedforward and/or feedback propagation using the ANN technique. Thanks to the independent variable parameters, ANN models with high prediction efficiency are created. In addition, data that will increase machining performances have become easily available [19–21]. Thanks to ANNs, machining performance can be improved and problems with different characteristics such as pattern classification, clustering, function modeling, prediction, optimization, and combining control can be solved [22]. It is integrated with algorithms such as genetic algorithm (GA), grey relational analysis (GRA), particle swarm optimization (PSO), adaptive neuro-fuzzy inference system (ANFIS), and nondominated sorting genetic algorithm (NSGA) to improve the performance of ANN. The prediction performance of the ANN is improved, and better results can be obtained by using these algorithms. An example of the integrated ANN–GA approach is given in Figure 11.4. This example consists of experimental data, regression modeling, GA optimization, ANN–GA optimization, and ANN modeling stages. Flow charts of ANN–PSO and ANN–NSGA-II algorithms are given in Figures 11.5 and 11.6, respectively. The relationship between the processes

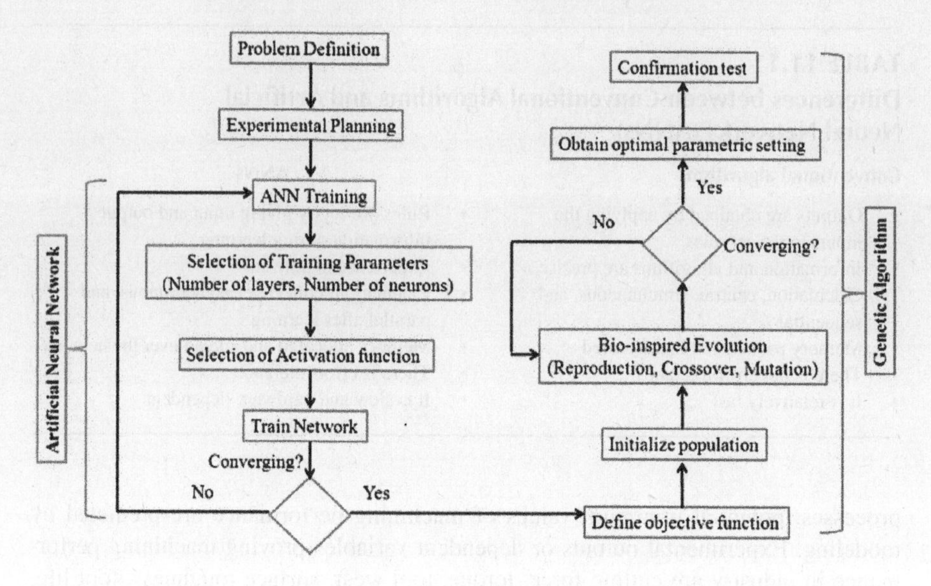

FIGURE 11.4 Flow chart of the integrated artificial neural network–genetic algorithm [23].

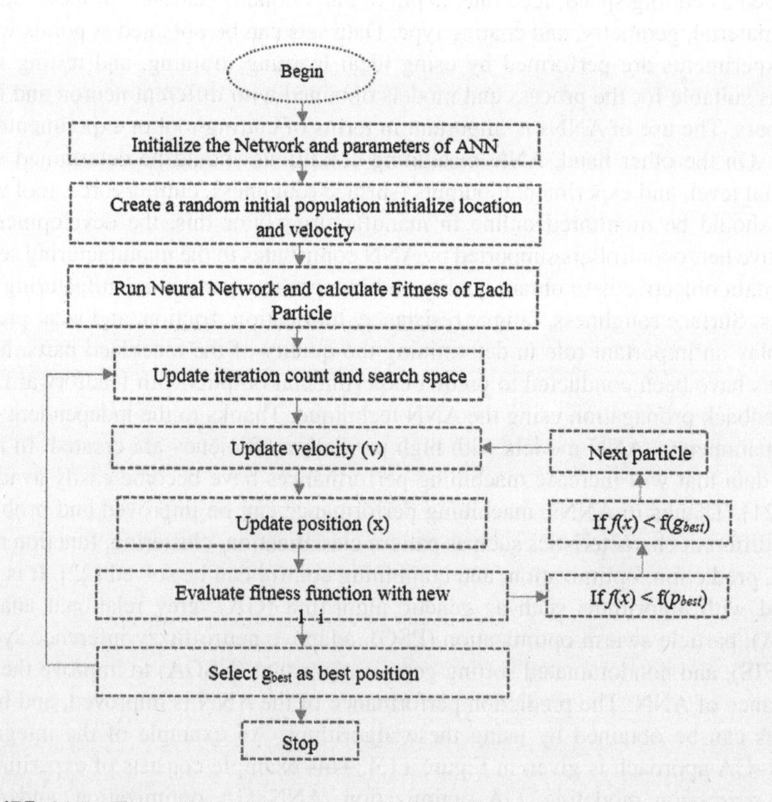

FIGURE 11.5 Flow chart of the integrated artificial neural network–particle swarming optimization [24].

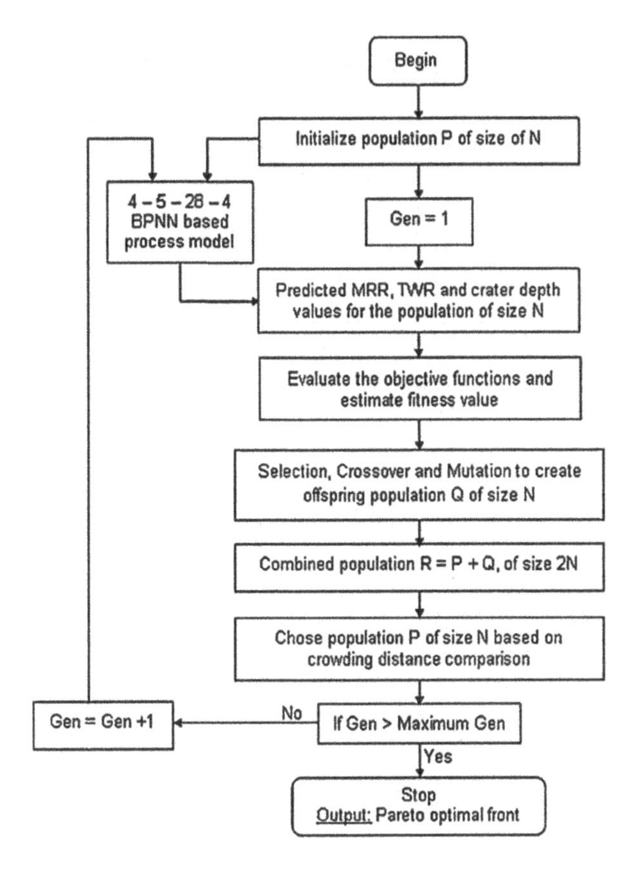

FIGURE 11.6 Flow chart of the integrated artificial neural network– nondominated sorting genetic algorithm-II [25].

and the sequence of applications are explained in detail on these charts. Different studies showing the success of ANNs in problems, such as demand prediction, engineering design, storage design, stock control, quality control, supply chain management, monitoring, diagnosis, and process selection, in the literature have been presented in the field of production management [26].

11.4 ANN APPLICATIONS IN MACHINING

The use of ANNs depending on the development of technology emerges in many areas of human lives. It has started to play an active role in the fields of mechanics, automotive, electronics, energy, space sciences, banking, finance, health, and military. The aim of the development of an ANN is to imitate the human brain to perform functions such as association, self-organization, and generalization. In addition, because it has the ability to accurately predict the independent variable in line with the data in the systems, it facilitates all machining processes. Therefore, it is more suitable for use in model development for nonlinear machining processes. Thanks to their ability to learn and generalize, they can easily create nonlinear and complex input–output relationships [27]. The main purposes of applying ANN in machining methods are predicting input and output values, classifying input values (errors, characters, etc.), associating all data with each other, analyzing input data, and editing input signals. Thus, it is important to quickly determine the quality index of the machined parts in the industry and to improve the quality of the workpiece. There are many recent studies in the literature regarding the application of ANN in machining processes. Yusoff et al. developed predictive models for performance outputs using ANN and orthoANN approaches in the machining of Inconel 718 with wire electrical discharge machining (WEDM). Performance outputs were determined according to cutting speed, sparking gap, MRR, and Ra. The independent variables were selected as pulse-on time, pulse-off time, peak current, servo voltage, and flushing pressure. It was observed that cascade-forward backpropagation neural network (CFNN) was the best network for the dataset. The single hidden layer and the 5-14-4 CFNN were revealed to be the most precise and generalized network structure with high prediction accuracy. It was determined that the error between the experimental results and the predicted results was 5.16% [28]. Prathik et al. investigated the acoustic emission and surface roughness in the WEDM process of titanium grade 2 material using an ANN. It was observed that the neural network trained with 70% of the data in the training set using the neural network fitting tool network structure for three neurons determined good prediction results when compared to 50–60% of the data in the training set with minimum regression coefficients [29]. Agrawal and Kamble optimized the minimum undercut, maximum MRR, and etch factor in the photochemical machining of Al/SiC composite using the GRA and ANN integrated approach and the independent variables of concentration, time, and temperature. Optimum conditions were determined as temperature of 40 °C, concentration of 400 gr/liter and etching time of 2 minutes. It was observed that good prediction performance was revealed using ANN at the optimum points obtained using GRA [30]. Abbas et al. optimized the surface roughness with ANN-based Edgeworth–Pareto by using different cutting speed, feed rate, and depth of cut parameters in turning of AZ61 magnesium alloy with minimum machining time (Tm) and cost (C). Optimum values were determined as Ra = 0.087 µm, Tm = 0.358 min/cm^3, and C = US$8.2973 at a cutting speed of 250 m/min, a depth of cut of 1 mm, and a feed rate of 0.08 mm/rev conditions. Surface roughness prediction using an ANN was revealed to be reliable with an error rate of ±1.35% [31]. Badiger et al. the cutting force, surface roughness, and tool wear modeled using a ANN and PSO in turning of MDN431 steel. The optimization of process parameters was carried out using desirability and MOPSO (multi-objective particle swarm optimization) approach. Optimum parameters for cutting force and surface roughness were determined as a cutting speed of 59 m/min, a feed rate of 0.063 mm/rev, and a depth of cut of 0.2 mm. It was observed that TiN/AlN-coated inserts reduced the cutting force and surface roughness by 9% and tool wear by 105% when compared to the uncoated tool. As a result of the regression and ANN models developed close to 1, it was revealed that the model developed for TiN/AlN coated tools is sufficient [32]. Zerti et al. predicted the cutting force and surface roughness using RSM and ANN models in the dry turning of AISI 420 martensitic stainless steel with ceramic inserts. According to the analysis of variance (ANOVA) results, it was revealed that while the feed rate was effective on the surface roughness, the depth of cut was effective on the cutting force. It was found that ANN models gave better results than RSM and the optimum parameters for multi-objective optimization were a cutting speed of 80 m/min, a feed rate of 0.08 mm/rev, and a depth of cut of 0.141 mm [33]. Sada and Ikpeseni compared ANN and ANFIS techniques to predict the machining performance of AISI 1050 steel. The effects of cutting speed, depth of cut, and feed parameters on MRR and tool wear were investigated in the turning process. The predicted error in ANN and ANFIS was calculated 6.1% and 11.5% for MRR and 4.1% and 7.2% for tool wear, respectively. In addition, it was found that ANN models give better prediction results than ANFIS [34]. Bagga et al. modeled tool wear with ANNs using cutting force and vibration outputs in dry turning of EN-8 steel with cemented carbide inserts. Cutting speed, feed, and depth of cut parameters were used as independent variables or control factors. It was determined that the cutting speed and depth of cut were the most important parameters to minimize tool wear. It showed that there was a 3.48% error rate for tool wear between the developed model and the experimental result. Thus, a very good agreement between the manual measurement results and the ANN results was obtained [35]. Xu et al. measured the cutting force and surface roughness in dry milling of graphite iron with Al_2O_3+Ti (C, N)+TiN-coated cutting tool using cutting speed, feed, and depth-of-cut parameters. They performed modeling on experimental outputs with improved case-based reasoning (ICBR), support vector regression machine (SVR), novel particle swarm optimization, and ANN techniques. It was stated that the ICBR method was the ability to improve itself. As a result of the trained ANN model, it was found that the cutting speed had a significant effect on the surface roughness, while the most important factor affecting the depth of cut was the cutting force. It was found that ICBR was better estimation accuracy and MSE (mean square error) than ANFIS, SVR, and standard CBR methods [36]. Paturi et al. modeled tool wear with ANN in turning of AISI 52100 under dry, wet and MQL (minimum quantity lubricant) conditions using different cutting speed, feed rate and depth of cut parameters. Levenberg–Marquardt (trainlm) and tangent sigmoid (tansig) were used as training and testing functions, respectively. An experimental design was performed by Taguchi orthogonal array. It was determined that the ANN had high accuracy in terms of the relationship between the predicted value and the targeted value. In addition, it was revealed that the wear was more in machining using the MQL technique [37]. Quarto et al. predicted the performance of the process parameters in the micro-electrical-discharge machining drilling process of AISI 304 stainless steel material using ANN–PSO and FEM (finite element model) techniques. Six drilling tests were performed considering two different electrode materials and geometry types for comparison between two model applications. It was stated that the ANN-PSO method for dimensional deviation and MRR was a low error rate and that more data were required for an effectively trained ANN model. It was revealed that a lot of time was needed to run the FEM simulation, process the data, and calculate the information [38]. Trinath et al. the effects of graphite particle reinforcement on the machinability properties of self-lubricating Al/SiC/graphite-hybrid composites prepared with mix stir casting using ANN were investigated. It was stated to improve machinability as the graphite particles act as solid lubricants and chip breakers. It was revealed that the addition of 4% graphite in AA2024/SiCp composite reduced up to approximately 25% of the feed force, while the surface roughness of the hybrid composite was reduced compared to the SiC particles-added composite. Hyperbolic tangent sigmoid function (transig) and linear transfer function (purelin) were used as activation transfer functions for the ANN architecture. The Levenberg–Marcendet training algorithm was preferred in order to update the weight and bias values in the back propagation function. It was observed that there was a very good agreement between the experimental results and the ANN results [39]. Lalwani et al. predicted the machining performance of Inconel 718 in the WEDM process using ANN and RSM-based models. Central composite design (CCD) was used for WEDM experiments. While selecting the test inputs pulse-on time (T_{ON}), pulse-off time (T_{OFF}), servo-voltage (SV), peak current (IP), and wire tension (WT), the performance outputs were kerf width, surface roughness and MRR was determined. The double hidden layer ANN model and 5-13-15-3 feedforward back-propagation training algorithm were used. According to the ANOVA results, it was determined that T_{on} was the most important parameter in WEDM and that the kerf width, surface roughness, and the MRR increased with the increase in T_{on}. It was observed that the results were in agreement with the error rate of less than ±12% between the RSM models and the experimental models. It was stated that the prediction accuracy of the ANN model was higher than the RSM and the NSGA-II could be preferred to determine the optimum WEDM conditions [40]. Kaviarasan et al. modeled the surface roughness in machining of titanium alloys using GA and ANN. Twenty-seven experiments were performed with different combinations of cutting speed, feed rate, and depth of cut. RSM was applied using the experimental data in the first stage, while GA was applied later to minimize the surface roughness. Validation experiments showed that the error rate was 4.27%. The ANN data were used to improve the revised RSM model. It was determined that the predictive model and the optimization for the ideal surface roughness resulted in the best cutting parameters, and the optimum surface roughness in GA was determined as Ra of 1.42 μm for cutting speed of 148 m/min, feed rate of 0.1 mm/rev and depth of cut of 0.5 mm. The error rate between prediction and experimental results was found to be 2.27% [41]. Lin et al. predicted the surface roughness in milling of Al6061 alloy using multiple regression and ANN techniques. It was determined that the composite effect of spindle speed and depth of cut significantly affected the surface roughness. It was revealed that ANN models had very high prediction accuracy compared to regression models for all training data [42]. Nagaraj et al. predicted the MRR and Ra using regression and ANN techniques for milling soda-lime-silica glass material. The error rate in model prediction of the regression analysis for the MRR and Ra was determined to be 8.24% and 7.70%, respectively, while it was determined to be 1.89% and 1.70%, respectively, in the ANN. It was stated that the deviation in the prediction and experimental results was within the allowable range in both models. Thus, it was revealed that the ANN model was higher predictive ability [43].

11.5 CONCLUDING REMARKS

An ANN is a deep learning algorithm that is developed as a result of the human brain being affected by the structure of biological neural networks (neurons). It is an improved version of machine learning techniques. It has an important place today because it is used in areas such as diagnosis, classification, prediction, control, data association, data filtering, and interpretation. ANNs can be used as process models because they can process a large number of parameters and incomplete information. They are able to adapt themselves to changes in the manufacturing environment depending on their internal learning abilities. They can also be used when precise information about the relationships between the various parameters of manufacturing cannot be obtained. The structure of the ANN, which is one of the nonconventional algorithm methods, and the current studies on the prediction performance are researched in detail and presented together in this chapter. Many theoretical and experimental studies on manufacturing processes have been performed in the literature. Machining parameters for different cutting methods were predicted using ANN in most of these studies. Different neural network models have been developed to improve experimental outputs such as cutting force, surface roughness, tool wear, and MRR that occur during machining. Cutting speed, feed and depth of cut, pulse-on time, pulse-off time, servo-voltage, peak current and wire tension parameters in the improved models are generally used as independent variables or control factors. Experimental outputs or dependent variables have been predicted during machining using these parameters. Accuracy levels have been determined by comparing predictions with experimental results. It has been tried to improve the performances of an ANN by integrating it into PSO, GRA, GA, ANFIS, and NSGA-II processes. The performance of ANNs has been revealed in comparison with Taguchi, RSM, and regression techniques. It has been observed that better prediction performance is obtained using ANN models compared to other techniques in most of the studies. As it is understood from these studies obtained, it is foreseen that the ANN technologies used today will be developed with a high probability in the future. Thus, it is envisaged that the use of ANN technologies will be used in autonomous cars and robots that can talk and recognize the environment. In addition, it will also become increasingly common in diagnosing medical problems and solving more complex problems due to developing technology.

REFERENCES

1. Eberbach, E., Goldin, D. and Wegner, P. (2004). Turing's ideas and models of computation. In *Alan Turing: Life and Legacy of a Great Thinker*. Berlin: Springer. ISBN 978-3-642-05744-1.
2. Naresh, C., Bose, P.S.C. and Rao, C.S.P. (2020). Artificial neural networks and adaptiand neuro-fuzzy models for predicting WEDM machining responses of Nitinol alloy: comparatiand study. *SN Applied Sciences*, 2, 314.
3. McCulloch, W.S. and Pitts, W.A. (1943). A logical calculus of the ideas immanent in nervous activity. *Buttetin of Mathematics and Biophysics*, 5, 115–133.

4. McClelland, J.L. and Rumelhart, D.E. (1986). *Parallel Distributed Processing, Explorations in the Microstructure of Cognition*. Psychological and biological models. Cambridge: MIT Press.
5. Basheer, I.A. and Hajmeer, M. (2000). Artificial neural networks: fundamentals, computing, design, and application. *Journal of Microbiological Methods*, 43, 3–31.
6. Azimi, P. and Soofi, P. (2017). An ANN-based optimization model for facility layout problem using simulation technique. *Scientia Iranica E*, 24(1), 364–377.
7. Efendigil, T., Önüt, S. and Kahraman, C. (2009). A decision support system for demand forecasting with artificial neural networks and neuro-fuzzy models: A comparatiand analysis. *Expert Systems with Applications*, 36, 6697–6707.
8. Benardos, P.G. and Vosniakos, G.C. (2003). Predicting surface roughness in machining: A review. *International Journal of Machine Tools and Manufacture*, 43(8), 833–844.
9. Beltratti, A., Margarita, S. and Terna, P. (1996). *Neural Networks for Economic and Financial Modelling*. London: International Thomson Computer Press.
10. Davim, J.P., Gaitonde, V.N. and Karmik, S.R. (2008). Investigations into the effect of cutting conditions on surface roughness in turning of free machining steel by ANN models. *Journal of Material Processing Technology*, 205(1–3), 16–23.
11. Singh, A.K., Panda, S.S., Pal, S.K. and Chakraborty, D. (2006). Prediction drill wear using an artificial neural network. *International Journal of Advanced Manufacturing Technology*, 28, 456–462.
12. Wang, T., Gao, H. and Qiu, J. (2016). A combined adaptiand neural network and nonlinear model predictiand control for multirate networked industrial process control. *IEEE Transactions on Neural Networks and Learning Systems*, 27(2), 416–425.
13. Mitrea, C.A., Lee, C.K.M. and Wu, Z. (2009). A Comparison between Neural Networks and Traditional Forecasting Methods: A Case Study. *International Journal of Engineering Business Management*, 1, 11.
14. Janikova, D. and Bezak, P. (2016). Prediction of production line performance using neural Networks. In: *3rd International Conference on Artificial Intelligence and Pattern Recognition (AIPR)* (pp. 1–5).
15. Ding, L. and Matthews, J. (2009). A contemporary study into the application of neural network techniques employed to automate CAD/CAM integration for die manufacture. *Computers and Industrial Engineering*, 57(4), 1457–1471.
16. Fast, M. and Palme, T. (2010). Application of artificial neural networks to the condition monitoring and diagnosis of a combined heat and power plant. *Energy*, 35(2), 1114–1120.
17. Mukherjee, I. and Ray, P.K. (2006). A review of optimization techniques in metal cutting processes. *Computers & Industrial Engineering*, 50, 15–34.
18. Jawahir, I.S., Balaji, A.K., Rouch, K.E. and Baker, J.R. (2003). Towards integration of hybrid models for optimized machining performance in intelligent manufacturing systems. *Journal of Materials Processing Technology*, 139 (1–3), 488–498.
19. Zain, A.M., Haron, H. and Sharif, S. (2010). Prediction of surface roughness in the end milling machining using artificial neural network. *Expert Systems with Applications*, 37(2), 1755–1768.
20. Rashid, M.F.A. and Abdul Lani, M.R. (2010). Surface roughness prediction for CNC milling process using artificial neural network. In: *Proceedings of the World Congress on Engineering (WCE '10)*, London, United Kingdom.
21. Razfar, M.R., Zinati, R.F. and Haghshenas, M. (2011). Optimum surface roughness prediction in face milling by using neural network and harmony search algorithm. *International Journal of Advanced Manufacturing Technology*, 52(5–8), 487–495.

22. Levin, A.U. and Narendra, K.S. (1993). Control of nonlinear dynamical systems using neural networks: Controllability and Stabilization. *IEEE Transactions on Neural Networks*, 4(2), 192–206.
23. Mukhopadhyay, A., Barman, T.K., Sahoo, P. and Davim, J.P. (2019). Modeling and optimization of fractal dimension in wire electrical discharge machining of EN 31 steel using the ANN-GA approach. *Materials*, *12*(3), 454.
24. Khatir, S., Tiachacht, S., Thanh, C.L., Bui, T.Q. and Wahab, M.A. (2019). Damage assessment in composite laminates using ANN-PSO-IGA and Cornwell indicator. *Composite Structures*, 230, 111509.
25. Joshi, S.N. and Pande, S.S. (2011). Intelligent process modeling and optimization of die-sinking electric discharge machining. *Applied soft computing*, *11*(2), 2743–2755.
26. Hakimpoor, H., Arshad, K.A.B., Tat, H.H., Khani, N. and Rahmandoust, M. (2011). Artificial Neural Networks' Applications in Management. *World Applied Sciences Journal*, 14(7), 1008–1019.
27. Kosko, B. (1994). *Neural Networks and Fuzzy Systems*. Prentice-Hall of India. New Delhi: MIT Press.
28. Yusoff, Y., Zain, M.A., Sharif, S., Sallehuddin R. and Ngadiman, M.S. (2016). Potential ANN prediction model for multiperformances WEDM on Inconel 718. *Neural Computing and Applications*, 30, 2113–2127.
29. Jain, S.P., Ravindra, H.V., Ugrasen, G., Prakash, G.V.N. and Rammohan, Y.S. (2017). Study of Surface Roughness and AE Signals while Machining Titanium Grade-2 Material using ANN in WEDM. *Materials Today: Proceedings*, 4(9), 9557–9560.
30. Agrawal, D.P. and Kamble D.N. (2017). GRA and ANN Integrated Approach for Photochemical Machining of Al/SiC Composite. *Materials Today: Proceedings*, 4(8), 7177–7188.
31. Abbas, A. T., Pimenov, D. Y., Erdakov, I. N., Taha, M. A., Soliman, M. S. and El Rayes, M. M. (2018). ANN Surface Roughness Optimization of AZ61 Magnesium Alloy Finish Turning: Minimum Machining Times at Prime Machining Costs. *Materials*, 11(5), 808.
32. Badiger, P.V., Desai, V., Ramesh, M.R., Prajwala B.K. and Raveendra K. (2019). Cutting Forces, Surface Roughness and Tool Wear Quality Assessment Using ANN and PSO Approach During Machining of MDN431 with TiN/AlN-Coated Cutting Tool. *Arabian Journal for Science and Engineering*, 44, 7465–7477.
33. Zerti, A., Yallese M. A., Zerti O., Nouioua M. and Khettabi R. (2019). Prediction of machining performance using RSM and ANN models in hard turning of martensitic stainless steel AISI 420. *SAGE Journals*, 233(13), 4439–4462.
34. Sada, S.O. and Ikpeseni, S.C. (2021). Evaluation of ANN and ANFIS modeling ability in the prediction of AISI 1050 steel machining performance. *Heliyon*, 7(2), e06136.
35. Bagga, P. J., Makhesana, M.A., Patel, H.D. and Patel, K.M. (2021). Indirect method of tool wear measurement and prediction using ANN network in machining process. *Materials Today: Proceedings*, 44(1), 1549–1554.
36. Xu, L., Huang, C., Niu, J., Li, C., Wang, J., Liu, H. and Wang, X. (2021). An improandd case-based reasoning method and its application to predict machining performance. *Soft Computing*, 25, 5683–5697.
37. Paturi, U.M.R., Cheruku, S., Pasunuri, V.P.K. and Salike, S. (2021). Modeling of tool wear in machining of AISI 52100 steel using artificial neural networks. *Materials Today: Proceedings*, 38(5), 2358–2365.
38. Quarto, M., D'Urso, G., Giardini, C., Maccarini, G. and Carminati, M. (2021). A Comparison between Finite Element Model (FEM) Simulation and an Integrated Artificial Neural Network (ANN)-Particle Swarm Optimization (PSO) Approach to

Forecast Performances of Micro Electro Discharge Machining (Micro-EDM) Drilling. *Micromachines*, 12(6), 667.

39. Trinath, K., Aepuru, R., Biswas, A., Viswanathan, M.R. and Manu, R. (2021). Study of self lubrication property of Al/SiC/Graphite hybrid composite during Machining by using artificial neural networks (ANN). *Materials Today: Proceedings*, 44, 3881–3887.
40. Lalwani, V., Sharma, P., Pruncu, C.I. and Unune, D.R. (2020). Response surface methodology and artificial neural network-based models for predicting performance of wire electrical discharge machining of inconel 718 alloy. *Journal of Manufacturing and Materials Processing*, 4(2), 44.
41. Kaviarasan, V., Elango, S., Gnanamuthu, E.M.A. and Durairaj, R. (2021). *Smart Machining of Titanium Alloy Using ANN Encompassed Prediction Model and GA Optimization*. In Futuristic Trends in Intelligent Manufacturing. Cham: Springer.
42. Lin, Y.C., Wu, K.D., Shih, W.C., Hsu, P.K. and Hung, J.P. (2020). Prediction of surface roughness based on cutting parameters and machining vibration in end milling using regression method and artificial neural network. *Applied Sciences*, 10(11), 3941.
43. Nagaraj, Y., Jagannatha, N., Sathisha, N. and Niranjana, S.J. (2020). Prediction of material removal rate and surface roughness in hot air assisted hybrid machining on soda-lime-silica glass using regression analysis and artificial neural network. *Silicon*, 13 (11), 4163–4175.

12 Machining and Vibration Behavior of Ti-TiB Composites Processed through Powder Metallurgy Techniques

M. Selvakumar and T. Ramkumar

Dr. Mahalingam College of Engineering and Technology, Pollachi, India

S. Vinoth Kumar

Shandong University, Ji'nan, China

P. Chandramohan

Sri Ramakrishna Engineering College, Coimbatore, India

R. Ganesh

Dr. Mahalingam College of Engineering and Technology, Pollachi, India

CONTENTS

DOI: 10.1201/9781003284574-13

12.1 INTRODUCTION

For the last two decades, titanium composites have been used in industrial applications, such as medical, automobile, and others, because of their distinctive properties, namely, high strength, low density, and enriched corrosion behavior [1–2]. During machining, selecting a tool with various grades is very difficult for ductile and brittle materials. The accuracy of the surface quality is mainly based on the tool wear. Many researchers found that the surface quality of the machined surface mostly depended on the burrs and residual stress [3]. Carbon-rich working oil is used for protecting the tool and the specimen surface during experimentation, and it reduces erosion. There are many manufacturing methods; however, electrical-discharge machining (EDM) possessed fewer environmental hazards owing to the pyrolysis and ease of EDM sinking. Much research has been carried out on titanium (Ti)–titanium boride (TiB) machining. Pecas and Henriques [4] investigated the surface quality of composites with respect to electrode area; the results showed that while increasing the electrode area, it directly influences the surface roughness and crater. Lee and Li [5] reported that the material removal rate (MRR) is mainly reliant on the electrode, and the experimental outcomes revealed that a graphite electrode possessed a higher MRR compared to a copper tungsten electrode. Khan et al. [6] predicted the optimal parameters for Ti-5Al-2.5Sn using EDM followed by the response surface methodology (RSM) technique. Servo-voltage and pulse-on and -off time are taken as the factors for this investigation. The outcomes display that the acquired model achieved remarkable accuracy, demonstrated better properties, and was more cost-effective. None of the researchers performed research in the area of titanium powder metallurgy composites and followed EDM and vibration analysis. Hence, the current study focused on assessing the machining and vibration behavior of Ti-TiB composites fabricated through a powder metallurgy (P/M) technique. Ti composites reinforced with 20 and 40 vol% of TiB were fabricated using spark plasma sintering (SPS), hot isostatic pressing (HIP), and vacuum sintering (VS). The composites were characterized through scanning electron microscopy (SEM). The machining performance of the sintered composites was analyzed using EDM. Damping measurements of all the specimens were taken by a dynamic mechanical analyzer (DMA) at different frequencies.

12.2 MATERIALS AND METHODS

Ti with 20% and 40% of TiB volume fractions were selected and the sintering processes, namely, SPS, HIP, and VS, were used in this study. The processing techniques are already elaborated in our previous published work elsewhere [7–8]. After processing, the distribution of the secondary particles is assessed using SEM. Figures 12.1a–c show the scanning electron image of sintered composites for 40 vol.% sample, and it is clearly evident that secondary particle such as TiB is evenly distributed in the matrix medium. The machining performances of the sintered composites were analyzed using EDM. Peak current, pulse-on, and -off time were selected as input factors. MRR, tool wear rate (TWR), and surface roughness were considered as performance indicators. Furthermore, the damping behavior of the composites was also

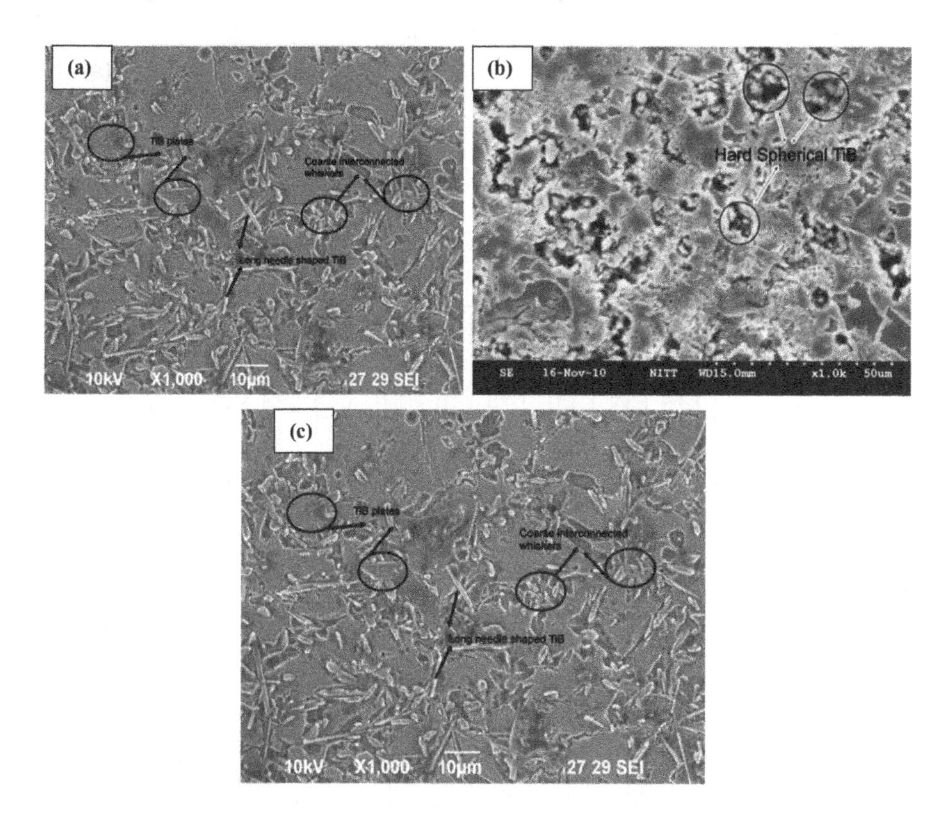

FIGURE 12.1 Scanning electron image of Ti-40% TiB (a) spark plasma sintering, (b) hot isostatic pressing, and (c) vacuum sintering.

analyzed. Damping measurements of all the specimens were attained by a DMA at different frequencies.

12.3 RESULTS AND DISCUSSION

12.3.1 Variation of the MRR with Respect to Current and Gap Voltage

Figure 12.2a–b shows the MRR of various composites for varying the current (I_p) and gap voltage (V). Three different types of current and gap voltage were considered, such as I_p = 12 A and I_p = 13 A and gap voltage = 50 V, 75 V, and 100 V. While increasing the gap voltage and current, the MRR is considerably enhanced, irrespective of pulse-on and -off time [9–10]. During experimentation, this was attributed to discharge energy, resulting in the vaporization and melting of the work piece. It led to an increase in the MRR.

12.3.2 Variation of the TWR with Respect to Current and Gap Voltage

Figures 12.3a–b show the TWR of various composites for varying the current (I_p) and gap voltage (V). Three different types of current and gap voltage were considered,

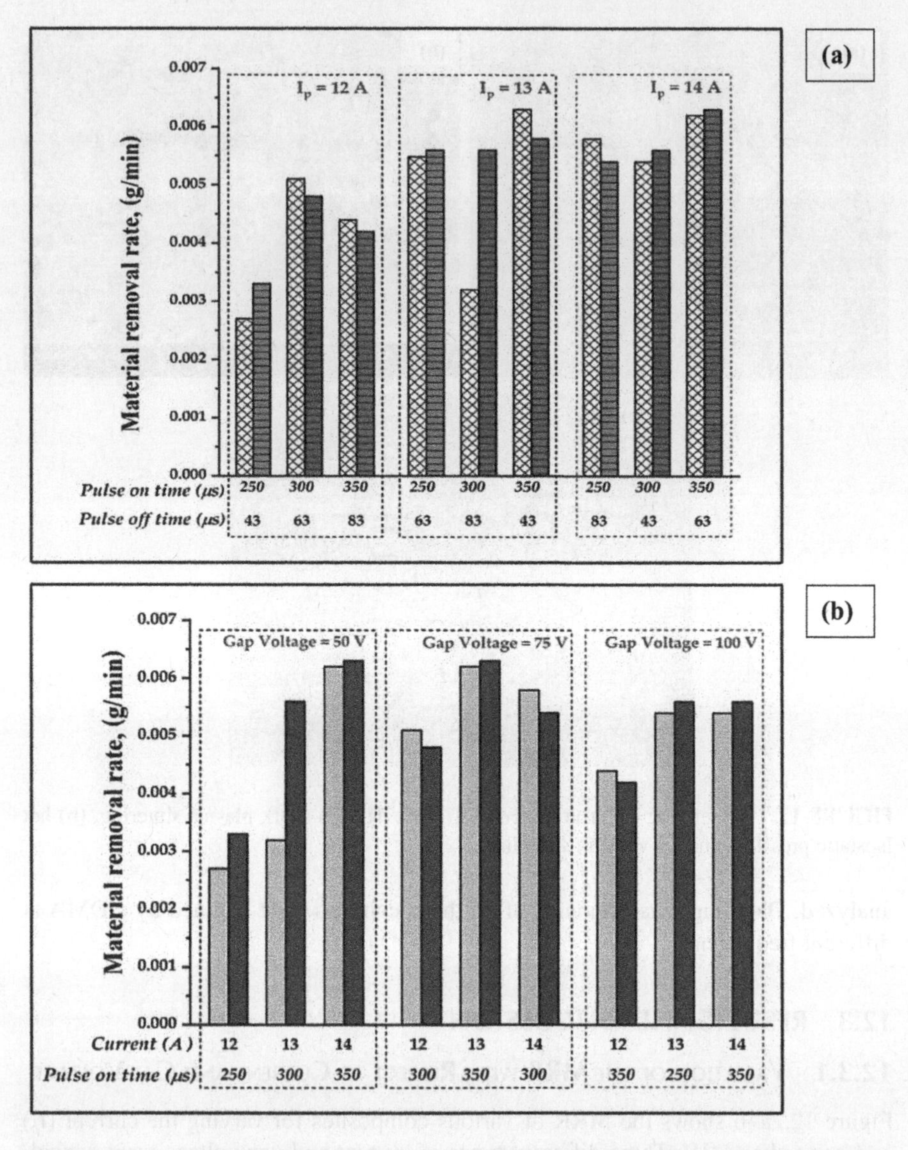

FIGURE 12.2 Variation of the material removal rate with respect to (a) current and (b) gap voltage.

such as I_p = 12 A and I_p = 13 A and gap voltage = 50 V, 75 V, and 100 V. Increasing the discharge energy led to an increase in the TWR. During experimentation, a higher discharge energy increased the frictional coefficient of tool and the work piece, and hence, the TWR increased [11–12].

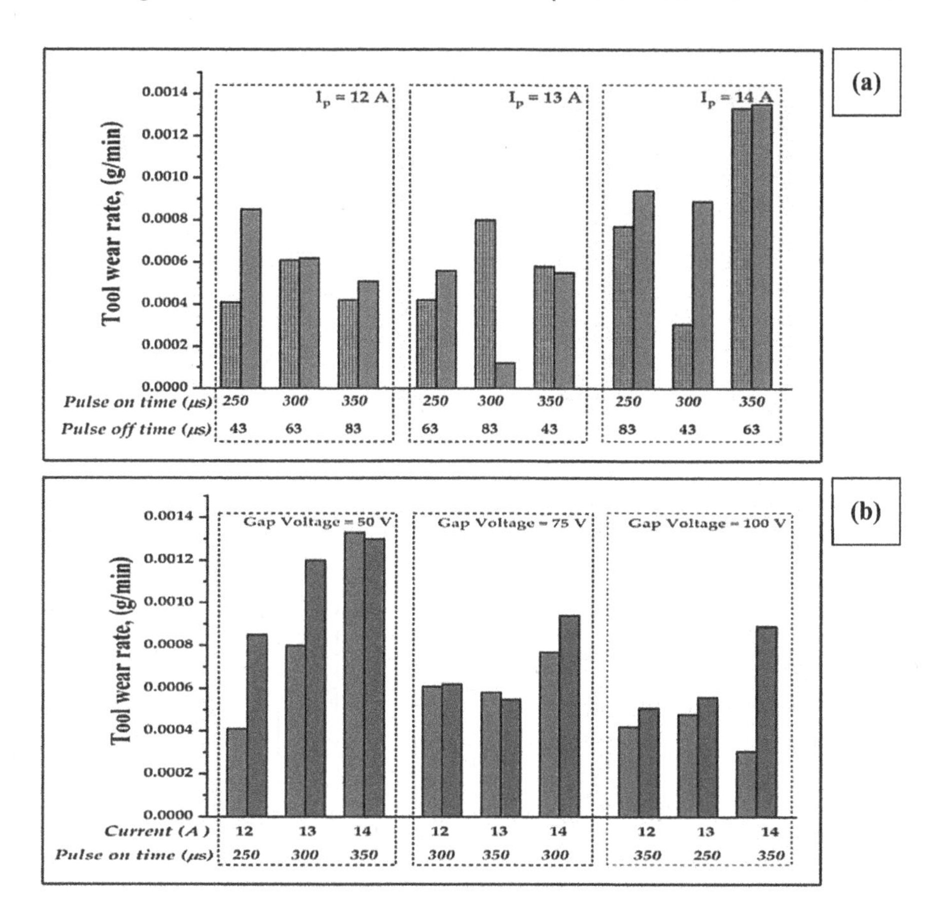

FIGURE 12.3 Variation of tool wear rate with respect to (a) current and (b) gap voltage.

12.3.3 Variation of Machining Time with Respect to Current and Gap Voltage

Figures 12.4a–b show the machining time of composites for varying the current (I_p) and gap voltage (V). Three different types of current and gap voltage were considered, such as I_p = 12 A and I_p = 13 A and gap voltage = 50 V, 75 V, and 100 V. Increasing the gap voltage and current led to a decrease in machining time. A higher machining time was obtained for 12 A and 50 V. Increasing the pulse-on and -off time will lead to a decrease in machining time for all the gap voltage and current due to higher volume percentage of composite [13]. It exhibits more needle-like structures that reduce the machining time.

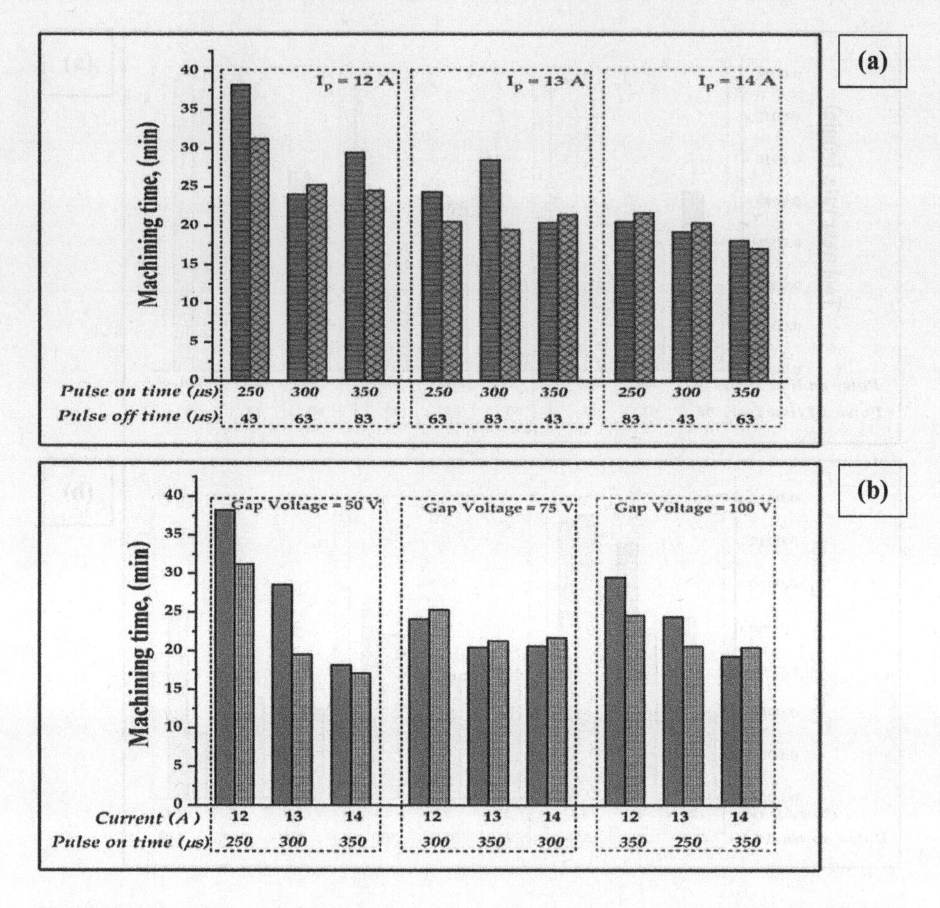

FIGURE 12.4 Variation of machining time with respect to (a) current and (b) gap voltage.

12.3.4 Damping Analysis

The samples were machined and sized (10 × 5 × 0.3 mm) for experimentation. Testing parameters of frequency = 0.5 Hz, T = 30 °C–600 °C, static load of 100–250 grams, strain rate of 0.5–1%, and ramp rate of 2.0 °C/min were applied. The elastic modulus of the composites was determined using DMA analysis. For accuracy, six samples for each system were considered for dynamic testing. Primarily, the static load was taken, and the strain rate was 0.5–1%. The storage modulus, loss modulus, and tangent of delta for the composite Ti-40% TiB processed through three techniques are shown in Table 12.1. It is clearly evident that storage and loss modulus were quite low for HIP-processed samples.

Figures 12.5a–c show the dynamic test results for the composites with Ti-40% TiB fabricated through HIP, SPS, and VS. From Figures 12.5a–c, it is observed that the SPS-processing samples show a better damping force than the other methods. When this technique was used, better bonding was obtained for the matrix and the reinforced medium. However, the nonlinearity of intensity was observed in the

TABLE 12.1
Storage and Loss Modulus of Ti-40% TiB for Three Processing Techniques

Ti Composition	Frequency Hz	Time Min	Temperature °C	Storage Modulus MPa	Loss Modulus MPa	Tan Delta
Ti-40% TiB (HiP)	1	6.81	144.11	718314	718314	3.264
Ti-40% TiB (SPS)	1	13.15	271.44	788679	786258	8.178
Ti-40% TiB (VS)	1	6.81	144.11	1444460	1444460	1

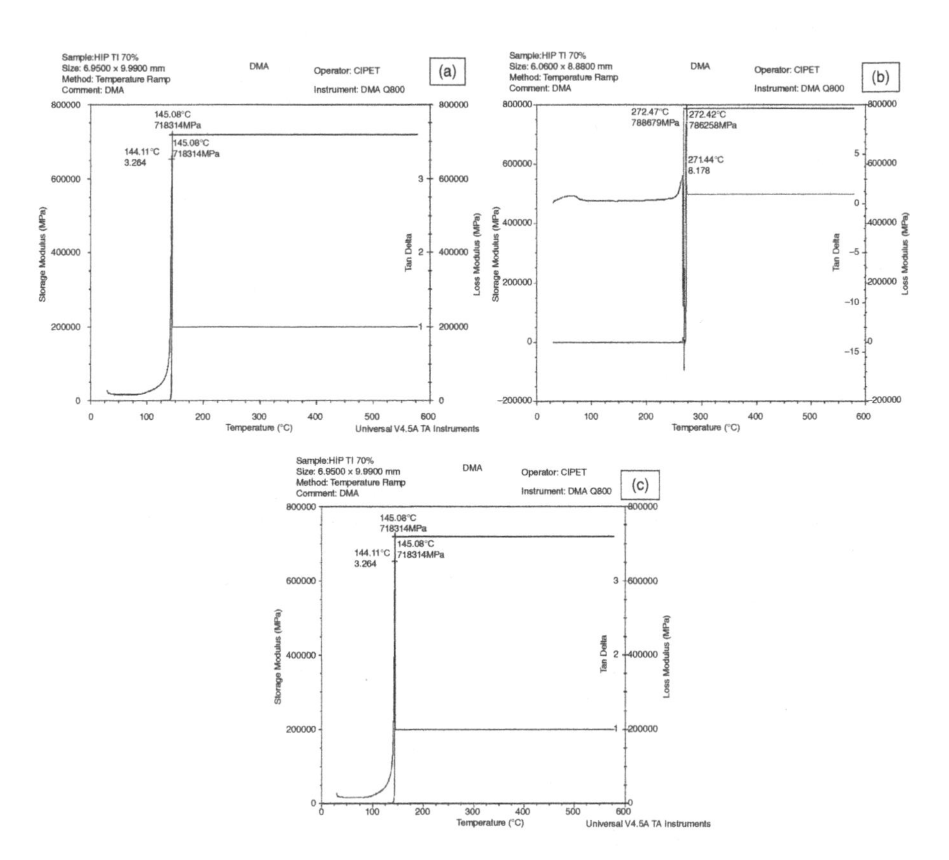

FIGURE 12.5 DMA analysis Ti-40% TiB (a) hot isostatic pressing, (b) spark plasma sintering, and (c) vacuum sintering.

mid-range of temperature and frequency for all the composites. This is due to the presence of hard asperities [14–15]. The frequency influences the damping capability of the composite for 230 °C and above; its value reaches 975 × 10^{-3} N m/s. for the frequency of 0.5 Hz, which makes the titanium composites, a suitable candidate for damping applications at different temperature ranges with good mechanical strength and stiffness properties.

12.4 CONCLUSION

The machining and vibration behavior of Titanium composites of 20% and 40% TiB fabricated by the P/M route. The following are the outcomes of this study:

- Increasing the gap voltage and current led to a considerable increase of the MRR irrespective of pulse-on and -off time.
- Increasing the discharge energy led to an increase in the TWR. More heat is produced during the experimentation, which increases the frictional coefficient and the TWR.
- Higher machining time was obtained for 12 A and 50 V. Increasing the pulse-on and -off time decreases the machining time for all the discharge current and gap voltage values.
- The frequency influences the damping capability of the composite for the 230 °C and above; its value reaches 975×10^{-3} N m/s for the frequency of 0.5 Hz.

REFERENCES

1. Klocke, F., Mohammadnejad, M., Holsten, M., Ehle, L., Zeis, M., Klink, A. (2018). A Comparative study of polarity-related effects in single discharge EDM of Titanium and Iron alloys, *Procedia CIRP*, 68, 52–57.
2. Kou, Z., Han, F. (2018). On sustainable manufacturing titanium alloy by high-speed EDM milling with moving electric arcs while using water-based dielectric, *Journal of Cleaner Production*, 189, 78–87.
3. Bhaumik, M., Maity, K. (2018). Effect of different tool materials during EDM performance of titanium grade 6 alloy, *Engineering Science and Technology, an International Journal*, 21 (3), 507–516.
4. Pecas, P., Henriques, E. (2003). Influence of silicon powder-mixed dielectric on conventional electrical discharge machining, *Internatioanl Journal of Machine Tools Manufacture*, 43, 1465–1471.
5. Lee, S.H., Li, X.P., (2001). Study of the effect of machining parameters on the machining characteristics in electrical discharge machining of tungsten carbide, *Journal of Materials Processing and Technology*, 115, 344–358.
6. Boujelbene, M., Bayraktar, E., Tebni, W., Salemn, S.B. (2009). Influence of machining parameters on the surface integrity in electrical discharge machining, *Archives of Materials Science and Engineering*, 37, 110–116.
7. Selvakumar, M., Ramkumar, T., Mohanraj, M., Chandramohan, P., Narayanasamy, P. (2020). Experimental investigations of reciprocating wear behavior of metal matrix (Ti/TiB) composites, *Archives of Civil and Mechanical Engineering*, 20, 1–9.
8. Ramkumar, T., Selvakumar, M., Mohanraj, M., Chandrasekar, P. (2019). Experimental investigation and analysis of drilling parameters of metal matrix (Ti/TiB) composites, *Journal of the Brazilian Society of Mechanical Sciences and Engineering*, 41, 1–12.
9. Torres, A., Luis, C.J., Puertas, I. (2017). EDM machinability and surface roughness analysis of TiB_2 using copper electrodes, *Journal of Alloys and Compounds*, 690, 337–347.
10. Sales, W.F., Oliveira, A.R.F., Raslan, A.A. (2016). Titanium perovskite (CaTiO3) formation in Ti6Al4V alloy using the electrical discharge machining process for biomedical applications, *Surface and Coatings Technology*, 307, 1011–1015.

11. Rao, P.S., Ramji, K., Satyanarayana, B. (2016). Effect of wire EDM conditions on generation of residual stresses in machining of aluminum 2014 T6 alloy, *Alexandria Engineering Journal*, 55, 1077–1084.
12. Li, L., Feng, L., Bai, X., Li, Z.Y. (2016). Surface characteristics of Ti–6Al–4V alloy by EDM with Cu–SiC composite electrode, *Applied Surface Science*, 388, 546–550.
13. Urso, G.D., Ravasio, C. (2017). Material-Technology Index to evaluate micro-EDM drilling process, *Journal of Manufacturing Processes*, 26, 13–21.
14. Hou, J., Zhao, Z., Fu, Y., Qian, N. (2021). Machining stability enhancement in multi-axis milling of titanium hollow blade by introducing multiple damping and rigid supporters, *Journal of Manufacturing Processes*, 64, 198–208.
15. Sajadifar, S.V., Atli, C., Yapici, G.G. (2019), Effect of severe plastic deformation on the damping behavior of titanium, *Materials Letters*, 244, 100–103.

13 Numerical Analysis of Machining Forces and Shear Angle during Dry Hard Turning

Anupam Alok and Amit Kumar
National Institute of Technology Patna, Patna, India

Manjesh Kumar and Manas Das
Indian Institute of Technology Guwahati, Assam, India

Kishor Kumar Gajrani
Indian Institute of Information Technology, Design and Manufacturing, Kancheepuram, India

CONTENTS

13.1 INTRODUCTION

Manufacturing is continually changing these days because of certain key variables, such as asset impediment, competition in the market, and expanding assumptions. Machining hard material involves extraordinary consideration for ongoing modern

DOI: 10.1201/9781003284574-14

development, and logical exploration manufacturing is continually changing these days. Machining hard material involves extraordinary consideration for mechanical creation and logical exploration [1, 2]. The turning of hard material having a hardness greater than 45 HRC is known as hard turning. Machining is preferred for its high dimensional accuracy and high productivity [3]. AISI 4340 steel alloy of medium carbon steel is known for its hardness and endurance limits and is used for a wide variety of applications. AISI 4340 serves as a viable substitute for AISI 4140 steel due to its high tensile and yield strength.

Nowadays, finite element method (FEM) is widely used for simulation and analysis of material removal processes, supported by highly developed computers and software packages. Along with minimizing the need for high-cost and time-consuming experimental setups, it also predicts the variables that are difficult to evaluate, such as stress, strain, and machining temperature. It shows how effectively the selection of the input variables is reflected in the deformation process involving chip formation. Parihar et al. analyzed the cutting force during the turning of AISI H13 with a ceramic insert. They have compared simulation results and experimental results. For the simulation, they used Deform 3D software [4]. FEM is capable of providing both qualitative and quantitative aspects of the machining process and can also correctly predict the difficult-to-measure variables associated with the hard-turning process, such as shear strains, deformation, and others. Numerous finite element (FE) codes such as ABAQUS®, DEFORM, Ansys/LS-DYNA have come up that are currently being used by researchers for hard turning simulation. ABAQUS®, finite element software is used for simulating the chip formation process [5]. It is employed mainly to generate chip of a two-dimensional (2D) FE model under orthogonal cutting conditions. Furthermore, the arbitrary Lagrangian Eulerian (ALE) adaptive meshing technique is generally applicable in ABAQUS/Explicit. It also provides mesh distortion control, which influences the cutting process undergoing large deformations.

In this work, hard machining AISI 4340 steel alloy is carried out using Al_2O_3-coated carbide tool insert. Numerical analysis of cutting forces and variation of shear angle were investigated using ABAQUS®, and cutting force results were compared with experiments. The numerically obtained variation of the shear angle was compared with existing literature.

13.2 EXPERIMENTAL PROCEDURES

In this study, an HMT make and NH-26 model lathe were used for the experiments. Experiments were conducted with an Al_2O_3-coated carbide insert during the machining of the AISI 4340 steel of a hardness 52 HRC. The Al_2O_3-coating thickness is 5 μm. A 9272B Kistler piezoelectric dynamometer was used to measure the cutting forces. Hard-machining experiments were performed at a constant feed of 0.2 mm/rev and a depth of cut of 0.1 mm. Cutting speeds were varied in the range of 110–200 m/min. Table 13.1 shows the process parameter for turning operation.

TABLE 13.1
Machining Experimental Input Parameters

Parameters	Values				
Cutting speed, V_c (m/min)	110	130	150	180	200
Depth of cut, t (mm)			0.1		
Feed, f (mm/rev)			0.2		

13.2.1 2D FEM Formulation of Orthogonal Cutting

A 2D metal-cutting process was simulated using the ABAQUS® 6.13 explicit package. The workpiece and the tool are considered to be deformable, with the workpiece having a rectangular geometry and the tool having a 0° clearance and rake angle. The workpiece block was 2 × 0.4 mm for length and height, respectively. Figure 13.1 shows that the geometry of the workpiece as well as the tool. The geometry is taken as per the model given by Akbar et al. [6]. The carbide insert had a coating of 5 μm. Figure 13.1 shows the model.

13.2.2 Boundary Conditions

During the simulations, the workpiece was considered fixed at the base in all directions, and cutting velocity was given to the tool in the negative horizontal direction. The steady-state conditions will be achieved during the cutting. Figure 13.2 shows the boundary conditions.

13.2.3 Element Formulation

A four-noded plane-strain bilinear displacement and temperature quadrilateral feature (CPE4RT) with a decreased hourglass control and integration scheme was

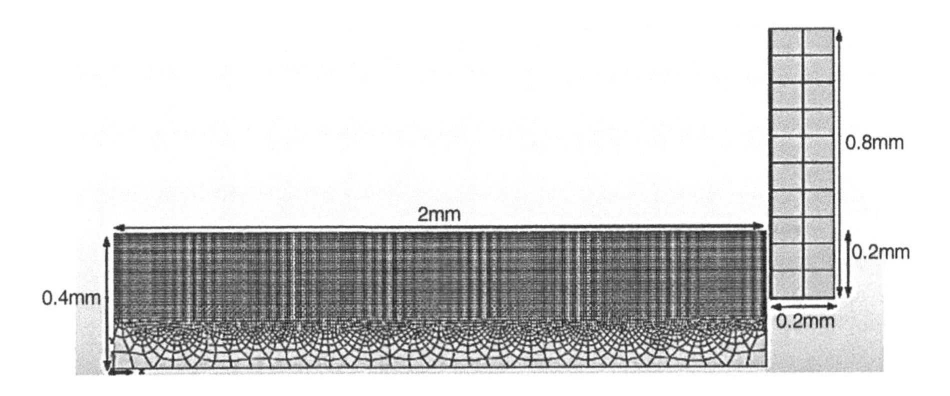

FIGURE 13.1 Mesh configurations with dimensions in the computational domain.

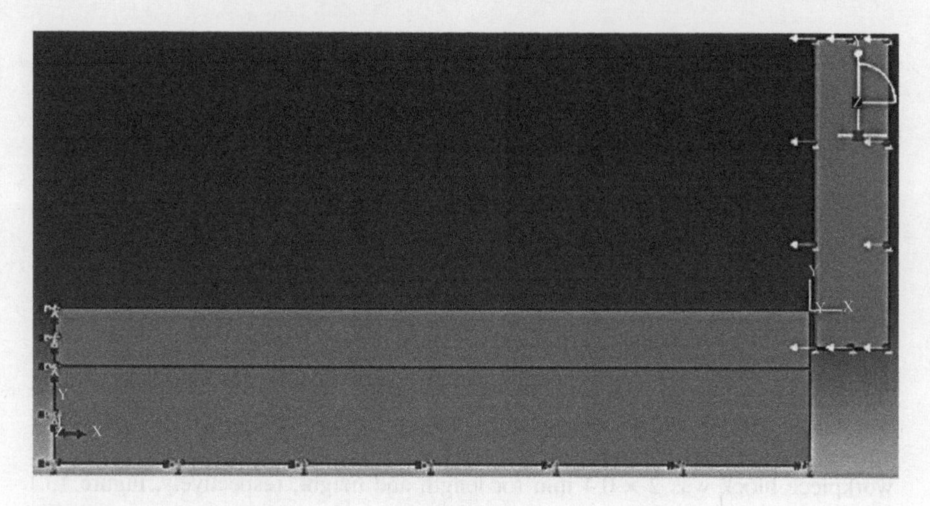

FIGURE 13.2 Boundary conditions on cutting tool and workpiece.

considered for the cutting tool and the workpiece. The workpiece has 10,332 nodes and 10,010 elements; however, the tool has 200 nodes and 168 elements. The stress analysis along with heat transfer has been coupled in the combined thermo-displacement components. Temperature, as well as displacement and degrees of freedom, have these components. At the time of machining, heat generation increases, which influences the mechanical behavior of the workpiece. Therefore, the coupled analysis was considered.

13.2.4 Material Model

Johnson–Cook constitutive equation (1983) is used for thermo-mechanical analysis of the workpiece, which reflects the effect of strain rate, strain, and temperature on the materials flow stress behavior [7, 8]. This model is followed by most of the researchers. This model requires the value of the parameters to calculate the equivalent flow stress. The model is represented as given in Equation 13.1:

$$\bar{\sigma} = \left(A + B\bar{\varepsilon}^{n}\right)\left[1 + C\ln\left(\frac{\dot{\bar{\varepsilon}}}{\dot{\bar{\varepsilon}}_{0}}\right)\right]\left[1 - \left(\frac{\theta - \theta_{room}}{\theta_{melting} - \theta_{room}}\right)^{m}\right], \tag{13.1}$$

where θ represents process temperature, $\theta_{melting}$ denotes melting temperature, $\dot{\bar{\varepsilon}}_0$ is the strain rate (reference), $\bar{\varepsilon}$ denotes the plastic strain (equivalent), $\dot{\bar{\varepsilon}}$ denotes the strain rate (plastic), m denotes the thermal softening coefficient, n denotes the work-hardening coefficient, C denotes the strain rate dependency coefficient, B denotes hardening modulus, and A denotes the yield stress.

For chip formulation, an equivalent plastic strain standard was instigated. The critical value of the equivalent plastic strain dictates the material failure. This cumulative law by Johnson and Cook (1985) is used and is given in Equation 13.2:

$$D = \sum \left(\frac{\Delta \bar{\varepsilon}}{\bar{\varepsilon}_f} \right), \quad (13.2)$$

where $\bar{\varepsilon}_f$ is the equivalent strain at failure, $\Delta \bar{\varepsilon}$ is the increment of the equivalent plastic strain, and D is the damage parameter. The equivalent plastic strain is given in Equation 13.3:

$$\bar{\varepsilon}_f = \left[D_1 + D_2 \exp\left(D_3 \frac{P}{\bar{\sigma}} \right) \right] \left[1 + D_4 \ln \left(\frac{\dot{\bar{\varepsilon}}}{\dot{\bar{\varepsilon}}_0} \right) \right] \left[1 + D_5 \left(\frac{\theta - \theta_{room}}{\theta_{melt} - \theta_{room}} \right) \right], \quad (13.3)$$

where $\Delta \bar{\varepsilon}$ gets updated at every load step; P is the hydrostatic pressure; and D_1, D_2, D_3, D_4, and D_5 are the experimentally obtained failure constants. Material stiffness begins to degrade after the element satisfies the criterion of initiation of harm. It is presumed that failure occurs when the value of D reaches 1. The element deletion in the ABAQUS program completely deletes the element after the element is completely degraded. The parameters as per the Johnson–Cook equation are presented for the workpiece in Table 13.2 [7].

For both the workpiece and the tool, the thermo-mechanical properties are specified. Table 13.3 tabulates the various properties of the workpiece and the tool materials. Temperature-dependent thermal conductivity (K) values are given in Table 13.4.

TABLE 13.2
Parameters as per Johnson–Cook for AISI 4340 [7]

A (MPa)	*B* (MPa)	*n*	*C*	*m*	D_1	D_2	D_3	D_4	D_5
950	725	0.375	0.015	0.625	−0.8	2.1	−0.5	0.002	0.61

TABLE 13.3
Material Properties of the AISI 4340 Workpiece, the Carbide Tool, and the Al_2O_3 Coating [9]

Material	Carbide tool	AISI 4340	Al_2O_3
C_p (J/kg-°C)	126	475	904
ρ (kg/m^3)	11900	7850	3780
E (N/m^2)	650 E9	205 E9	415E9
ν	0.25	0.3	0.22
α (μm/m-°C)	–	13.7	–
θ_{melt} (°C)	–	1520	–
Θ_{room} (°C)	25	25	25

TABLE 13.4
Thermal Conductivity of the Carbide Tool and the Al_2O_3 Coating [7]

Material	Property						
Al_2O_3	Temperature (°C)	50	90	300	500	1000	1300
	K* of (W/m-°C)	33	28	19	13	7	7
Carbide tool	Temperature (°C)	30	100	300	500	1000	1300
	K of (W/m-°C)	30	32	34	37	44	47.5

* K – thermal conductivity.

13.2.5 Contact Properties

Contact between the formed chip and the tool rake face was established using a kinematic contact algorithm. The frictional forces tangential behavior effect was also considered. A penalty contact with a constant coefficient of friction (0.25) is used to model the contact.

13.2.6 ALE Adaptive Meshing Technique

The ALE meshing method is a combination of both the Lagrangian and the Eulerian method. In the Lagrangian model, the mesh is locked onto the material. Depending on the need of the mesh at a point, ALE randomly switches between the Eulerian and the Lagrangian model. The ALE method prefers the Eulerian model for regions having huge plastic deformation as well as along the material boundary. Usually in Eulerian model, even though the material flows through mesh, the mesh elements remain fixed/locked.

13.3 RESULTS AND DISCUSSION

13.3.1 Cutting Force Model Validation

Figure 13.3 explains the obtained simulation results showing the variation of thrust and cutting force with time at a constant cutting speed of 110 m/min. The thrust and cutting forces achieved steady-state condition at 0.00015 s with a certain variation as

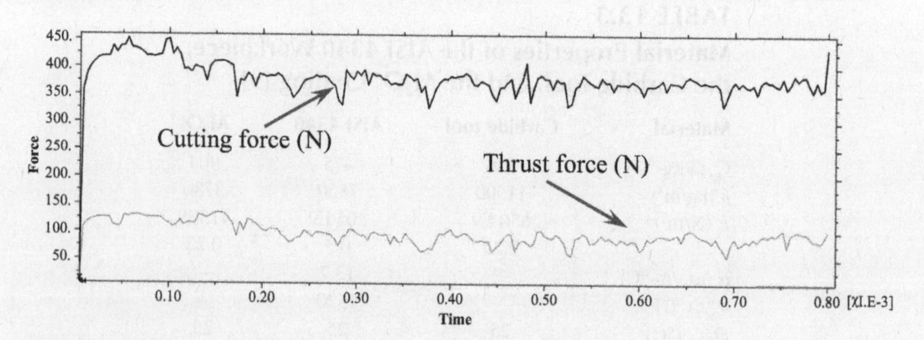

FIGURE 13.3 Force plot for a single cutting tool at a speed of 110 m/min.

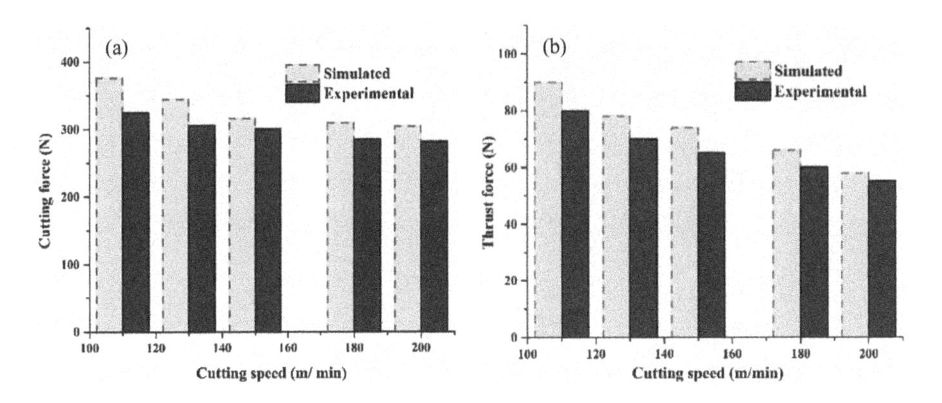

FIGURE 13.4 Effect of cutting speed on (a) cutting force and (b) thrust force.

shown in Figure 13.3. The assessment between predicted and experimental results of cutting force is shown in Figure 13.4, where a good agreement between the measured and predicted forces at all cutting speeds is observed. As the cutting speed increases from 110 to 200 m/min, the machining forces decrease gradually from an average value of 325–281 N experimentally and 376–305 N in simulation results.

From Figure 13.4, it is observed that the thrust and the cutting force reduce at higher cutting speeds. Also, the temperatures are pointedly higher at higher cutting speeds. At higher cutting speeds, the material becomes softer due to thermal softening [10]. Hence, the machining force is reduced.

13.3.2 Variation of Shear Angle

This section illustrates the simulated influence on shear angle with varying cutting speed. Figure 13.5 shows the simulated shear angle at a constant cutting speed of 110 m/min. From Figure 13.6, it is observed that with an increase in cutting speed, the shear angle increases. At high speeds, the cutting temperature increases. Hence, the chip thickness is reduced, which leads to an increased shear angle.

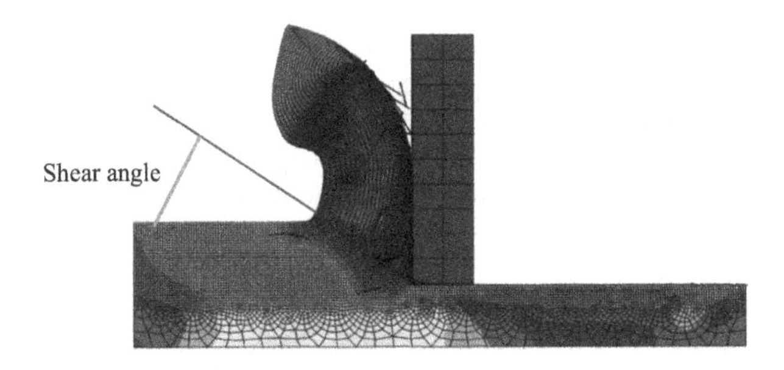

FIGURE 13.5 Shear angle measured at a 110 m/min cutting speed and a 0.2 mm/rev feed rate.

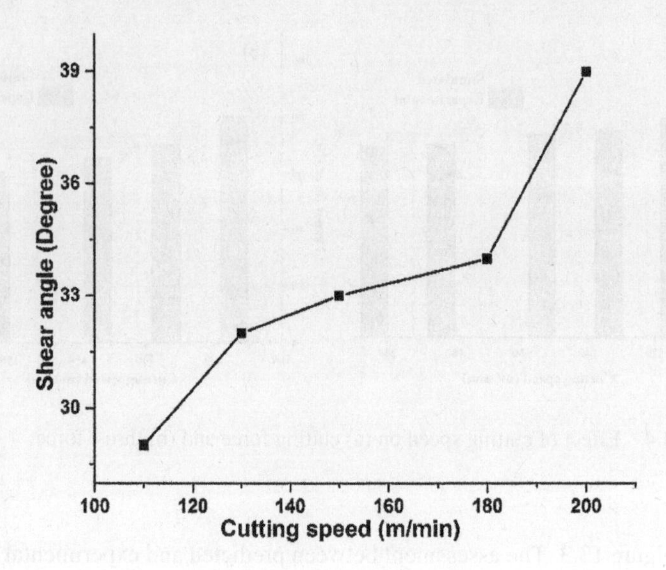

FIGURE 13.6 Simulation results of the effect of cutting speed on shear angle.

13.4 CONCLUSION

For the orthogonal machining of hard AISI 4340 steel with an Al_2O_3-coated carbide tool, a FE-based simulation was accomplished. The consequence of coating on the thrust force, cutting force, and shear angle during hard machining was studied. From the results, it can be concluded that at higher cutting speeds, both forces (i.e., cutting and thrust forces) are reduced; hence, if the cutting tool permits, then a higher cutting speed is good for cutting. The experimental results are very close to the simulated results for thrust force and cutting force. The pattern of shear angle closely matches with the results available in the literature. Hence, from the preceding results, it is clear that ABAQUS® can be used for to predict the forces and shear angle for 2D orthogonal hard-turning operations.

REFERENCES

1. Bouacha, K., Yallese, M.A., Khamel, S., Belhadi, S. (2014). Analysis and optimization of hard turning operation using cubic boron nitride tool, *International Journal of Refractory Metals and Hard Materials* 45, 160–178.
2. Gajrani, K.K., Suvin, P.S., Kailas, S.V., Rajurkar, K.P., Sankar, M.R. (2021). Machining of hard materials using textured tool with minimum quantity nano-green cutting fluid, *CIRP Journal of Manufacturing Science and Technology* 35, 410–421.
3. Bartarya, G., Choudhury, S.K. (2012). Effect of Cutting Parameters on Cutting Force and Surface Roughness during Finish Hard Turning AISI52100 Grade Steel, *Procedia CIRP* 1, 651–656.
4. Alok, A., Das, M. (2018). Cost-effective way of hard turning with newly developed HSN^2-coated tool, *Material and Manufacturing Process* 33, 1003–1010.

5. Zhang, Q., Zhang, S., Li, J. (2017). Three dimensional finite element simulation of cut-ting forces and cutting temperature in hard milling of AISI H13 steel, *Procedia Manufacturing* 10, 37–47.
6. Gajrani, K.K., Divse, V., Joshi, S.S. (2021). Burr reduction in drilling titanium using drills with peripheral slits, *Transactions of the Indian Institute of Metals* 74(5), 1155–1172.
7. Akbar, F., Mativenga, P.T., Sheikh, M.A. (2010). An experimental and coupled thermo-mechanical finite element study of heat partition effects in machining, *The International Journal of Advanced Manufacturing Technology* 46(5–8), 491–507.
8. Zerilli, F.J. (2004). Dislocation mechanics-based constitutive equations, *Metallurgical and Materials Transactions A* 35(9), 2547–2555.
9. Johnson, G.R., Cook, W.H. (1983). A Constitutive Model and Data for Metals Subjected to Large Strains, High Strain Rates, and High Temperatures Proceedings 7th International Symposium on Ballistics, *The Hague*, 19–21, April 1983, 541–547. - References - Scientific Research Publish, (n.d.).
10. Ucun, İ., Aslantas, K. (2010). Numerical simulation of orthogonal machining process using multilayer and single-layer coated tools, *The International Journal of Advanced Manufacturing Technology* 46(5–8), 491–507.

14 Machining Performance Evaluation of Titanium Biomaterial, Ti6Al4V in CNC cylindrical turning Using CBN Insert

Raju Pawade

Dr. Babasaheb Ambedkar Technological University, Lonere, Raigad, India

Nikhil Khatekar

PCT's A. P. Shah Institute of Technology, Thane, India

Sameer Ghanvat and Nitesh Gothe

Dr. Babasaheb Ambedkar Technological University, Lonere, India

CONTENTS

DOI: 10.1201/9781003284574-15

14.1 INTRODUCTION

The requirement for materials with higher machinability are less cutting power, a faster rate of cutting, and achieving a desired surface finish easily with less tool wear. The machinability is often degraded by the factors that usually improve the performance of the material. In this regard, metal-cutting scientists, as well as manufacturing engineers, are working on ways to improve the material's ability, keeping the economy and productivity intact [1]. It may not be easy to predict machinability as it has many variables. The various aspects, such as work material type and their physical properties, working conditions, and cutting tool geometry, and the machining control factors have an influence on the machining performance [2].

Aerospace, biomedical, and chemical industries employ the use of titanium biomaterial, Ti6Al4V due to its light weight and elevated temperature corrosion resistance. During machining, the heat generated is not able to dissipate quickly from the tool edge because of the lower thermal diffusivity of these alloys. The cubic boron nitride (CBN) cutting inserts are nowadays increasingly being used for machining various components in high-end industries. Recently, the turning of titanium alloy was carried out using polycrystalline cubic boron nitride (PCBN) and polycrystalline diamond (PCD) tools and better performance of PCD tools was reported [3]. However, the PCD tools are expensive. Therefore, CBN tools are preferred for machining to cut costs. The main contents of CBN tools are boron and nitrogen. A relatively flexible and smooth hexagonal structure of boron nitride (BN), like graphite, appears, which can be easily broken into small pieces. The conversion of hexagonal boron nitride (HBN) into CBN can be achieved using high temperatures and pressures, similar to the conversion of graphite into diamond. The CBN grains, once grown by breaking down unwanted foreign particles, are released and returned to a normal state for further processing. The multiple CBN crystals are bound together by using ceramic bonds and a large mass of polycrystalline CBN is formed [4].

The PCBN, obtained by the sintering process, is an isotropic compound that consists of randomly oriented anisotropic crystals. In fact, there are two types of PCBN tools. The first one is known as high-volume CBN content cutting tools and has a 0.9 volume CBN content with metal connectors (e.g., cobalt). The second one is known as low-volume CBN content tools [4] with 0.5–0.7% volume particles of CBN content having ceramic binders such as titanium nitride and titanium carbide.

Additionally, the CBN inserts are coated with special coatings that make them possible to sustain sudden impact and resist wear due to rubbing and extreme heat. Some of the major advantages are increased productivity, extremely good capacity to absorb heat, high impact and wear resistance, increased tool life, and high material removal rate. PCBN provides natural benefits by recycling unwanted chips and removing grinding swarf. Not only this, but the time required for machining can be also reduced because the metals can be machined in a solid state as an alternative to grinding.

14.2 LITERATURE REVIEW

It is revealed from the literature that factors such as CBN percentage, cutting edge type, coating material, binder used, grain size of the CBN, the use of lubrication methods, and cutting conditions have significant effect on the CBN cutting tool efficiency. Titanium biomaterial Ti6Al4V has been studied by many researchers in the recent past and still is a topic of importance for many researchers worldwide. Although studies report on machining titanium biomaterial Ti6Al4V using plain and coated carbide tools, very few attempts have been found using CBN cutting tools.

Research on slot milling titanium biomaterial, Ti6Al4V has been conducted by Zareena [5] using CBN, PCD, and binderless cubic boron nitride (BCBN) cutting tools. They observed that better performance was exhibited by CBN tools when the following cutting conditions are used as compared to other cutting tools: cutting speed (400 m/min), feed rate (0.05 mm/tooth), and depth of cut (0.05 mm). Hirosaki et al. [6] tried machining of vanadium-free titanium alloy Ti-6Al-2Nb-1Ta with a binder-less PCBN tool. They found that low flank wear and retained sharp cutting edge were exhibited by the PCBN tool under the application of a high-pressure coolant.

High-speed slot milling was performed on Ti-6Al-4V by Wang et al. [7] using BCBN tools. The working and the wear pattern of the BCBN tool with respect to machining forces, tool life, and tool wear have been studied. It is found that the uneven wear on the flank face dominated the wear behavior of BCBN tools.

Wang et al. [8] in another study investigated that tool life at higher cutting speed is more than at low cutting speed. A diffusion bond was found at the tool–work interface, and the welded work constituents to the tool flank face caused a reduction in tool wear. Based on the energy dispersive X-ray analysis of chips, the alloying elements of the tool diffused into the nearby region of the chip. This led to the development of diffusion-activated dissolution wear mechanism for BCBN tools.

Ezugwu et al. [9] tested the machining performance of various grades of CBN in fine turning titanium biomaterial Ti6Al4V at a higher speed of 250 m/min with various coolants. They found that at higher cutting speed, a low-CBN-content (50% CBN) tool shows a reduction in tool life. A similar result, with severe notching and cutting-edge chipping found, when the higher CBN-content (90% CBN) tool was used. It was also seen that there was an accelerated notch wear rate with higher CBN content which diminished the tool life consequently for the cutting conditions that were investigated.

Multilayer coated inserts were evaluated by Ozel et al. [10] while investigating the turning of Ti-6Al-4V. A new modified material model for titanium biomaterial Ti6Al4V titanium alloy is established where the softness (flow), hardness, and temperature effects are combined. A serrated chip model using the adiabatic shear phenomenon is applied to validate the model with elastoviscous plastic FE simulation while machining titanium biomaterial Ti6Al4V. Measured cutting forces and chip morphology are used to compare simulation predictions with orthogonal cutting test results. Bhaumik et al. [11] found that for machining titanium alloys, a composite tool of WBN-CBN (wurtzite boron nitride–cubic boron nitride) can be used economically. The results showed that a steady state of flank wear is reached after a machining time of 5 minutes. The magnitude of wear rate tends to increase after 50 minutes.

Burhanuddin et al. [12] investigated critical factors that influenced tool life, tool flank wear rate, and different modes of tool wear patterns of the CBN cutting tool while machining of titanium biomaterial Ti6Al4V. It was investigated that cutting speed and feed rate critically influenced the life of the tool. The thorough analysis of a worn-out tool using scanning electron microscopy found that the wear took place on the flank face and rake face of the tool's cutting edge. The parameter such as feed rate influences the tool life most; in addition, cutting speed, depth of cut used, and CBN grades also affect much.

Zoya and Krishnamurthy [13] employed CBN tools for high-speed turning of titanium alloys and investigated their machining performance. They concluded that machining titanium alloy is a thermally dominant process. The surface finish is highly influenced by the replication of the tool nose on the work surface during the turning operation.

Liang et al. [14] investigated surface topography and its deterioration resulting from tool wear evolution while dry turning titanium alloy Ti-6Al-4V. They analyzed that the amplitude and functional parameters increased with the increasing tool wear. The experiments were performed by Li et al. [15] demonstrating the effect of very high speed and found that the chip configuration rapidly varied at higher cutting speed. As a result, the tool wore out at a rapid pace. Polini et al. [16] found that higher values of cutting forces were achieved and better surface quality was produced at higher cutting speeds. This can also be obtained with an increase in radial depth of cut thus maintaining the cutting force value. Guo et al. [17] analyzed the wear of the tool and found coating delamination and adhesion are the common tool wear failure that is observed when coated carbide tools were used.

Tool wear mechanisms were analyzed by Jaffery et al. [18] to improve the coating on the cutting tool for machining of titanium alloys. It was recommended that the life of the tool can be improved by employing materials that have coatings with lower thermal diffusivity compared to other material. Dorlin et al. [19] compared a proposed model with experimental data of cutting forces and confirmed that the increased cutting force causes a rise in the nose radius of the tool. Celik et al. [20] analyzed the tool wear and surface quality when turning was performed dry using physical vapor deposition (PVD) and chemical vapor deposition (CVD) cutting tools. It was presumed that a common rise in tool wear in both types when machining factors became much difficult. This gave rise to a worsening of the surface finish.

Highly satisfactory results were obtained when the V_c value reached 300–400 m/min in high-speed machining employing cubic boron nitride tools without a binder (BCBN) [21]. Based on the literature review on past research work on machining titanium alloy using a CBN insert in machining, the following summary is presented.

The performance of CBN insert depends on cutting edge geometry, coating type, CBN content, type of binder, the grain size of CBN, use of cooling methods, and variation in cutting parameters. Titanium alloy has low thermal conductivity and chemical reactivity with other materials. In recent years, the titanium alloy was turned using PCBN tools and PCD tools and better performance was reported using PCD tools. The cutting performance of CBN tools was relatively dull compared to uncoated carbide tools with respect to tool life at a cutting speed of 250 m/min. The optimum cutting speed condition for machining titanium alloy with CBN insert is 185–225 m/min. The wear mechanisms, such as mechanical abrasion, adhesion, diffusion–dissolution, and fracture were seen during the machining of titanium alloy with a CBN insert. There are chances of better machinability by changing feed conditions and cutting edge geometry of the insert.

Thus, it is revealed that none of the studies has reported using AE signals to correlate the CBN tool insert for evaluating the machining performance of Ti-6Al-4V alloy. Besides, the effect of machining parameters on machinability is less understood. In this study, indigenously developed new CBN inserts were employed to assess the machinability of Ti-6Al-4V alloy in terms of cutting forces, tool wear, and AE signal characteristics.

Therefore, the objectives of the current study are to

- Analyze the effect of low CBN content cutting tool on the machinability of titanium alloy,
- Decide the optimum cutting conditions for higher tool life and good surface finish, and
- Explore the effect of cutting parameters with the AE signal characteristics during machining.

14.3 MATERIALS AND METHODS

The experimental plan involves the selection of process parameters, response variables, experimental method and measurement performed to assess the machinability of Ti-6Al-4V Alloy. It also describes the plan of experimentation, machining facilities, and equipment used. The schematic of experimental setup and the p-diagram of parallel turning is depicted in Figures 14.1 and 14.2, respectively.

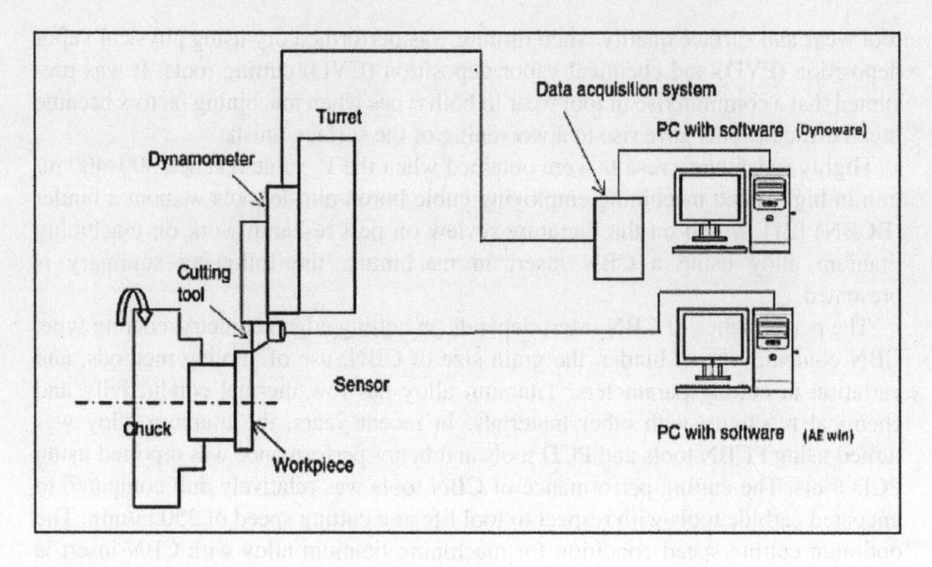

FIGURE 14.1 Schematic of experimental setup.

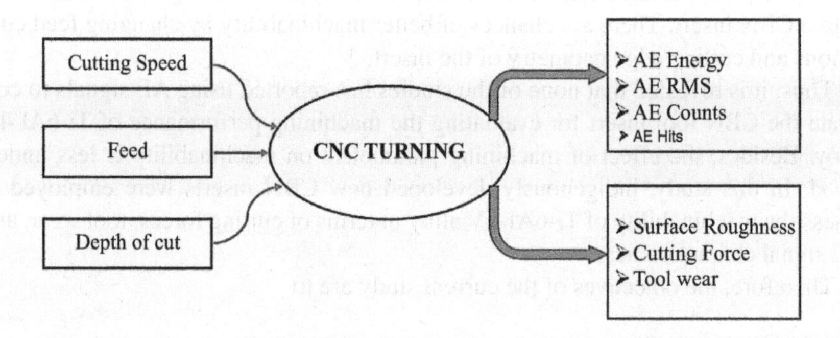

FIGURE 14.2 p-diagram of CNC turning.

14.3.1 Input Machining Parameters

14.3.1.1 Cutting Speed

Cutting speed is the relative speed at which the tool passes through the work material and removes metal. It is normally expressed in meters per minute and is evaluated by Equation 14.1:

$$v = \frac{\pi dn}{1000}, \quad (14.1)$$

where cutting speed (v) is in m/min, spindle speed (n) is in rpm, and the diameter of the workpiece (d) is in mm.

14.3.1.2 Depth of Cut

It is the thickness of the material removed in one pass when the tool penetrates the workpiece (in mm).

14.3.1.3 Feed

It is the distance traveled by the cutting tool. In case of turning, it is given by feed per revolution, stating the advancement of the tool during one workpiece revolution [22].

14.3.2 Selection of Response Variables

The machining performance of titanium alloy was assessed in terms of output variables that include cutting forces, tool wear, and AE signal characteristics. These are defined as follows:

- *Cutting forces*
  One of the most promising techniques for evaluating the machinability involves the measurement of cutting forces. Cutting forces influence the dimensional accuracy and machining system stability and therefore were chosen as one of the machinability parameters.
- *Tool wear*
  Cutting tools are subjected to an extremely severe rubbing process. Usually, it is a gradual process. Tool wear can be defined as progressive loss of material. The basic mechanism of wear between the sharp cutting tool edge and the workpiece surface is due to abrasion and adhesion, diffusion, localized galvanic action, surface cracks, and so on. It depends on various factors including tool and workpiece material, tool shape, cutting parameters, and cutting fluid. Tool wear affects the machining performance in a number of ways. These include increased cutting forces, increased cutting temperature, decreased accuracy of produced parts, decreased tool life, poor surface finish, and the economics of machining.
- *AE*
  AE is the phenomenon of radiating elastic waves for a short time interval by the rapid release of energy when a material undergoes deformation due to stresses. AE is one of the most significant indirect methods for monitoring tool wear and surface damages. AE signals are captured using an AE sensor. AE can monitor tool condition very precisely because the frequency range of the AE signal is quite higher than that of the machine tool vibrations and environmental noises [23].

The response variables selected are the cutting force components, surface roughness, tool life, and AE parameters, which include AE energy, counts, root mean square (RMS), and the AE hits. The machinability is assessed in terms of the previously described performance indicators also. These are elaborated in the following.

14.3.2.1 Machining Forces

The three components of cutting forces are illustrated as follows:

- The **cutting or tangential force** (F_t) acting in downward direction on the tool tip, allowing the upward deflection of the workpiece. It delivers the energy needed for the turning process.
- The **axial or feed force** (F_f) acting in the longitudinal direction. It is also known as feed force because it is parallel to the axis of the workpiece. The tool is being pushed away from the chuck due to this force.
- The **radial or thrust force** (F_r) acting in the radial direction, which tends to push the tool away from the workpiece.

14.3.2.2 Surface Roughness

Surface roughness is defined as the shorter frequency of real surfaces relative to the troughs. Surface roughness not only affects the appearance, but it also produces texture or tactile differences. Surface roughness is given as the arithmetic mean value for a randomly sampled area. The theoretical value of the arithmetic mean surface roughness [24] is given in Equation 14.2:

$$R_a = f^2/8R_n, \tag{14.2}$$

where f is the feed in mm/rev and R_n is the nose radius in mm.

14.3.2.3 Acoustic Emission Signal Parameters

The electrical signal identified as an AE signal is generated by the elastic deformation and fracture. Therefore, AE signal variables are studied to get insight into the physical phenomena during machining [25]. The AE parameters are defined graphically as shown in Figure 14.3.

Hit: It is a signal exceeding the limit which makes the channel suitable for data collection. This usually shows the AE function by calculated time (average). In Figure 14.3, a single waveform is corresponding to one "hit."

Count: When the AE signal crosses the preset limit within an interval, it is called as count. It is a function of amount of work done and threshold value.

Often, the counts in between the trigger time above threshold value, as well as peak amplitude, is known as counts to peak (see Figure 14.3).

Energy: A rated area under the adjusted envelope of the signal. It is the value of source event power, is sensitive to the magnitude and length of time, and is highly dependent on the frequency threshold of operating frequency.

RMS: A variable proportional to the square root of the energy carried by AE waves. It is the most widely used term to quantify the AE signals. It is determined from the rate of energy dissipation and the count rate.

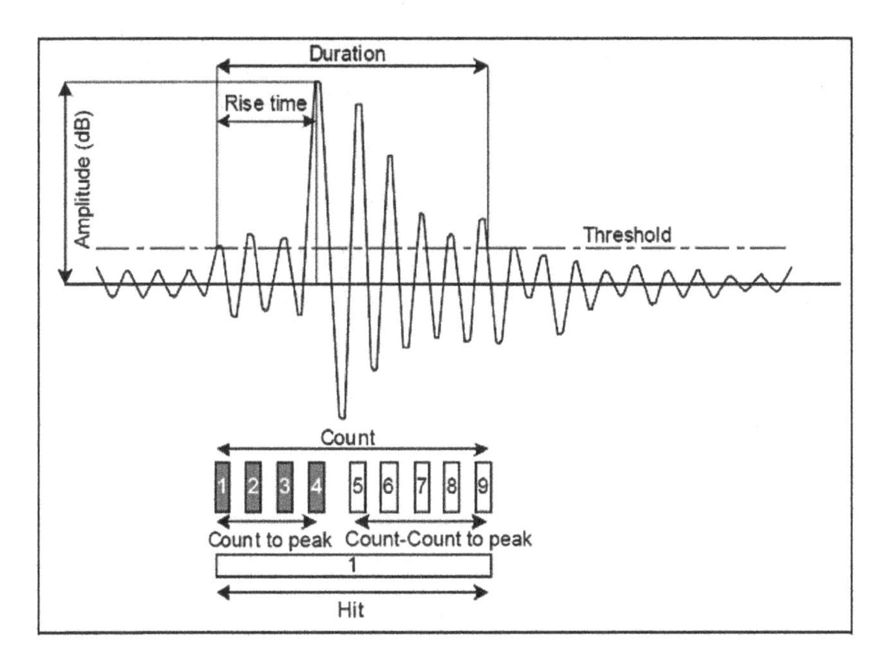

FIGURE 14.3 Acoustic emissions parameters [26].

14.3.3 Selection of Workpiece Material

Ti-6Al-4V (UNS designated R56400), also known as TC4, is an alpha–beta titanium alloy having a high strength-to-weight ratio and excellent resistance to corrosion, was selected as work material for this research (see Figure 14.4). It is one of the most widely used titanium alloys and is applied where low density and extremely good corrosion resistance are necessary, such as in the aircraft industries and in biomedical applications (implants and prostheses).

14.3.4 Tool Material

The cutting tools used for the experimental trials were CBN inserts. Table 14.1 is showing the chemical composition of CBN insert of Grade (2NK-CNGA120408) indigenously manufactured which is obtained from the EDAX results, performed with PHI 710 Scanning Auger Nanoprobe (high-performance Auger electron spectroscopy) at ICO Analytical lab at Worli, Mumbai (see Figure 14.5).

14.3.5 Cutting Tool and Tool Holder

Cylindrical turning on Ti-6Al-4V workpiece was performed using CBN inserts (2NK-CNGA120408), as shown in Figure 14.6a. The typical insert–holder (PCLNL2525M12) combination is shown in Figure 14.6b. It was a right-handed tool holder because the tool's cutting edge points to the left when viewing the holder from the top as shown in Figure 14.6c. Figure 14.7 shows the chemical composition of the CBN tool insert.

FIGURE 14.4 Ti-6Al-4V workpiece.

TABLE 14.1
Chemical Composition of CBN Tool Insert

Element	C	Ti	N	O	Na	Al	Si	Cl
Weight (%)	10.7575	51.63	15.7125	11.76	1.1425	0.2225	8.545	0.2325

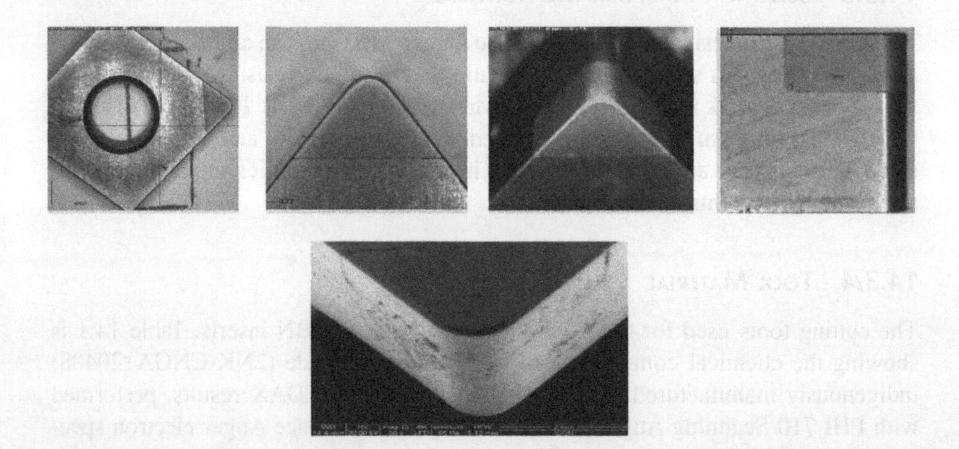

FIGURE 14.5 Digital microscope and scanning electron microscopy images of different views of a CBN tool insert.

14.3.6 Experimental Procedure

Experiments were performed using CBN insert (2NK-CNGA120408) with two cutting edges. The tool holder was PCLNL2525M12. The experiments were performed on a CNC lathe (micromatic—Ace, Jobber XL) with a dry condition. A round bar of

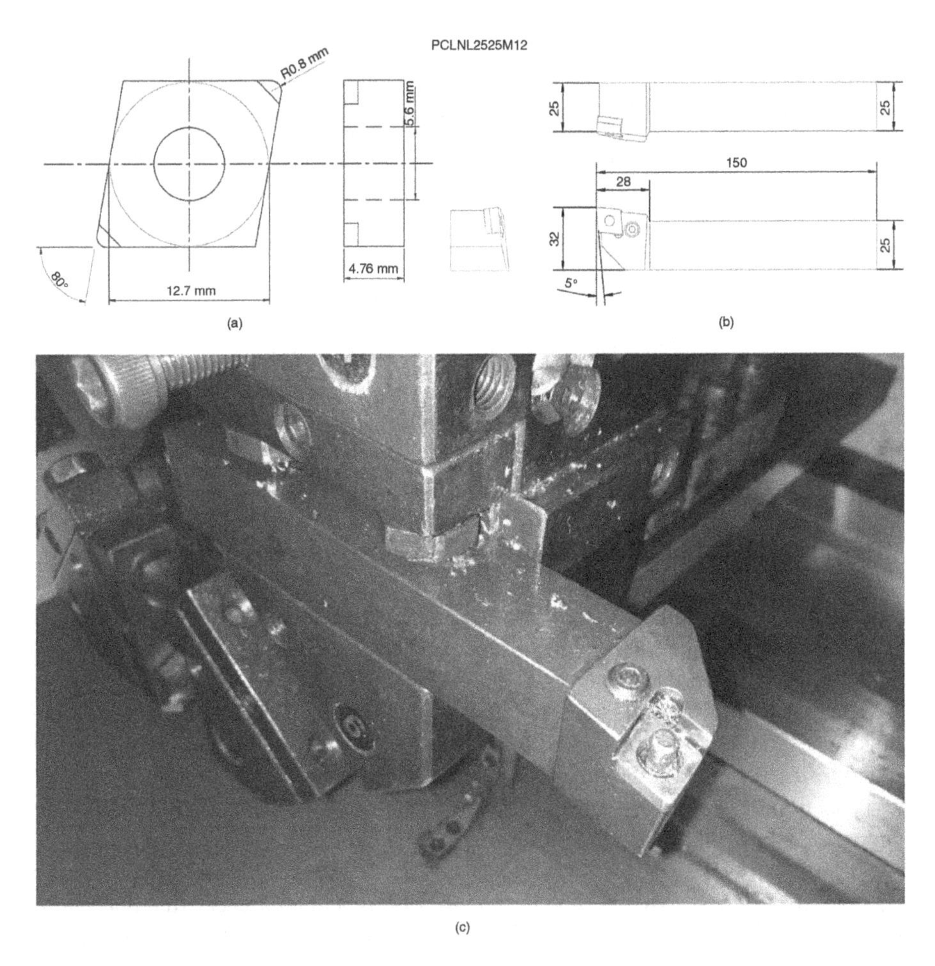

FIGURE 14.6 (a) CBN Insert (2NK-CNGA120408), (b) tool holder geometry, and (c) tool holder used.

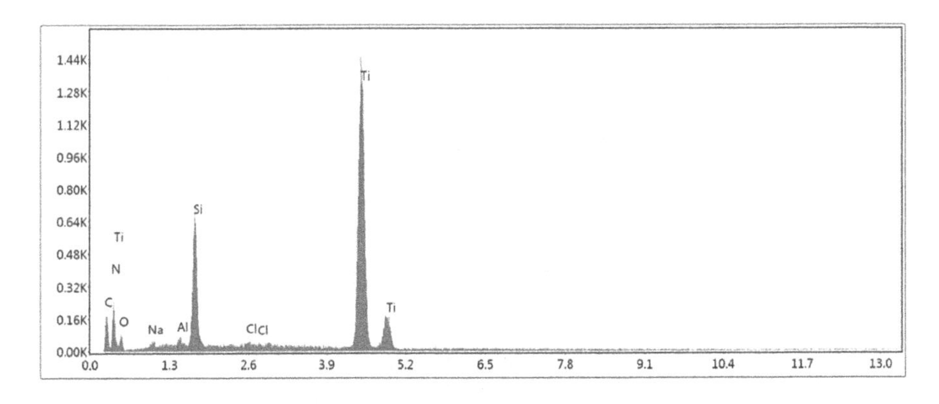

FIGURE 14.7 Chemical composition of CBN and the EDAX profile.

TABLE 14.2
Cutting Conditions for the First Set of Experiments

Experiment No.	Cutting Speed (m/min)	Feed Rate (mm/rev)	Depth of Cut (mm)
1	185	0.15	0.25
2	200	0.15	0.25
3	220	0.15	0.25
4	185	0.1	0.25
5	150	0.15	0.25

Ti-6Al-4V biomaterial with a hardness of 36 HRC was selected for the experiments. It was ϕ 65-mm × 150-mm-long cylinder.

The first set of experiments was performed to select the best set of cutting parameters to determine the minimum cutting forces and lower surface roughness and to correlate the AE signal characteristics with cutting parameters. This set of experiments were performed using the first cutting edge of the CBN insert. The optimum cutting parameters obtained from the first set of experiments were further used to assess the tool life. These optimum parameters were considered as the cutting parameters for the second set of experiments.

The first set of experiment was conducted to measure the surface roughness, cutting forces and AE parameters. The following parameters are identified based on the literature review and the manufacturer's catalogue. Cutting speed—150, 185, 200, and 220 m/min; feed rate—0.1 and 0.15 mm/rev; and depth of cut—0.25 mm. The experimental parameter combinations are given in Table 14.2.

The second set of experiments was performed using the second cutting edge of the CBN insert. In the second set of experiments, during and after every run, the output parameter was recorded. While machining, the cutting force components, as well as AE signals, were recorded, and after each run, the values for surface roughness and tool wear were recorded by roughness tester and digital microscope, respectively. A second set of experiments were conducted to monitor tool life for the selected optimum cutting parameters based on the first set of experiments. The parameters used for these experiments follow:

Cutting Speed—200 m/min, feed rate—0.15 mm/rev, depth of cut—0.25 mm

For each experiment, the cutting length considered was 15 mm for analyzing the effect of cutting length on the cutting force components. These experiments were performed until the value of V_{Bavg} reached the critical value of 0.6 mm.

14.3.7 Machine Tool and Measuring Instruments

14.3.7.1 CNC Lathe

All the experiments were performed on CNC lathe. The CNC lathe is a micromatic Ace model Jobber XL (see Figure 14.8). Tables 14.3 and 14.4 show the specifications of the cutting inserts and tool holder, respectively.

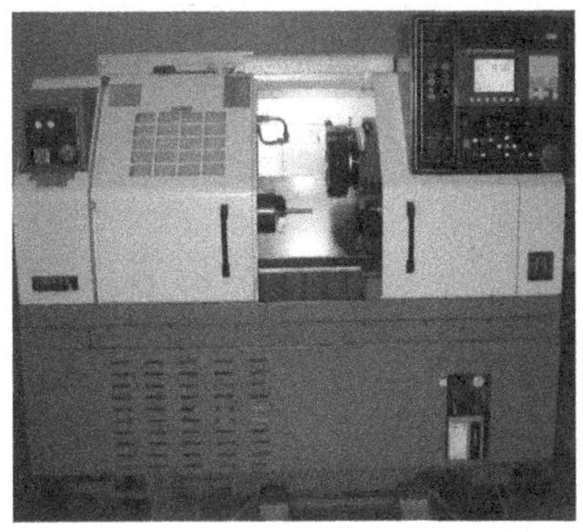

Make and model	Micromatic – Ace, Jobber XL
Distance between centres	425 mm
Maximum turning diameter	270 mm
Maximum turning length	400 mm
Spindle motor rated power	5.5/7.5kw
Speed range	50–5000 rpm

FIGURE 14.8 CNC lathe machine with specifications.

Make and Model	Kistler 5233A1
Manufacturer	Kistler, Switzerland
Operating Temperature Range	0 to 60°C
Force Range	–5 to 5 kN for F_x and F_y –5 to 10 kN for F_z
Sensitivity	10 mV/N for F_x and F_y 5 mV/N for F_z

FIGURE 14.9 Setup of cutting force dynamometer with specifications.

TABLE 14.3
Specification of Cutting Insert

Alphabet/Number	Specification	Insert's Shape or Size
2	Number of cutting edges	2
NK	Chip breaker geometry	No chip breaker
C	Shape	80° Rhombus
N	Clearance angle	0°
08	Corner/nose radius	0.8 mm
CBN content and binder		50–60%
CBN grain size		0.5–1.0
Coating type		Uncoated
Cutting edge geometry		Chamfer

TABLE 14.4
Tool Holder Specifications

Alphabet/Number	Specification	Size
P	Insert holding	Hole type
C	Insert shape	80^0
L	Tool holder style	5^0 side and end cut, offset shank
L	Hand	Left hand

14.3.7.2 Cutting Force Dynamometer

A cutting force dynamometer recorded the following forces, which are acting on the tool–workpiece interface while oblique machining:

- F_t: Cutting force along the direction of the cutting velocity vector, known as a tangential force component
- F_f: Feed force along the direction of the tool travel, known as ab axial force component
- F_r: Thrust force in the direction normal to the produced surface, known as a radial force component

14.3.7.3 Surface Roughness Tester

The stylus of the roughness tester unit traces the minute irregularities of the machined surface. Surface roughness is determined from the vertical stylus displacement produced during the detector traversing over surface irregularities of workpiece material (see Figure 14.10).

14.3.7.4 Digital Microscope

The most common instrument used for measuring tool wear, chip thickness, and width is a Dino-Lite digital microscope. This microscope helps with taking a photograph of tool wear and the chip thickness and width, and the Dino-Lite software is used

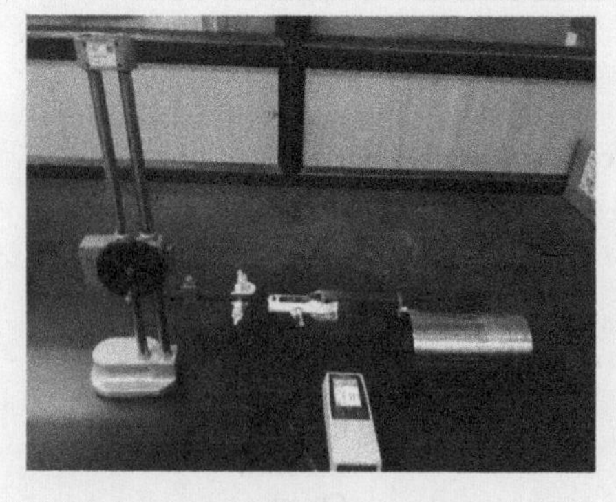

Parameters	*Ra, Rz*
Make	Mitutoyo
Measuring Range	Ra:0.05-10.0μm Rz:0.1-50.0μm
Accuracy	+/– 10% of actual value
Dimensions	125 x 73 x 26mm
Operating temperature	0°C - 40°C

FIGURE 14.10 Surface roughness tester with specifications.

for measuring tool wear and chip thickness and width in photographs. Figure 14.11 shows the overview of digital microscope and its specifications.

For selecting the optimum cutting parameters that were to be used for minimum cutting forces and lower surface roughness, the cutting force piezoelectric dynamometer and a surface roughness tester were used (see Figures 14.12 and 14.13 and Table 14.5).

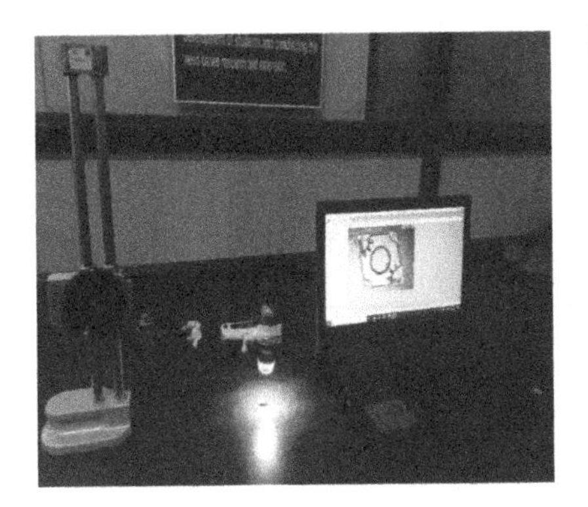

Make	Dinolite
Optical magnification power	10x-220x
Resolution	1280 x 1024 pixels
Lighting	8 white LED lights switched on/off by software
Dimension	10.5cm x 3.2cm

FIGURE 14.11 Digital microscope setup with specifications.

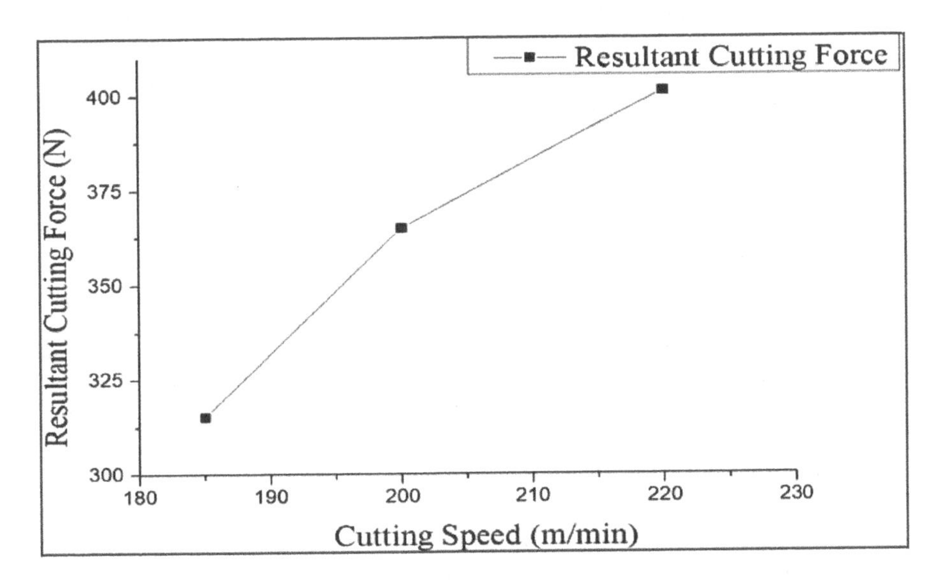

FIGURE 14.12 Resultant cutting force versus cutting speed.

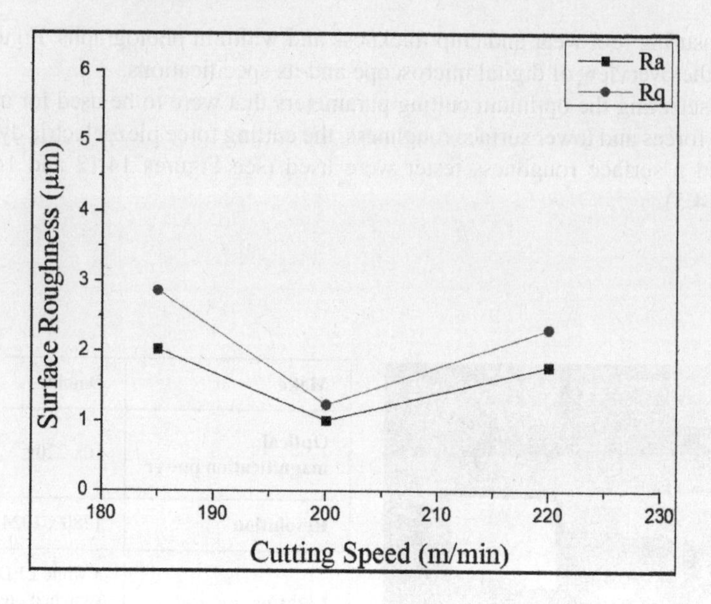

FIGURE 14.13 Surface roughness (R_a and R_q) versus cutting speed.

TABLE 14.5
Cutting Force and Surface Roughness Observations

Expt No.	Cutting speed (m/min)	Feed (mm/rev)	Depth of cut (mm)	F_t (N)	F_r (N)	F_f (N)	Resultant cutting force (N)	R_a (μm)	R_q (μm)
1	185	0.15	0.25	162.73	269.83	10.5	315.2769	2.028	2.86
2	200	0.15	0.25	109.53	348.11	12.5	365.1488	1.017	1.24
3	220	0.15	0.25	104.02	387.279	14	401.2495	1.78	2.31
4	185	0.1	0.25	107.5	376.661	14.4	391.9657	1.124	1.41
5	150	0.15	0.25	115.66	328.67	9.3	348.5508	2.88	3.787

14.4 RESULTS AND DISCUSSION

This deals with the analysis of experimental data, that is, cutting forces, surface roughness, tool wear, and AE signal characteristics recorded during the second set of experiments.

14.4.1 Cutting Force Analysis

It is observed from the graph in Figure 14.14 that an increasing trend is observed for all the force components with respect to cutting length. The radial force is found to be maximum compared to the other two cutting forces. Initially, the cutting edge is

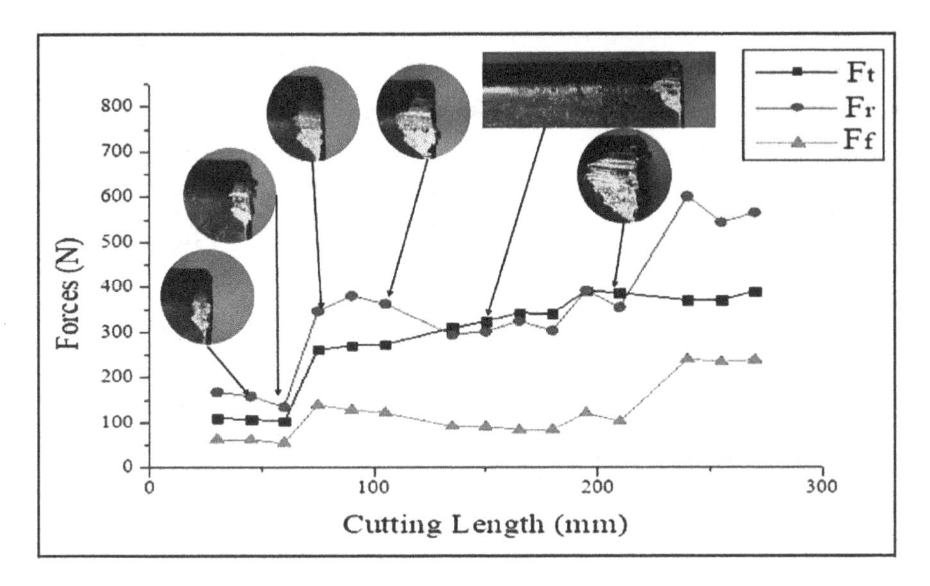

FIGURE 14.14 Cutting force component versus cutting length.

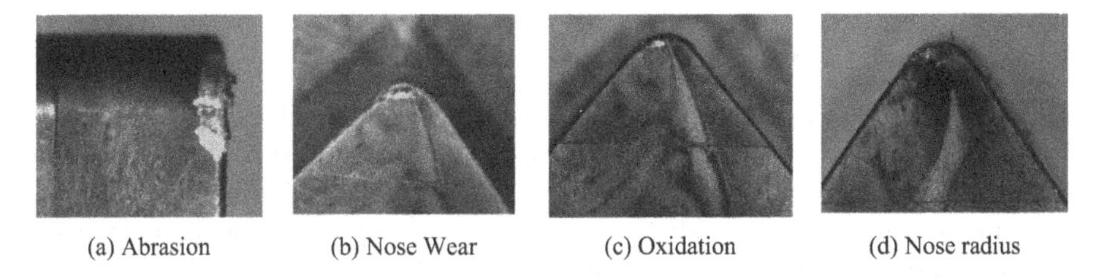

(a) Abrasion (b) Nose Wear (c) Oxidation (d) Nose radius

FIGURE 14.15 Wear pattern and wear mechanism.

new, and there is a point contact as seen in Figure 14.14. Therefore, cutting forces are low, which were constant until a 60-mm cutting length. Abrasion, oxidation, and adhesion wear mechanisms were observed at a 60-mm cutting length. After that, the forces remain almost constant up to 210 mm. At 210 mm, severe nose radius wear and chipping were observed due to which cutting forces increased drastically.

As the cutting length increases, a steady rise in R_a values because of an increase in tool flank wear was found. During machining, built-up edge (BUE) forms, which leads to an increase in roughness values. Figure 14.15 illustrated the wear pattern and mechanism on tool surface after machining. Table 14.6 presents the magnitude of the cutting force component during each pass of 15 mm.

TABLE 14.6
Values of Cutting Force Components for Varying Cutting Length

Expt No.	Cutting length (mm)	*Ft* (N)	*Fr* (N)	*Ff* (N)	*Ra* (μm)	*Rq* (μm)
1	15	115.36	172.42	68.26	0.924	1.1445
2	30	109.67	167.8	62.1	0.8555	1.0605
3	45	107.73	158.76	61.82	1.2385	1.5185
4	60	102.43	135.2	54.76	1.559	1.919
5	75	262.01	346.47	139.62	1.786	2.196
6	90	270.51	380.75	128.23	1.6435	2.0555
7	105	272.94	362.42	121.75	1.922	2.341
8	120	278.17	332.43	98.28	1.6905	2.1065
9	135	309.4	294.34	91.14	1.7795	2.2065
10	150	324.37	301.4	90.64	1.8735	2.1965
11	165	342.22	324.56	83.25	1.714	2.0575
12	180	340.4	303.18	84.18	2.2765	2.7775
13	195	391.3	391.63	121.21	1.8365	2.157
14	210	386.23	354.56	102.26	1.895	2.2145
15	225	385.43	788.7	278.24	1.782	2.114
16	240	370.51	600.84	239.88	1.9335	2.1535
17	255	370.83	543.4	234.29	1.4895	1.8245
18	270	389.67	564.58	238.4	1.276	1.5685

14.4.2 Surface Roughness Analysis

The surface roughness in terms of R_a and R_q was measured for all the specimens and is presented in Table 14.6. From Figure 14.16, the graph followed the same trend for surface roughness parameters R_a and R_q. It was observed that with an increase in cutting length, there was a slight increase in surface roughness at approximately 180-mm cutting length. This may be due to the rapid formation of oxide layers due to temperature rise and their removal, continuously leading to tool wear. However, after 250 mm of cutting length, the surface roughness was reduced. This may be due to the BUE, causing chip weld, protecting the cutting edge of the tool, and thus not affecting the machined surface.

14.4.3 Tool Wear Analysis

Tool wear was measured in terms of V_{Bmax} and V_{Bavg}, and its variation was shown with the cutting length. Figure 14.17 shows the graphical variation of average values for the flank wear of the CBN tool with respect to an increase in cutting length. The variation is normally classified into three categories: the accelerated beginning stage, intermediate stage, and the rapid failure stage [12]. The tool worn out mostly due to adhesion, attrition, oxidation as shown in Figure 14.18.

Figure 14.19 shows the different tool wear profiles. The increase in wear rate tends to reduce when the turning is employed for more than a 90-mm cutting length.

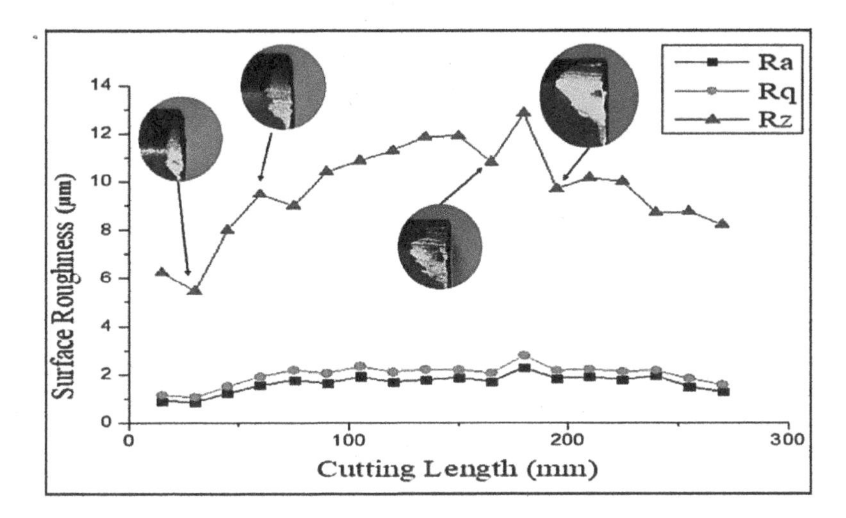

FIGURE 14.16 Surface roughness versus cutting length.

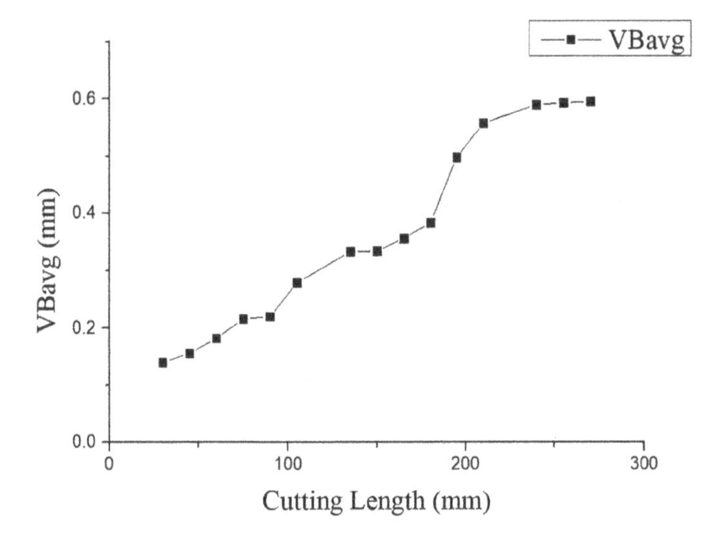

FIGURE 14.17 V_{Bavg} versus cutting length.

The chip flowing on the tool rake face gets partially welded at the tool edge. This sticking or welding of chip is attributed to the adhesive wear on the rake face of the tool. Under dry cutting conditions, temperature increases quickly while turning at high rate of feed and/or high rate of cutting speed. This high temperature welds the work material to the tool and forms the BUE-like layer. The welded material causes the wear growth.

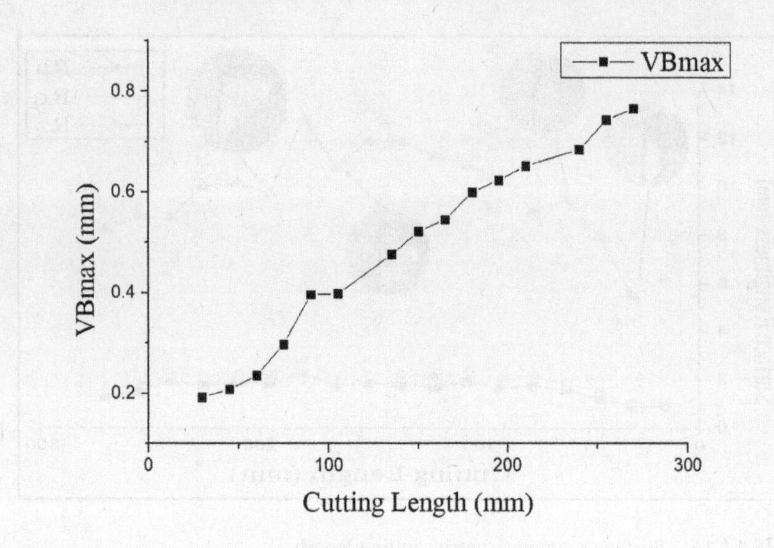

FIGURE 14.18 V_{Bmax} versus cutting length.

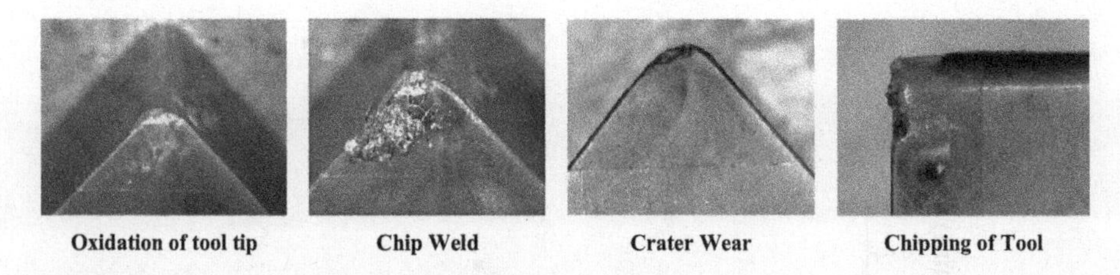

FIGURE 14.19 Different tool wear profiles.

The blunt tool, as well as the BUE, increases the cutting force. Despite of the mechanism procedures outlined above, CBN tool was diagnosed with crater wear. The breakdown and dispersion phenomena took place at the interface of the cutting tool and the workpiece being welded.

14.4.4 AE Analysis

Analysis of the AE parameter is done according to the number of passes. After every seven experiments, the diameter of the workpiece is decreased by 0.4 mm. Variation in the AE parameters, such as hit, counts, RMS voltage, and absolute energy, is shown in Figures 14.20 and 14.21.

From Figure 14.20a, it is observed that absolute energy increases with an increase in number of passes. However, after four passes, the absolute energy reduces because of severe tool wear. A similar trend is observed for AE counts also. From Figure 14.21a, it is seen that the RMS value increases steadily with the number of passes. This may be attributed to the fact that the RMS value increases with an increase in tool wear. However, the opposite trend is followed in AE hits from the second pass to the fourth pass.

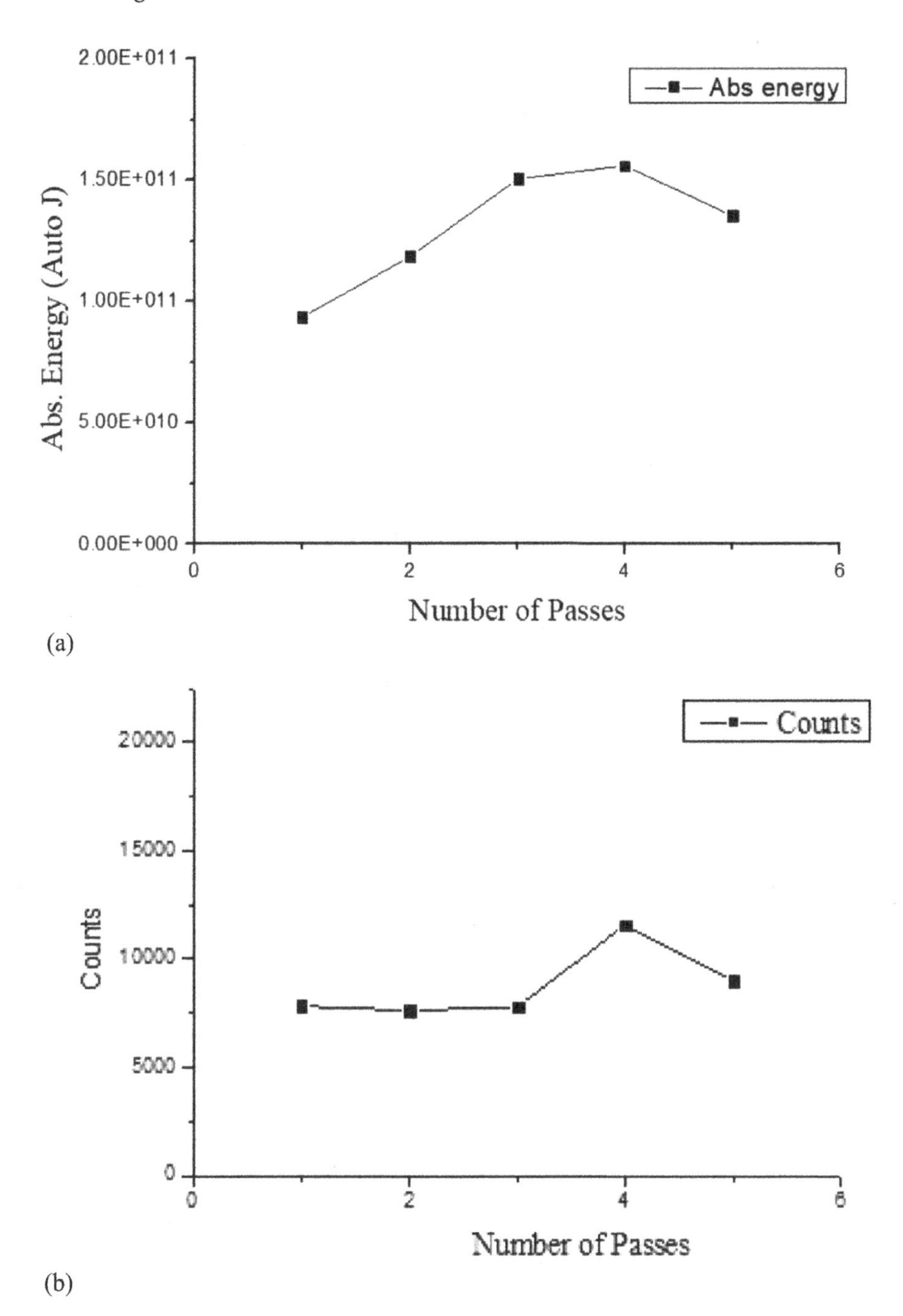

FIGURE 14.20 (a) Absolute energy versus number of passes; (b) counts versus number of passes.

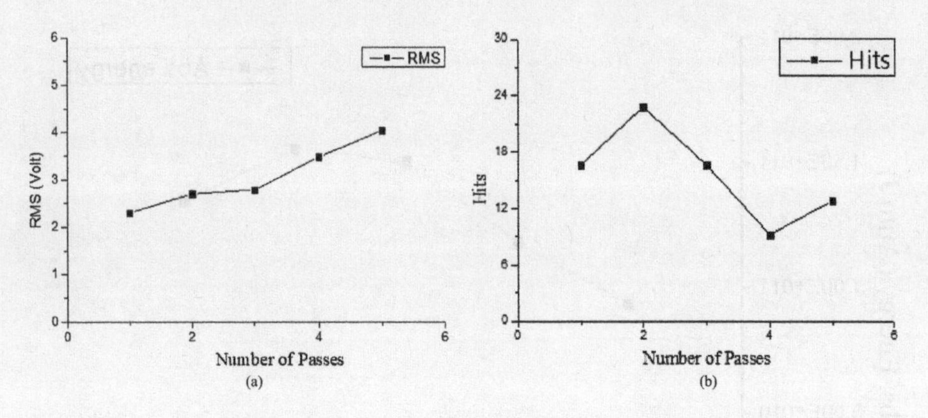

FIGURE 14.21 (a) RMS voltage versus number of passes; (b) hits versus number of passes.

14.5 CONCLUSION

The experimental research was done to analyze the effect of process parameters on machinability in turning of Ti-6Al-4V with a CBN insert. In order to study the tool wear mechanism, variation of the cutting forces, AE signal trends, and surface roughness has been observed with respect to cutting length. The following conclusions are drawn on the basis of exhaustive research work carried out:

- Lower cutting forces (150–400 N) are observed at 200 m/min cutting speed and 0.15 mm/rev feed.
- Thrust force is higher than the cutting force and feed force due to lower depth of cut, feed rate, and negative rake angle (0.25 mm).
- Cutting force value increases drastically, with nose radius wear, abrasion, and chipping wear becoming severe.
- Low cutting forces have been observed with CBN (2NK-CNGA120408). It is in accordance with the past research related to the study of cutting forces during the turning of Ti-6Al-4V using low-CBN content.
- As the thermal conductivity of titanium alloy is low, the temperature is concentrated at the machining zone, due to which sparks are seen during machining. Due to an increase in temperature of the workpiece material and, after that, cooling in the atmosphere, it may get hardened, and force values might increase.
- Lowest surface roughness (R_a – 0.9 μm) is observed. As tool wear progresses, roughness values increase.
- Due to a higher nose radius (0.8 mm), surface roughness values are low.
- With an increase in number of passes, AE absolute energy increases. However, after four passes, the absolute energy reduces due to a rise in tool wear.
- The AE RMS value increases with an increase in tool wear.
- With the increase in cutting length, BUE formation has been observed, and it may cause to increase in roughness value up to 1.6 μm R_a.

- Because there is no chip breaker geometry, chips accumulate in the machining zone as the tool progresses and may damage the machined surface.
- The 0.6-mm V_{Bavg} criterion is considered for experimentation for that tool life is 1.628 min.
- Abrasion, nose radius wear, oxidation, adhesion, and flank wear are observed during the study.
- The tool wear progression curve divides into three categories: the accelerated beginning stage, the intermediate stage, and the rapid failure stage.

14.6 FUTURE SCOPE

The present investigation was carried out with certain limited resources. However, it has a scope in the future so as to enhance the understanding of machining characteristics of titanium alloy with a CBN insert.

In this context, the following aspects will be useful:

- All these experiments are performed in dry machining conditions. Therefore, other cutting environments, like MQL conditions with different lubricant, can be explored.
- Different CBN tool geometry such as a round cutting edge and a different nose radius can be studied.
- Characterization like scanning electron microscopy and EDAX of tool wear can be explored to get more understanding of wear mechanism.
- The modeling of machining forces, roughness parameter R_a, and tool flank wear can be studied.
- In-depth correlation of process parameters with AE signals can also be done.

REFERENCES

1. Degarmo, E. P., Black, J. T. (2003). *Materials and Processes in Manufacturing* (9th ed.), Wiley, ISBN:0-471-65653-4.
2. Schneider, G. (2002). *Cutting Tool Applications, Tooling and Production*, Chapter 3, Nelson Publishing LLC.
3. Bhaumik, S. K., Divakar, C., Singh, A. K. (1995). Machining Ti-6Al-4V alloy with a WBN-CBN composite tool. *Materials & Design*, 16, 221–226.
4. Huang, Y., Chou, Y. K., Steven Y. L. (2007). CBN tool wear in hard turning: A survey on research progresses. *The International Journal of Advanced Manufacturing Technology*, 35, 443–453.
5. Zareena, A. R. (2002). High-speed machining of titanium alloys. *Master Thesis, National University of Singapore.*
6. Hirosaki, K., Kazuhiro S., Hideharu K. (2004). High speed machining of bio-titanium alloy with a binder-less PCBN tool. *JSME International Journal Series C Mechanical Systems, Machine Elements and Manufacturing*, 47, 14–20.
7. Wang, Z. G., Wong, Y. S., Rahman M. (2005). High-speed milling of titanium alloys using binderless CBN tools. *International Journal of Machine Tools and Manufacture*, 45, 105–114.

8. Wang, Z. G. (2005). Modelling of Cutting Forces during Machining of Ti-6Al-4V with Different Coolant Strategies. In *8th CIRP International Workshop on Modelling in Machining Operations, Chemnitz, Germany*.
9. Ezugwu, E. O., Da Silva R. B., Machado A. R. (2005). The effect of argon-enriched environment in high-speed machining of titanium alloy. *Tribology Transactions*, 48, 18–23.
10. Ozel, T., Sima M., Srivastava, A. K., Kaftanoglu, B. (2010). Investigations on the effects of multi-layered coated inserts in machining Ti–6Al–4V alloy with experiments and finite element simulations. *CIRP Annals*, 59, 77–82.
11. Bhaumik, S. K., Divakar, C., Singh, A. K. (1995). Machining Ti6Al4V alloy with a WBN-CBN composite tool. *Materials & Design*, 16, 221–226.
12. Burhanuddin, Y., Ghani J. A., Ariffin, A. K. (2008). The effects of CBN cutting tool grades on the tool life and wear mechanism when dry turning of titanium alloy. *Asian Int. Journal of Science and Technology*, 105–110.
13. Zoya, Z. A., Krishnamurthy, R. (2000). The performance of CBN tools in the machining of titanium alloys. *Journal of Materials Processing Technology*, 100, 80–86.
14. Liang, X., Wang, B. (2019). Investigation of surface topography and its deterioration resulting from tool wear evolution when dry turning of Titanium alloy Ti-6Al-4V. *Tribology International*, 135, 2019.
15. Li, A., Zhao, J., Hou, G. (2017). Effect of cutting speed on chip formation and wear mechanisms of coated carbide tools when ultra-high-speed face milling titanium alloy Ti-6Al-4V. *Advances in Mechanical Engineering*, 9(7), 1–13.
16. Polini, W., Turchetta, S. (2016). Cutting force, tool life and surface integrity in milling of titanium alloy Ti-6Al-4V with coated carbide tools. *Proceedings of the Institution of Mechanical Engineers, Part B*, 230, 694–700.
17. Guo, K., Yang, B., Sivalingam, V. (2018). Investigation on the tool wear model and equivalent tool life in end milling titanium alloy TI6AL4V. *ASME 13th International Manufacturing Science and Engineering Conference MSEC*, Vol. 4, 1–8.
18. Jaffery, S. H. I., Mativenga, P. T. (2012). Wear mechanisms analysis for turning Ti-6Al-4V-towards the development of suitable tool coatings. *The International Journal of Advanced Manufacturing Technology*, 58, 479–493.
19. Dorlin, T., Fromentin, G., Costes, J. P. (2016). Generalised cutting force model including contact radius effect for turning operations on Ti6Al4V titanium alloy. *The International Journal of Advanced Manufacturing Technology*, 86, 3297–3313.
20. Celik, Y. H., Kilickap, E., Guney, M. (2017). Investigation of cutting parameters affecting on tool wear and surface roughness in dry turning of Ti-6Al-4V using CVD and PVD coated tools. *The Journal of the Brazilian Society of Mechanical Sciences and Engineering*, 39, 2085–2093.
21. Kamaruddin, A. M. N. A., Hosokawa, A., Ueda, T., Furumoto, T. (2016). Cutting characteristics of binderless diamond tools in high-speed turning of Ti-6Al-4V: Availability of single-crystal and nano-polycrystalline diamond. *International Journal of Automotive Technology*, 10, 411–419.
22. Wang, Z., Rahman, M., Wong, Y. S. (2005). Tool wear characteristics of binderless CBN tools used in high-speed milling of titanium alloys. *Wear* 258, 752–758.
23. Bhaumik, S. K., Divakar, C. (1995). Machining Ti-6Al-4V alloy with a WBN-CBN composite tool. *Materials & Design* 16, 221–226.
24. Shaw, M. C. (1997). *Metal Cutting Principles (Oxford Series on Advanced Manufacturing)*, Oxford University Press.
25. Christian, U. (2008). *Acoustic Emission Testing*, Springer.
26. Pawade, R. S., Joshi, S. S. (2012). Analysis of Acoustic Emission Signals and Surface Integrity in High-speed Turning of Inconel 718. *Journal of Engineering Manufacture, Institution of Engineers, Part B*, 126(1), 3–27.

Part II

Manufacturing Processes

15 Industrial Internet of Things in Manufacturing

Hridayjit Kalita and Kaushik Kumar
Birla Institute of Technology, Ranchi, India

CONTENTS

15.1 INTRODUCTION

With the introduction of the Internet of Things (IOT) in 1998 by Kevin Ashton and developments in wireless technologies, a new automation paradigm to connect and communicate different devices through the internet has come into existence. Communication between electronic devices through the internet have led to the establishment of concepts such as smart homes, smart transportation, smart health-care facilities, environment concerns, energy efficiency, and effective resource utilization.

The Industrial Internet of Things (IIOT) is a subset of IOT that focuses on building an efficient manufacturing system with reduced operational cost and capital expenditure, optimized monitoring and controlling, and improved performance by collecting and analyzing big data efficiently. The IIOT is different from the IOT in that its emphasis is on higher levels of safety, reliability, security, and real-time operations. It also attempts to automate predictive maintenance, operational, and asset management.

It is estimated that the number of internet-connected devices is expected to increase to nearly 70 billion by 2025 with the global market share of the IIOT at 14.2

DOI: 10.1201/9781003284574-17

trillion US dollars by 2023. Cloud service platforms integrated with IIOT systems have the potential to alter the way of designing, maintaining, and deploying processes in industries [1–3]. Cloud-based IIOT systems promote greater flexibility, agility, device intelligence, adaptability, scalability of the data communication, virtualization, and data visualization. The deployment of a cloud-based IIOT system [4–6],however, is a challenge due to a lack of engineering approaches or tools for solving the integration and limitations of the IIOT component hardware as of the real-time needs, safety, and reliability.

Different studies concerning challenges in deploying IIOT, issues in connectivity, energy awareness issues, IIOT application deployment strategies, IIOT commercialization, big data analysis, and cyber-physical systems have been carried out in the literature [7–13].

In this chapter, a review describes the overall basic strategy in implementing IIOT in industries based on the previous research work in detail. Section 15.2 gives an account of the architecture, communication protocols, and data management techniques available for IIOT integration in industries. Section 15.3 gives an account of the software design methodologies or models in the industry setup, the major models such as the component-based model, multi-agent-based and service-oriented architecture (SOA), and model-driven engineering (MDE) have been described in detail.

15.2 IIOT ARCHITECTURE, COMMUNICATION PROTOCOLS, AND DATA MANAGEMENT

The current research on the implementation of IIOT architecture, communication protocols, and data management systems in an industry automation scenario is described in detail in this section as given in Khan et al.[14].

15.2.1 IIOT Architectures

IIOT architecture has been extensively described by the industrial internet consortium [15] consisting of different stages of data collection, processing, and deployment using the IOT as the main driving force of the technology. Data are collected continuously by the IIOT sources and devices (sensors and actuators) in the first layer, which are empowered by the edge and cloud computing systems for IIOT applications in layers 2 and 3, respectively. These data and information are finally employed for their application in enterprise, which is placed in the layer 4. Based on the variations in the design of location, computational assignments, safety, privacy, communication, resource, execution, resilience, and addressability different views of this architecture have been perceived by the researchers.

Wireless evolution for automation (WEVA) employing a wireless sensor network (WSN) as a graphical user interface for maintenance and network setup had been proposed by Campobello et al. [16], based on open-source communication and software protocols. This implementation of practical design for general industries employs IRIS, TelosB, TP-Link, and MR3020 as hardware and TinyOS, OpenWrt, and EasyWSN as software, with specifically IPv6 protocols for flexibility in IIOT performance. The difficulty level in implementing these network technologies has been realized in the case of lower latency and higher security. Lee et al. [17] proposed a conceptual design of the IIOT architecture for the manufacturing industry in the form of an IIOT suite that has two major components: a smart hub and a cloud platform. The smart hub is used for communicating and data transfer between the IIOT devices, providing secure channels for data transfer between the IIOT devices and the cloud platform, providing scaling solutions while introducing new IIOT devices to the system. The cloud platform helps in optimizing and decision-making in terms of IIOT device location/configuration, load balancing, routing, and monitoring and can be considered as the brain of the IIOT suite. The architecture uses Wi-Fi, Bluetooth, Z-Wave, MQTT/CoAP, and ZigBee as the communication protocols and JSON/XML and RESTful web Services as software. Martinez et al. [18] employed a prototyping development design of IIOT architecture for general industry. This design employed I3Mote as an open hardware platform consisting of the components, such as processors, multisource power features, sensors, and wireless radio interface, with the aim to provide connectivity and sensing support for prototyping and production of final IIOT products. I3Mote provides a software package suite for faster IIOT application development and two separate processors as a unique application feature for applications and communications, thereby promoting faster adoption of automation in industries. Tao et al. [19] employed IIHub (based on an IIOT hub) for a test-bed design in the manufacturing industry, consisting of three modules. In the first module, heterogeneous IIOT devices are connected by a customized access module (CA-module) via a few communication protocols. The second module connects the CA-module with the smart terminals and the labor workforce using an access hub (A-Hub) via. Constraint application protocols (CoAP), Ethernet, and Wi-Fi. The smart terminal comes under the third module, which is also called the local pool service (LPS), and performs decision-making, processing, and data collection and storage. Three different library packages for each CA-module, A-hub, and LPS are present for communication protocols, multidimensional model deployment, and data processing, respectively, which collectively operate the integration between heterogeneous IIOT devices and humans.

15.2.2 Communication Protocols

A zero-message queuing (ZMQ) messaging protocol of high latency had been proposed by Meng et al. [20] that enables a machine-to-machine communication mechanism for data sharing, including commands and events, machine discovery, presence, and connectivity. This protocol solves the heterogeneity and complex structure of the IIOT applications and facilitates the implementation of devices and powerful computers with Matlab and Visual C as the hardware component. Katsikeas et al. [21] employed a security protocol named message queue telemetry transport (MQTT) in the domain of industry for evaluating and comparing different security mechanisms using link layers and payload encryption. Publisher and subscriber are the two nodes taken and compared for their latency, energy consumption, and memory use.

Publisher was used to emulate the IIOT sensors and data encryption while subscriber was used to emulate the IIOT actuators and data decryption. Kiran et al. [22] introduced an analytical/theoretical model (Markov chain) that is validated by a real-time test bed and Monte Carlo simulation, with an error percentage of less than 5% and higher reliability. The slotted and unslotted prioritized contention access (PCA) and carrier-sense multiple access with collision avoidance (CSMA/CA) are compared for analysis in terms of reduction in power consumption and delay. In both cases, slotted and the unslotted PCA achieved better results than the slotted and the unslotted CSMA/CA. The hardware used for the model were the IITH (Indian Institute of Technology, Hyderabad) motes and Contiki 3.0. A round-trip time (RTT) latency of transferring data to the cloud and back had been investigated for MQTT protocol by Ferrari et al. [23] considering the intercontinental cloud transfer and local transfer. It was found that the former transfer took less than 300 ms while the latter one achieved an RTT of less than 50 ms. The hardware employed for the experimentation was the IOT2040 from Siemens, Yocto Linux, Intel i3-5000, Windows 7, Intel quark x1020, 1GB RAM, battery-backed RTC, and two ethernet ports.

15.2.3 Data Management in the IIOT

A distributed data management layer (DML) has been adopted by Theofanis Raptis et al. [24], which works along with the network layer to identify the nodes generating data, thereby requesting the data and storing them. It is kept separately from the routing operation so that lower latency could be maintained and improved performance of the network achieved. Proxy nodes or the intermediate nodes transfer data from the generating (source) nodes to the requesting (destination) nodes using cooperation algorithms and maintains a fast and efficient data delivery system. In another work by Raptis et al. [25], data distribution based on edge computing has been proposed for the IIOT network and proxy nodes or edge nodes introduced as intermediate nodes for receiving data from the IIOT devices and storing. A centralized algorithm based on heuristic approach has been adopted to search requests from the consumer nodes, locate and select the distributed edge nodes for data follow-up, and store data. An association and aggregation scheme had been proposed by Rao et al. [26] employing intermediate devices called user equipment (UE) to connect IOT devices to the base station (BS). The associated UEs receive the sensed data from the IOT devices to aggregate and forward the data to the BS, which uploads themto the cloud servers for processing and storage.

Data management in IOT-cloud systems has always been a centralized one in which the decision-making or the big data analytics of the large volume of data from the devices remained with the cloud end, resulting in a high-cost transfer of data, security and privacy threat, incremental processing cost of duplicate data, and the utilization of bandwidth. In order to tackle these problems in a conventional system, researchers have switched to a concentric computing model (CCM), in which the computational tasks would be based on overall application objectives [27] in a multilayer processing format to enhance the data-processing collaboration between the IIOT devices and the system. Attempts have been made to process the data at an early stage by catering the devices and systems into five layers, which are sensing systems, outer and inner gateway processing, and outer and inner central processing. The sensing system includes the data from the machine, people, and IOT devices; inner gateway processing includes the micro clouds and cloudlets; outer gateway processing includes the smart routers, smart switches, and application servers; outer central processing includes the cloud controllers, VPN (virtual rivate network) servers and virtual servers; and finally, the inner central processing includes the data centers and WAN (wide area network) servers.

15.3 INDUSTRIAL AUTOMATION SOFTWARE DESIGN METHODOLOGIES

The methodology for software design in an engineering industrial automation are categorized in the literature on the basis of different models such as component-based models, multi-agent-based models, model-based engineering (MBE), formal-based and SOA-based, and design pattern–based models [28]. With these model applications by industry designers, classification awareness has increased, which has lowered the cost of design, validation, and verification. Moreover, the costly affairs in post-validation and upgrades have been realized. Previous research has been performed with the implementation of these methods in different concerned problems, integrations, and interactions [29–34]. In this chapter, the methodologies such as component-based, multi-agent-based, SOA models, and MBE-based models are discussed in detail, with more emphasis given to the MBE model. MBE, which employs models for performing software design and component testing, has been found to be a promising solution among all other. This is due to its inherent capability to auto-generate codes in industry automation systems [35], which enhanced handling of complexities and multi-domain entanglements and interactions [36].

15.3.1 Component-Based Software Systems

This methodology enables individual software components to be reused for different software systems without any modification just as in the case of hardware components such as an integrated circuit. These independent components are defined, implemented, and integrated into systems toperform different tasks.

Component-based industrial automation involves two different aspects: implementation-level issues and application-level issues. In the implementation-level issues, any details regarding component locations are masked, which might represent any remote servers connected via networks and operated by an operating system very different from the client system. Component-based implementation models such as CORBA [37] and DCOM [38] come under the middleware implementation mechanism and are commonly used in information and communication technology. Other models that provide quality of service based on time relation include data distribution services and object management facilities[39]. The major challenges in implementation models involve the execution of real-time situations and determinism properties in the middleware mechanism.

At the application level, the modularity planning is involved in the component-based models. For reuse purposes, the design and development of components and maintenance are some of the complex situations that need to be modeled to ensure better functionalities and flexibilities. The granularity of the models for reusability is an important aspect, which can be fine-, coarse-, or medium-grained, determined by the purpose, dependence, and domain considered in the model. The authors recommend the allotment of different levels of granularity at each hierarchical position of the model based on the functionality. Szen-Ming et al. [40] proposed a control system composed of intelligent and autonomous components capable of operating without any master controller and facilitating distributed automation. This network of components would enhance agility and reusability, which reduces engineering costs as it has been adopted to solve complexity in automation using graphical programming languages by Strasser et al. [41].

15.3.2 Multi-Agent-Based Models

Woodbridgc [42] defined *multi-agent system* (MAS) as an alternate decision-making strategy in which the functions are decentralized and allotted over some distributed system entities. This system enhances flexibility scalability and modularity of the software design, which facilitates local preprocessing and collection of distributed data. This system, in the absence of any control command through the network channels, is able to handle scenarios and can be remotely controlled due to intelligence and in situ reactivity built on them. The control and monitoring of the proprietary devices, along with their access, can be well abstracted using this model. An agent platform called AMES can be employed, which has been suggested by Theiss et al. [33] for building efficiency in resources and is implemented in C++ and JAVA. An integration for industrial control of an MAS and an SOA had been presented by Herrera et al. [43],in which the reconfigurable manufacturing is focused on and a strong synergy between the SOA and the MAS is realized.

15.3.3 SOA

Semantic web services are an important tool for taking logical and rapid decisions in the current customized and reconfigurable manufacturing environment. It enables process classifications by performing logical ontologies of machines to execute different manufacturing scenarios simultaneously and autonomously. Considering services as the primary means to build logic for strategic goals in service-oriented programming, SOA is able to enhance enterprise productivity, agility, and efficiency [44]. Some of the features of SOA described by the work of Cannata et al.[45],in which by employing SOA, the authors have attempted to relate components in both manufacturing and business management systems, thereby promoting vertical integration that includes loose coupling (software modules are generic-format design-specific and present asynchronous component communication), software component modularization (or individual component programming for distributed control systems), and possessing common communication protocols (for implementing hardware devices from both low-level to high-level abstraction in the case of service providers). Some of the qualitative measures of SOA are better interoperability, implementation speed, requirement of less skilled labor, and cost reduction.

15.3.4 MDE

MDE relies on a code generation/development strategy that promotes flexible automation by automating tasks that are error-prone and labor-intensive and can be upgraded/modified without much engineering effort, which, in turn, reduces the cost of design, builds efficiency in the exchange of data, and enhances system verifiability and reusability. As MDE facilitates MBD and model-based verification (MBV), the entire process of design, verification, and validation can be automated. Few other drawbacks of the automating industrial components include their complexities, heterogeneity, and involvement of multiple domains closely entangled to each other. A proper semantic relation is tough to be established between several domains due to a lack of formalism, tools, and semantics. Muthukumar et al. [46] investigated the multi-view approach to implement MDE with a cloud-based IIOT.

In a multi-view approach, viewpoints from stakeholders are usually captured as domain knowledge for solving complex systems [47]. Different domain views must be solved simultaneously which can create contradictions and overlaps in information exchange and must be dealt with a variety of approaches. The solution for these multiview engineering systems can be in the form of model transformation and mapping. In order to design an automation system based on IIOT, Muthukumar et al. [46] employed Automation ML as a meta-model that provides a topology view for integrating different models' view requirements in verification and design. Autogeneration of codes for IEC61131 has been studied in the literature [35] for determining MBD's role in process automation and its significance in designing MPC (model predictive control) for solving complex optimization problems and validating using MIL (model-in-the-loop), HIL (hardware-in-the-loop), and SIL (software-in-the-loop) designs. The MBV was performed using the Uppaal timed automata model [48], which recognizes the patterns of actions regarding time, thereby verifying the timing behavior or performance [49]. Plug-and-play control, on-the-fly verification, and smart manufacturing are a few of the flexible cloud-capability features in IIOT.

15.4 FUTURE SCOPE

Although the models incorporating IIOT capabilities have been thoroughly analyzed, made inclusive as much as possible, and planned according to industrial requirement, extensive deployment hasyet to be performed on wide-scale enterprise connectivity, along with plant optimization. The deployment strategies can be the future research work on this area, which alters with changing manufacturing-processing environments of the industries. The future trend of the research on this topic might also hold personalized manufacturing enablers in IIOT systems. Integrating IIOT systems into industrial automation suffers from certain challenges that need to be solved in domains such as heterogeneous system collaboration, data management strategy and its efficiency, flexibility and robustness in data analytics, decentralization strategy, public safety, and trust in IIOT systems, and merging protocols with wireless devices and operating systems.

15.5 CONCLUSION

The preceding discussion gave an overall description of all the architectural, communication, data management, and software aspects of implementing IIOT for industry automation. Different hardware platforms, sensors, actuators, and protocols involved in establishing an efficient IIOT system in industries and their latency, security, energy efficiency, and compatibility were compared and discussed in the first section of this chapter. In the second section, different software design methodologies and approaches commonly employed in the literature, such as component-based models, multi-agent-based models, SOA, and MDE were discussed. Among all the approaches, MDE has been found to be more promising in designing software, automating tests, and validating and verifying industry automation components. The multidomain complexities involving heterogeneity and entanglement in opinions of stakeholders can also be solved by establishing proper semantic relationships and auto–code generation.

REFERENCES

1. Botta, A., De Donato, W., Persico, V., Pescapé, A. (2016). Integration of cloud computing and internet of things: A survey. *Future Gener. Comput. Syst.* 56:684–700.
2. Gubbi, J., Buyya, R., Marusic, S., Palaniswami, M. (2013). Internet of things (IoT): A vision, architectural elements, and future directions. *Future Gener. Comput. Syst.* 29(7): 1645–60.
3. Wang, C., Bi, Z., Da Xu, L. (2014). IoT and cloud computing in automation of assembly modeling systems. *IEEE Trans. Ind. Inf.* 10(2):1426–34.
4. Colombo, A.W., Karnouskos, S., Kaynak, O., Shi, Y., Yin, S. (2017). Industrial cyber-physical systems: A backbone of the fourth industrial revolution. *IEEE Ind. Electron. Mag.* 11(1): 6–16.
5. Da Xu, L., He, W., Li, S. (2014). Internet of things in industries: A survey. *IEEE Trans. Ind. Inf.* 10(4): 2233–43.
6. He, W., Da Xu, L: (2014). Integration of distributed enterprise applications: A survey. *IEEE Trans. Ind. Inf.* 10(1): 35–42.
7. Aazam, M., Zeadally, S., Harras, K.A. (2018). Deploying fog computing in industrial internet of things and industry 4.0. *IEEE Trans. Ind. Inf.* 14(10):4674–82. doi:10.1109/TII.2018.2855198.
8. Al-Gumaei, K., Schuba, K., Friesen, A., Heymann, S., Pieper, C., Pethig, F., et al. (2018). A survey of internet of things and big data integrated solutions for industrie 4.0. In: *2018 IEEE 23rd International Conference on Emerging Technologies and Factory Automation (ETFA)*, 1. IEEE; 2018, pp. 1417–24.
9. Long, N.B., Tran-Dang, H., Kim, D. (2018). Energy-aware real-time routing for large-scale industrial internet of things. *IEEE Internet Things J.* 5(3):2190–99. doi:10.1109/JIOT.2018.2827050.
10. Mumtaz, S., Alsohaily, A., Pang, Z., Rayes, A., Tsang, K.F., Rodriguez, J. (2017). Massive internet of things for industrial applications: addressing wireless IIoT connectivity challenges and ecosystem fragmentation. *IEEE Ind. Electron. Mag.* 11(1):28–33.
11. Perera, C., Liu, C.H., Jayawardena, S. (2015). The emerging internet of things marketplace from an industrial perspective: a survey. *IEEE Trans. Emerg. Top Comput.* 3(4):585–98. doi:10.1109/TETC.2015.2390034

12. Xu, H., Yu, W., Griffith, D., Golmie, N. (2018). A survey on industrial internet of things: a cyber-physical systems perspective. *IEEE Access* 6:78238–59. doi:10.1109/ACCESS.2018.2884906.
13. Zhu, C., Rodrigues, J.J.P.C., Leung, V.C.M., Shu, L., Yang, L.T. (2018). Trust-based communication for the industrial internet of things. *IEEE Commun. Mag.* 56(2):16–22. doi:10.1109/MCOM.2018.1700592.
14. Khan, W.Z., Rehman, M.H., Zangoti, H.M., Afzal, M.K., Armi, N., Salah, K. (2020). Industrial internet of things: Recent advances, enabling technologies and open challenges. *Comput. Elect.Eng.* 81:106–522.
15. Lin, S.W., Miller, B., Durand, J., Bleakley, G., Chigani, A., Martin, R., Murphy, B., Crawford, M. (2017). The industrial internet of things volume G1: Reference architecture. *Ind. Internet Consort.* 1:10–46.
16. Campobello, G, Castano, M, Fucile, A, Segreto, A. (2017).Weva: A complete solution for industrial internet of things. In: *International Conference on Ad-hoc Networks and Wireless*. Springer, pp. 231–8
17. Lee, C.K.M., Zhang, S.Z., Ng, K.K.H. (2017). Development of an industrial internet of things suite for smart factory towards re-industrialization. *Adv. Manuf.* 5(4):335–43. doi:10.1007/s40436-017-0197-2.
18. Martinez, B., Vilajosana, X., Kim, I., Zhou, J., Tuset-Peiró, P., Xhafa, A., Poissonnier, D., Lu, X. (2017). I3mote: an open development platform for the intelligent industrial internet. *Sensors* 17(5):986. doi:10.3390/s17050986.
19. Tao, F., Cheng, J., Qi, Q. (2018). IIHub: An industrial internet-of-things hub toward smart manufacturing based on cyber-physical system. *IEEE Trans. Ind. Inf.* 14(5):2271–80. doi:10.1109/tii.2017.2759178
20. Meng, Z., Wu, Z., Muvianto, C., Gray, J. (2017). A data-oriented M2M messaging mechanism for industrial IoT applications. *IEEE Internet Things J.* 4(1):236–46.
21. Katsikeas, S., Fysarakis, K., Miaoudakis, A., Van Bemten, A., Askoxylakis, I., Papaefstathiou, I., Plemenos, A. (2017).Lightweight & secure industrial IoT communications via the MQ telemetry transport protocol. In: *Computers and communications (ISCC), 2017 IEEE symposium on. IEEE*,pp. 1193–200.
22. Kiran, M., Rajalakshmi, P. (2018). Performance analysis of CSMA/CA and PCA for time critical industrial IoT applications. *IEEE Trans. Ind. Inf.* 14(5):2281–93.
23. Ferrari, P., Sisinni, E., Brandão, D., Rocha, M. (2017). Evaluation of communication latency in industrial IoT applications. In: *Measurement and networking (M&N), 2017 IEEE international workshop on. IEEE*; 2017,pp. 1–6.
24. Raptis, T.P., Passarella, A. (2017).A distributed data management scheme for industrial IoT environments. In: *Wireless and mobile computing, networking and communications (WiMob), IEEE*,pp. 196–203
25. Raptis, T.P., Passarella, A., Conti, M. (2018). Maximizing industrial IoT network lifetime under latency constraints through edge data distribution. In: *1st IEEE international conference on industrial cyber-physical systems, (ICPS)* (May 2018). Available at http://cnd.iit.cnr.it/traptis/2018-raptis-icps.pdf.
26. Rao, S., Shorey, R. (2017).Efficient device-to-device association and data aggregation in industrial IoT systems. In: *Communication systems and networks (COM-SNETS), 2017 9th international conference on. IEEE*, pp. 314–321.
27. ur Rehman, M.H., Yaqoob, I., Salah, K., Imran, M., Jayaraman, P.P., Perera, C. (2019). The role of big data analytics in industrial internet of things. *Fut. Gener. Comput. Syst.*; 99:247–59. doi:10.1016/j.future.2019.04.020
28. Vyatkin, V. (2013). Software Engineering in Industrial Automation: State-of-the-Art Review. *IEEE Trans. Ind. Inf.* 9(3): 1234–1249.

29. Calvo, I., Pérez, F., Etxeberria, I., Morán, G. (2010).Control communications with DDS using IEC61499 service interface function blocks.In: *Emerging Technologies and Factory Automation (ETFA), 2010 IEEE Conference on, IEEE*, pp. 1–4.
30. Dubinin, V.N., Vyatkin, V. (2012). Semantics-robust design patterns for IEC 61499. *IEEE Trans. Ind. Inf.* 8(2): 279–90.
31. Hanisch, H.M. (2004).Closed-loop modeling and related problems of embedded control systems in engineering.In: *International Workshop on Abstract State Machines*, Springer, pp. 6–19.
32. Jammes, F., Smit, H. (2005). Service-oriented paradigms in industrial automation. *IEEE Trans. Ind. Inf.* 1(1): 62–70.
33. Theiss, S., Vasyutynskyy, V., Kabitzsch, K. (2009). Software agents in industry: A customized framework in theory and praxis. *IEEE Trans.Ind. Inf.* 5(1):147–56.
34. Thramboulidis, K. (2005). Model-integrated mechatronics-toward a new paradigm in the development of manufacturing systems. *IEEE Trans. Ind. Inf.* 1(1): 54–61.
35. Obermeier, M., Braun, S., Vogel-Heuser, B.(2015). A model-driven approach on objec-toriented PLC programming for manufacturing systems with regard to usability. *IEEE Trans. Ind. Inf.* 11(3): 790–800.
36. Vogel-Heuser, B., Schütz, D., Frank, T., Legat, C.(2014). Model-driven engineering of manufacturing automation software projects–a SysML-based approach. *Mechatronics* 24(7): 883–97.
37. Vinoski, S. (1997). CORBA: Integrating diverse applications within distributed heterogeneous environments. *IEEE Commun. Mag.* 35: 46–55.
38. Sessions, R.(1997)*COM and DCOM: Microsoft's Vision for Distributed Objects*. New York, NY: Wiley.
39. Crnkovic, I.,Larsson, M. (2000). A case study: Demands on component- based development. *Proc. Int. Conf. Software Eng.*23–31. https://www.computer.org/csdl/proceedings-article/icse/2000/00870393/12OmNqzu6TR
40. Szer-Ming, L., Harrison, R., West, A.A. (2004).A component-based distributed control system for assembly automation.In *Proc. 2nd IEEEInt. Conf. Ind. Inf.*, pp. 33–38.
41. Strasser, T.et al. (2008). Structuring of large scale distributed control programs with IEC 61499 subapplications and a hierarchical plant structure model.In:*Proc. IEEE Int. Conf. Emerging Technol. Factory Autom.*,Sep. 15–18, pp. 934–41.
42. Wooldridge, M. (2002).*An Introduction to Multi-Agent Systems*. New York, NY: Wiley.
43. Herrera, V.V., Bepperling, A., Lobov, A., Smit, H., Colombo, A.W., Lastra, J. (2008). Integration of multi-agent systems and service-oriented architecture for industrial automation. *Proc. 6th IEEE Int. Conf.Ind. Inf.*, pp. 768–773
44. Erl, T. (2005).*Service-Oriented Architecture: Concepts, Technology, and Design*. Upper Saddle River, NJ: Prentice-Hall PTR.
45. Cannata, A., Gerosa, M., Taisch, M. (2008). SOCRADES: A framework for developing intelligent systems in manufacturing. *Proc. IEEEInt. Conf. Ind. Eng. Eng. Manag.*, pp. 1904–908.
46. Muthukumar, N., Srinivasan, S., Ramkumar, K., Pal, D., Vain, J., Ramaswamy, S. (2019). A model-based approach for design and verification of Industrial Internet of Things. *Fut. Gene. Comp. Syst.* 95: 354–363
47. Reineke, J., Tripakis, S. (2014).Basic problems in multi-view modeling.In: *International Conference on Tools and Algorithms for the Construction and Analysis of Systems*, Springer, pp. 217–32.

48. Balasubramaniyan, S., Srinivasan, S., Buonopane, F., Subathra, B., Vain, J., Ramaswamy, S. (2016). Design and verification of cyber-physical systems using truetime, evolutionary optimization and UPPAAL. *Microprocess. Microsyst.* 42: 37–48.
49. Srinivasan, S., Buonopane, F., Vain, J., Ramaswamy, S. (2015). Model checking response times in networked automation systems using jitter bounds. *Comput. Ind.* 74: 186–200.

16 Improvement in Forming Characteristics Resulted in Incremental Sheet Forming

Saurabh Thakur and Parnika Shrivastava
National Institute of Technology, Hamirpur, India

CONTENTS

16.1 INTRODUCTION

The incremental sheet forming (ISF) process continues to receive attention from researchers worldwide as a substitute to the traditional sheet metal–forming process in certain aspects. ISF has a clear advantage over the conventional forming process in batch-type manufacturing systems, prototypes, and customized products [1]. It has been established as a process that significantly increases material formability in comparison to conventional sheet metal forming operations [2–4]. In addition, the process also facilitates flexibility in the manufacturing of asymmetric complex and customized components.

During the last decade, intensive development and efforts to push the process toward applicability have been reported. To facilitate practicality and overcome limitations, it is necessary to thoroughly understand the complex-forming mechanism that material is subjected to during the forming process. Reviews done in the past mostly focus on the effect of various process and material parameters that influence

DOI: 10.1201/9781003284574-18

the forming characteristics, that is, formability, geometrical accuracy, and surface quality in ISF. However, the process has several challenges in terms of the quality of parts produced by the process. The geometrical accuracy of the part produced vary in comparison to the computer-aided design (CAD) model [5]. Major geometrical errors reported are pillow effect, springback, and unwanted plastic deformation [6].

Recent studies have reported numerous strategies that improve the formability and quality of the parts produced by the ISF process. The work aims to highlight and systematically review the recent strategies related to numerical techniques, such as FEA, CAD, tool path development, experimental setup, and hybrid techniques, and explore their consequences on the forming characteristics of the ISF-formed parts.

16.2 FORMING CHARACTERISTICS IN ISF

To quantify the ISF process formability, geometrical accuracy and surface finish have been readily used throughout the literature. Formability, in the conventional sheet metal–forming process, is the ability of the metal to deform without failure. In ISF, however, the plastic zone is restricted to the small area of contact between the tool and the workpiece, and formability is quantified using the maximum forming or draw angle; sometimes, the depth of forming has also been used to measure the formability [7]. Draw angle is the largest angle of deformation in one pass of the tool without tearing of the sheet metal [4]. Some of the early research aimed at obtaining forming-limit curves (FLCs) for the process to compare it to the traditional sheet metal–forming process. Figure 16.1 shows FLC for the ISF process in contrast to the

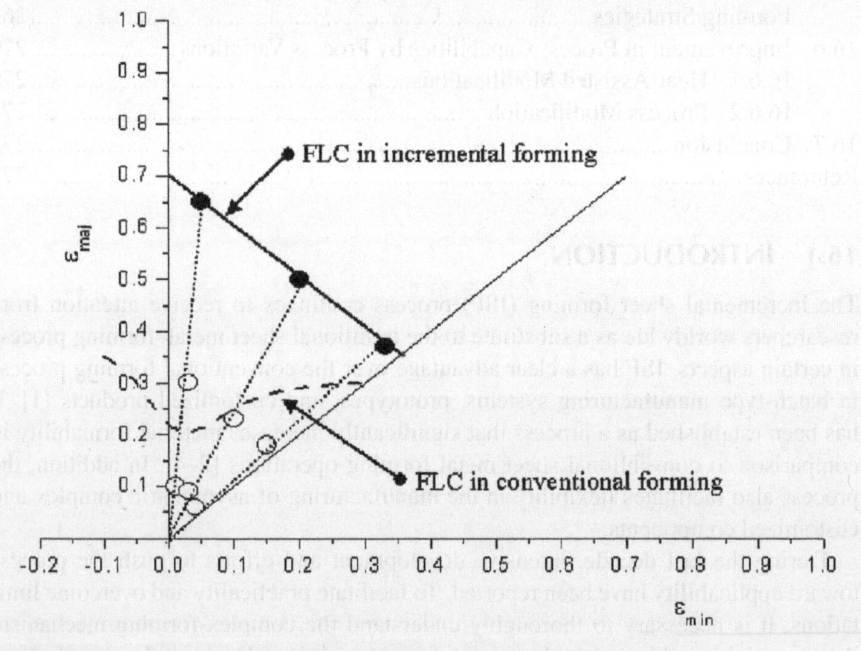

FIGURE 16.1 Conventional versus incremental sheet forming limit curve. **Reprinted from [8] with permission from Elsevier.**

conventional forming process in major and minor strain space for an aluminum sheet [8]. Formability, as defined by the formability limit curve, was a straight line with a negative slope.

Formability is influenced by parameters such as tool size and type, feed rate, and step size. As the process evolved, more process parameters, such as the curvature of generatrix [9], the reduction in area at tensile fracture, and forming speed [10], were analyzed to identify their effect on formability. Over time, various process parameters have been identified and extensively researched to report their effect on forming characteristics. Forming limit diagrams (FLD) and fracture-forming line (FFL) have been evaluated for different materials and different techniques as means for characterizing formability [10–15].

Although the process is simple and can be carried out on a CNC machine with slight modifications and a simple tool, this simplicity introduces inaccuracies in the final product. Geometrical inaccuracy can be the pillow effect, any unwanted plastic deformation, and springback [6]. Several methods are suggested in the literature for improving inaccuracies, but achieving industry-standard dimensional accuracy is still difficult. Surface roughness in industrial applications is always a concern as it is in ISF. Because of direct contact between the blank and the tool and high friction involvement, the surface quality is not good. Parametric approaches have tried to optimize conditions to obtain good surface quality, but it interferes with other parameters.

With the advent of manufacturing capabilities and advancement in computation capacity process, modifications and laborious numerical analyses are feasible. The purpose of this study, in addition to the already stated, is to report all other factors besides process parameters that are responsible for formability enhancement or improvement in the quality of parts produced using ISF.

16.3 DEFORMATION MECHANISM AND ITS INFLUENCE ON FORMING CHARACTERISTICS

Deformation in single-point incremental forming (SPIF) is a complex combination of bending, membrane stretching, and through-thickness shear. The resulting geometrical accuracy of the part produced is in direct relation with the underlying deformation mechanism [16]. Thus, the formability also depends on the dominant deformation mechanism. Filice, Fratini, and Micari fully characterized material formability by drawing FLD as shown in Figure 16.2. Wide ranges of deforming conditions ranging from pure uniaxial to biaxial stretching were used, which enabled the detection of critical strains and a complete investigation of formability. Later, to relate formability in ISF to the material properties, the strain hardening coefficient was chosen as the indirect measure of the capability of the material to undergo thinning without plastic failure. FLD was taken as a proper indicator for the material formability and a pure stretching deformation mechanism was pushed toward finding that strain hardening has a significant effect on the formability of a test specimen [17].

Allwood, Shouler, and Tekkaya attributed the increased forming limits to the presence of through-thickness shear. Considering through-thickness shear ensures that

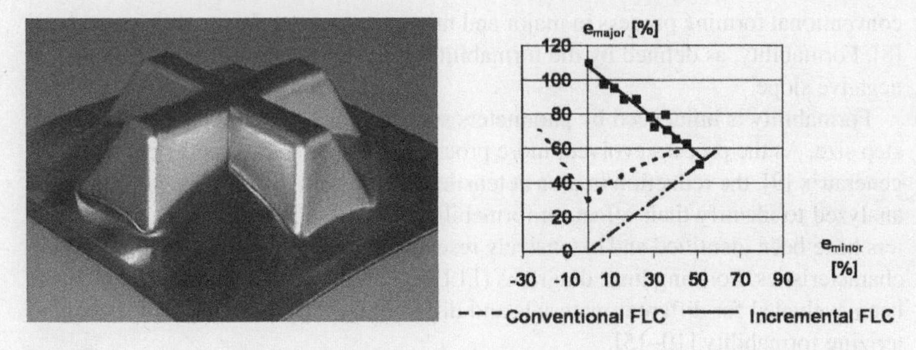

Figure 16.2 Part produced and FLD for ISF. Reprinted from [3] with permission from Elsevier.

the direct stresses are lower at the start of plastic deformation when effective stress is calculated, which, in turn, allows increased deformation and, thus, formability [12]. Silva et al. concluded that necking in SPIF is suppressed and FLDs constructed using necking strains are irrelevant in SPIF. A theoretical model based on ductile damage and membrane analysis was proposed to explain enhanced formability [18]. By taking into account the deformation mechanism coordinates of principal strain space, we obtained the FFL.

Emmens and van den Boogaard discussed every deformation mechanism, theoretical and experimental, that was found to enhance the formability of material in ISF. A total of six mechanisms were discussed, bending under tension, cyclic straining, contact stress, hydrostatic stress, shear, and geometrical inability to grow. In conclusion, the knowledge was found to be inadequate to assign enhanced formability mostly to only one kind of deformation mechanism [19].

Malhotra et al. stated that high local bending increases damage evolution which is greater than the enhanced formability effect produced by high levels of shear in SPIF. Even then, the plastic strain in SPIF is greater than the conventional forming, so greater shear does not fully explain the enhanced formability [20]. A noodle theory was proposed to explain the enhanced formability as a result of the local nature of deformation. Martínez-Donaire et al. studied the onset of ductile fracture due to stress triaxiality to have a better understanding of formability. Higher values of plastic strain and average stress triaxiality were obtained, explaining the enhanced formability [21].

16.4 EXPERIMENTAL AND NUMERICAL INVESTIGATIONS ON FORMING CHARACTERISTICS

ISF has certain limitations in terms of geometrical inaccuracies, surface finish, and formability. If volume consistency is considered, the sine law can be used to predict the final thickness of the sheet. But because of excessive thinning, there are limitations on the maximum wall angle, which limits the formability. Furthermore, because the deformation is not completely plastic, there is geometrical inaccuracy because of the plastic component of deformation. To address these limitations, various

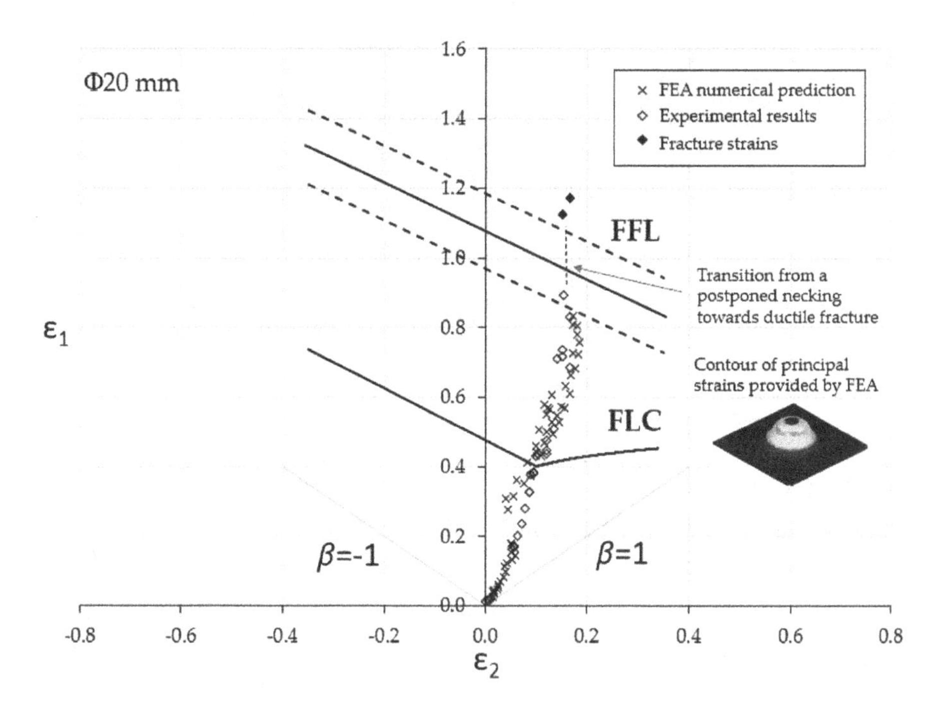

FIGURE 16.3 Numerical predictions versus experimental values of principal strains within the FLD [13].

experimental and numerical solutions have been researched throughout the history of the development of the process. Hirt et al. proposed a correction algorithm and multistage forming strategy to improve formability and geometrical accuracy with positive results [22] Figure 16.3.

Nguyen and Kim stated that to accurately estimate and understand the increased formability an understanding of the failure mechanism is necessary. They used a modified maximum force criterion to obtain FLD, which, in turn, was used to simulate the ISF process in order to compare the numerically estimated enhanced formability to the experimental value [23]. As it is already established that ISF results in increased formability, most of the simulation studies are based on comparing experimentally or theoretically obtained formability to simulated values.

Kurra and Regalla used thickness distribution as a predictor of formability and compared the simulation to the theoretical model and reported a better estimation of the actual thickness distribution [24]. To overcome excessive thinning, Mirnia and Shamsari proposed a finite element strategy using ductile fracture as the deformation mechanism. Fracture forming limit diagrams (FFLDs)in SPIF were constructed quite differently from the ones in the tension test. The reason for high formability was attributed to the severely nonproportional loading [25].

Said et al. used an elasto-plastic constitutive model with Hill yield criteria to simulate ISF. The efficiency of the finite element model to predict formability and deformation was evaluated by comparing it to the experimental data. Damage behavior

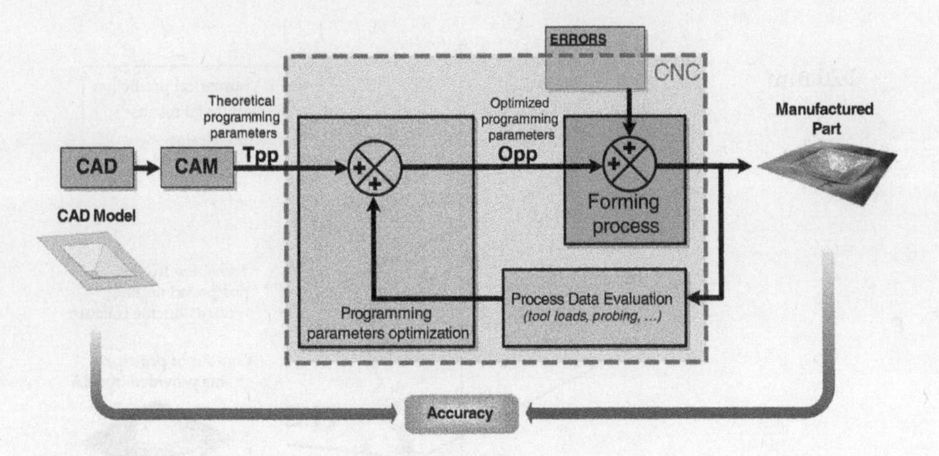

FIGURE 16.4 Intelligent computer-aided manufacturing approach.

Reprinted from [29] with permission from Elsevier.

was simulated considering the isotropic hardening model, and the value of the damage parameter was found to be increased. Still, the formability was found greater than the conventional forming process [26]. Centeno et al. presented a numerical model to analyze mechanisms involved in enhanced formability and future prediction of failure. Figure 16.4 shows that a good agreement was found between the numerically estimated and experimental values. A direct transition from stable plastic deformation was obtained using numerical simulation towards ductile fracture in the absence of necking.

Deokar, Jain, and Tandon focused on the formability of non-axis symmetric components using FLD. For plotting, FLD deformation produced on the sheet was simulated. FFL in ISF performed better when compared to deep drawing and stretch forming; hence, formability is increased [27]. Advancements in FEM software have made it possible to use FLD models for computational evaluation. Many such models have been used to compare the enhanced formability, fracture mechanisms, and geometrical accuracies in FEM simulations and experiments.

16.5 FORMING CHARACTERISTICS AS A FUNCTION OF TOOL PATH AND FORMING STRATEGIES

Despite all the advantages offered by ISF, its industrial use is still limited due to the already-stated limitations. To address these limitations, various tool paths and forming strategies have been suggested throughout the literature. T. J. Kim and Yang, in order to obtain a uniform final shape using ISF, proposed a double-pass forming method. An increase in formability was also reported. Bambach et al. identified the problem of inhomogeneous thickness distribution in worksheet with the conventional forming strategy in which the movement of CNC-controlled forming tool produces deformation layer by layer as the part is split into several two-dimensional layers [28]. Two new strategies, a contour and a radial strategy, were proposed, aiming for

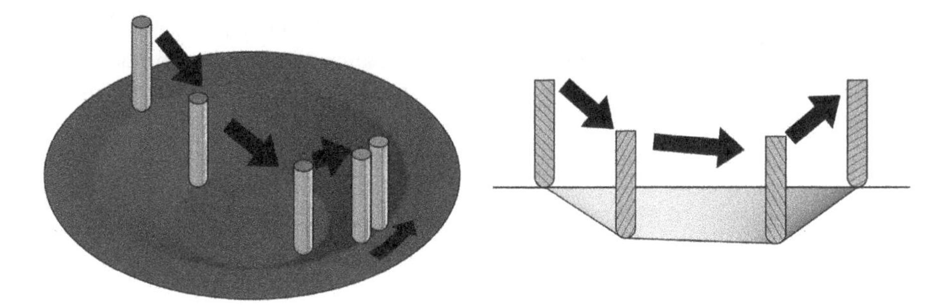

FIGURE 16.5 Isometric and profile view of radial toolpath [33].

a uniform thickness distribution. To identify an ideal strategy for computer-aided tool-path optimization FEA was used, and the thickness distribution for the proposed strategies was found better than the conventional forming strategies.

Rauch et al. listed sheet metal integrity and accuracy of the formed part as two major drawbacks and proposed a new approach (Figure 16.5), intelligent computer-aided manufacturing (ICAM), for tool-path control to overcome these drawbacks. For part-specific applications and correct tool-path programming compensations, the ICAM method was successfully carried out [29].

Liu et al., in 2014, proposed a multi-pass deformation strategy to avoid early forming failure and uniform thickness distribution. Later the formability was analyzed based on the outcome comparison of single-pass and multi-pass forming mechanisms [30]. Additionally, the quality of the final part was improved because of uniform thickness strain distribution [31].

In 2019, Wang et al., in an attempt to increase the formability, proposed an equal-diameter spiral tool-path strategy. The part manufactured by an equal diameter tool path showed enhanced formability regardless of the spiral direction when compared to the normal ISF [32]. The effect on surface quality was also reported by measuring surface roughness.

Grimm and Mears addressed the process limitation of low overall formability by presenting and experimentally verifying a novel tool path termed as radial toolpath as shown in Figure 16.6. Moving tool radially resulted in a consistent thickness of the

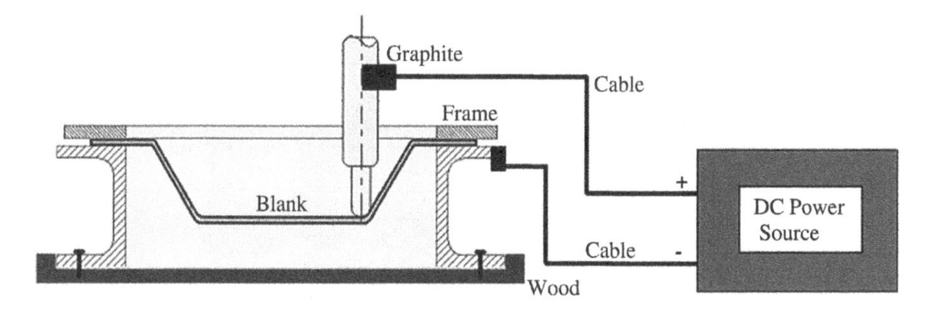

Figure 16.6 Heat-assisted incremental sheet forming. Reprinted from [35] with permission from Elsevier.

resulting part and a 20% increase in formability [33]. The unintentionally induced elastic deformation due to complex straining is a major disadvantage in traditional ISF. This is termed springback and is a major limitation in employing ISF in the industry environment. Various tool path and forming strategies are discussed here which may increase formability, surface conditions, and geometrical accuracy in SPIF.

16.6 IMPROVEMENT IN PROCESS CAPABILITIES BY PROCESS VARIATIONS

Several methods of tool path and forming strategies exist to improve forming characteristics. Still, it is quite difficult to achieve everything simultaneously by employing only these strategies. Research has proposed some modifications in the process or forming tool or external factors that coupled with forming strategies may increase the forming characteristics in the process. Modifications to the tool are required because of friction, which deteriorates the surface finish and other forming characteristics. Moreover, heat-assisted incremental forming, where heat is supplied either to the tool or the sheet metal to improve the ductility of the workpiece, has been found to increase the forming characteristics. Here, recently reported modifications in the process are being described by dividing them into two categories, heat-assisted modifications and process modifications.

16.6.1 Heat-Assisted Modifications

Duflou et al. increased the yield strength of sheet material and decreased the deforming forces using a dynamic temperature field localized to the area of contact of tool and workpiece set with the help of laser. The formability was significantly extended, and high dimensional accuracy was realized because of reduced springback [34]. Ambrogio, Filice, and Gagliardi used electric hot incremental forming (Figure 16.7)

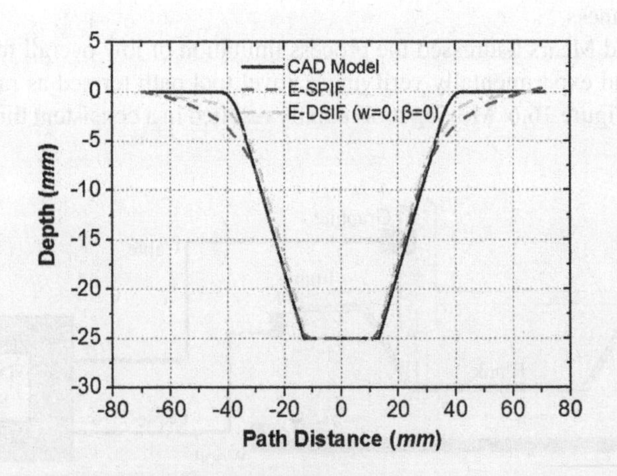

FIGURE 16.7 Geometrical accuracy comparison.

Reprinted from [38] with permission from Elsevier.

on lightweight aluminum alloys to specify a workability window. The formability was increased at the expense of the surface quality of the product [35]. Xu et al. compared friction stir and electric heat-assisted ISF to discuss the limitations and adaptability of the processes. For higher formability, electric hot-assisted ISF was concluded to be better [36].

Mohammadi et al. investigated the potential of laser-assisted ISF in improving the formability. Based on simulation and experimentation, different strategies for laser positioning were proposed and accuracy was measured [37]. D. K. Xu et al. compared electrically assisted SPIF with electrically assisted double-side incremental forming (EDSIF) in terms of geometrical accuracy and surface finish. The surface topography of both processes was compared, with EDSIF yielding better results [38]. The comparison in geometrical accuracies is shown in Figure 16.7.

Vahdani et al. evaluated the formability improvement of Ti-6Al-4V alloy using EISF and reported an 18% improvement in thickness distribution [39]. Xiao et al. performed ISF on AA7075 alloy by heating it at different temperatures and reported a remarkable increase in formability with temperature [40]. Zhang et al., in their 2020 research to overcome the non-uniform heating issue, proposed a novel warm ISF method. An oil bath was used to uniformly heat the hard to form Mg alloy sheet, thus overcoming the drawbacks of local heat accumulation present in simple heat-assisted ISF. The parametric approach reported temperature as the most influential parameter in increasing the formability [41].

16.6.2 Process Modification

The inherent process limitations in ISF have led to the development of modifications in the process to improve forming characteristics. Iseki developed water jet incremental forming (WJISF) to improve surface finish as it removed the need to use the tool [42]. Jurisevic, Kuzman, and Junkar 2006, found that WJISF has great potential in terms of surface finish but improvements in forming accuracy, time, and energy consumption are required [43]. Li et al. proposed a theoretical model based on WJISF based on work-energy theorem [44] to improve forming accuracy by controlling the water jet pressure. Malhotra et al., in 2011, developed double-side incremental forming (DSIF), in which a top and a bottom tool are used and the sheet is formed by squeezing it between the two tools. The deformation in DSIF stabilizes faster into a localized zone than in SPIF. Thus, geometrical accuracy was reported to be better [45]. Figure 16.8 represents a simple schematics diagram of the process. Smith et al. also used a supporting tool superimposing pressure in the forming zone which theoretically could lead to higher strain values and better geometrical accuracy. A partial increase in accuracy and increase in maximum formability was reported [46]. Smith et al. used accumulative DSIF, which removed the need for a shape-specific tool and explained the increased formability to SPIF. Ai, Dai, and Long, to study the increased formability in DSIF, used a new tension under the cyclic bending testing method [47].

P. Li et al., in 2017, introduced the advantages of ultrasonic vibrations in plastic forming to SPIF. The tool was required to vibrate and rotate as well. On visual inspection, it was confirmed that the surface quality increased [49]. Do et al. improved

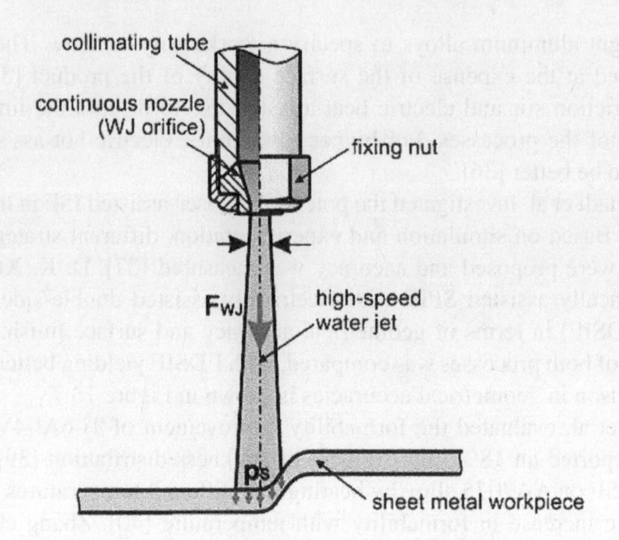

FIGURE 16.8 Schematics of double-side incremental forming [48].

the formability of the process by cutting holes on the shoulder area of the blank. The hole lancing increased the achievable forming angle considerably and the profile error was also reduced [50]. Araghi et al. combined ISF and stretch forming and termed it hybrid forming. The process has true potential as a uniform thickness distribution was reported and maximum thinning was also reduced. Lora et al. introduced a hybrid forming approach by performing using conventional forming and then using ISF. The pre-strains influenced the maximum deformations increasing the total formability [51].

16.7 CONCLUSION

The literature includes an investigation into the recent advances in the ISF process that may affect the forming characteristics. Geometrical accuracy, surface finish. and formability were the most important forming characteristics discussed in this study. Following conclusions were made regarding the study:

- FFL and FLD have been successfully used to interpret formability. Different deformation mechanisms have been used to access formability but to single out a single one as the reason for enhanced formability is not possible. Several process parameters, forming conditions, and material properties work hand in hand with deformation criteria in increasing the formability of ISF process.
- A reasonable prediction regarding deformation, surface quality, strain distribution, and formability are provided by numerical simulation. Also, numerical estimation of parameters in ISF allows checking the efficacy of mathematical models and comparison to actual experimental data. But the models are limited in their functionality because of computational

ineffectiveness. The cost associated and time involved is also high so a need to do sophisticated modeling is identified to reproduce the experimental procedure specifically.

- Tooling and forming strategies have significantly improved the forming characteristics. Hybrid ISF technology can be considered more of a forming strategy than a process modification. These strategies function well within predefined constants or a range of process parameters, thus limiting the overall applicability of the process. An application-based strategy would benefit in introducing the process to industrial applications.
- Introducing heat definitely eases the deformation of material and overall formability, which has been proved experimentally as reported in the literature. The problem lies in the overall equipment, which becomes complex. It is difficult to control the temperature using simple construction. Lasers are easier to control, but the process becomes costly and adds variables to the process, making it difficult to find optimized working conditions.
- Process modifications, like using slave and master tools in DSIF, have proved to increase geometrical accuracy as uniform thickness distribution is obtained. But it changes the deformation mechanisms and introduces unknown variables to the process and limited literature is available to make any useful comment.

To conclude, despite significant advancement in the ISF process, the industrial feasibility and applicability of the process are still obscure due to process limitations. The work aimed at identifying the forming characteristics that play important role in the industry environment and reporting any significant advancements aimed at improving those characteristics.

REFERENCES

1. S. Ai and H. Long, "A review on material fracture mechanism in incremental sheet forming," *Int. J. Adv. Manuf. Technol.*, vol. 104, no. 1–4, pp. 33–61, 2019. doi:10.1007/s00170-019-03682-6.
2. T. J. Kim and D. Y. Yang, "Improvement of formability for the incremental sheet metal forming process," *Int. J. Mech. Sci.*, vol. 42, no. 7, pp. 1271–1286, 2000. doi:10.1016/S0020-7403(99)00047-8.
3. L. Filice, L. Fratini, and F. Micari, "Analysis of material formability in incremental forming," *CIRP Ann. - Manuf. Technol.*, vol. 51, no. 1, pp. 199–202, 2002. doi:10.1016/S0007-8506(07)61499-1.
4. J. Jeswiet and D. Young, "Forming limit diagrams for single-point incremental forming of aluminium sheet," *Proc. Inst. Mech. Eng. Part B J. Eng. Manuf.*, vol. 219, no. 4, pp. 359–364, 2005. doi:10.1243/095440505X32210.
5. G. Ambrogio, V. Cozza, L. Filice, and F. Micari, "An analytical model for improving precision in single point incremental forming," *J. Mater. Process. Technol.*, vol. 191, no. 1–3, pp. 92–95, 2007. doi:10.1016/j.jmatprotec.2007.03.079.
6. A. Taherkhani, A. Basti, N. Nariman-Zadeh, and A. Jamali, "Achieving maximum dimensional accuracy and surface quality at the shortest possible time in single-point

incremental forming via multi-objective optimization," *Proc. Inst. Mech. Eng. Part B J. Eng. Manuf.*, vol. 233, no. 3, pp. 900–913, 2019. doi:10.1177/0954405418755822.
7. A. Attanasio, E. Ceretti, A. Fiorentino, L. Mazzoni, and C. Giardini, "Experimental tests to study feasibility and formability in incremental forming process," *Key Eng. Mater.*, vol. 410–411, 2009. doi:10.4028/www.scientific.net/KEM.410-411.391.
8. Y. H. Kim and J. J. Park, "Effect of process parameters on formability in incremental forming of sheet metal," *J. Mater. Process. Technol.*, vol. 130, no. 131, pp. 42–46, 2002. doi:10.1016/S0924-0136(02)00788-4.
9. G. Hussain, L. Gao, N. Hayat, and L. Qijian, "The effect of variation in the curvature of part on the formability in incremental forming: An experimental investigation," *Int. J. Mach. Tools Manuf.*, vol. 47, no. 14, pp. 2177–2181, 2007. doi:10.1016/j.ijmachtools.2007.05.001.
10. G. Hussain, L. Gao, N. Hayat, and N. U. Dar, "The formability of annealed and pre-aged AA-2024 sheets in single-point incremental forming," *Int. J. Adv. Manuf. Technol.*, vol. 46, no. 5–8, pp. 543–549, 2010. doi:10.1007/s00170-009-2120-x.
11. J. Jeswiet, F. Micari, G. Hirt, A. Bramley, J. Duflou, and J. Allwood, "Asymmetric single point incremental forming of sheet metal," *CIRP Ann. - Manuf. Technol.*, vol. 54, no. 2, pp. 88–114, 2005. doi:10.1016/s0007-8506(07)60021-3.
12. J. M. Allwood, D. R. Shouler, and A. E. Tekkaya, "The increased forming limits of incremental sheet forming processes," *Key Eng. Mater.*, vol. 344, pp. 621–628, 2007. doi:10.4028/www.scientific.net/kem.344.621.
13. G. Centeno, A. J. Martínez-Donaire, I. Bagudanch, D. Morales-Palma, M. L. Garcia-Romeu, and C. Vallellano, "Revisiting formability and failure of AISI304 sheets in SPIF: Experimental approach and numerical validation," *Metals (Basel)*, vol. 7, no. 12, pp. 1–14, 2017. doi:10.3390/met7120531.
14. K. Jawale, J. F. Duarte, A. Reis, and M. B. Silva, "Characterizing fracture forming limit and shear fracture forming limit for sheet metals," *J. Mater. Process. Technol.*, vol. 255, no. 2010, pp. 886–897, 2018. doi:10.1016/j.jmatprotec.2018.01.035.
15. S. Gatea, D. Xu, H. Ou, and G. McCartney, "Evaluation of formability and fracture of pure titanium in incremental sheet forming," *Int. J. Adv. Manuf. Technol.*, vol. 95, no. 1–4, pp. 625–641, 2018. doi:10.1007/s00170-017-1195-z.
16. B. T. Araghi, G. L. Manco, M. Bambach, and G. Hirt, "Investigation into a new hybrid forming process: Incremental sheet forming combined with stretch forming," *CIRP Ann. - Manuf. Technol.*, vol. 58, no. 1, pp. 225–228, 2009. doi:10.1016/j.cirp.2009.03.101.
17. L. Fratini, G. Ambrogio, R. Di Lorenzo, L. Filice, and F. Micari, "Influence of mechanical properties of the sheet material on formability in single point incremental forming applied to manufacturing of biocompatible polymer prostheses forming," *CIRP Ann. - Manuf. Technol.*, vol. 53, no. 1, pp. 207–210, 2004.
18. M. B. Silva, M. Skjoedr, A. G. Atkins, N. Bay, and P. A. F. Martins, "Single-point incremental forming and formability-failure diagrams," *J. Strain Anal. Eng. Des.*, vol. 43, no. 1, pp. 15–35, 2008. doi:10.1243/03093247JSA340.
19. W. C. Emmens and A. H. van den Boogaard, "An overview of stabilizing deformation mechanisms in incremental sheet forming," *J. Mater. Process. Technol.*, vol. 209, no. 8, pp. 3688–3695, 2009. doi:10.1016/j.jmatprotec.2008.10.003.
20. R. Malhotra, L. Xue, T. Belytschko, and J. Cao, "Mechanics of fracture in single point incremental forming," *J. Mater. Process. Technol.*, vol. 212, no. 7, pp. 1573–1590, 2012. doi:10.1016/j.jmatprotec.2012.02.021.
21. A. J. Martínez-Donaire, M. Borrego, D. Morales-Palma, G. Centeno, and C. Vallellano, "Analysis of the influence of stress triaxiality on formability of hole-flanging by single-stage SPIF," *Int. J. Mech. Sci.*, vol. 151, pp. 76–84, 2019. doi:10.1016/j.ijmecsci.2018.11.006.

22. G. Hirt, J. Ames, M. Bambach, R. Kopp, and R. Kopp, "Forming strategies and process modelling for CNC incremental sheet forming," *CIRP Ann. - Manuf. Technol.*, vol. 53, no. 1, pp. 203–206, 2004. doi:10.1016/S0007-8506(07)60679-9.
23. D. Nguyen and Y. Kim, "A Numerical Study on Establishing the Forming Limit Curve and Indicating the Formability of Complex Shape in Incremental Sheet Forming Process," *Int. J. Precis. Eng.*, vol. 14, no. 12, pp. 2087–2093, 2013. doi:10.1007/s12541-013-0283-8.
24. S. Kurra and S. P. Regalla, "Experimental and numerical studies on formability of extra-deep drawing steel in incremental sheet metal forming," *J. Mater. Res. Technol.*, vol. 3, no. 2, pp. 158–171, 2014. doi:10.1016/j.jmrt.2014.03.009.
25. M. J. Mirnia and M. Shamsari, "Numerical prediction of failure in single point incremental forming using a phenomenological ductile fracture criterion," *J. Mater. Process. Tech.*, 2017. doi:10.1016/j.jmatprotec.2017.01.029.
26. L. Ben Said, J. Mars, M. Wali, and F. Dammak, "Numerical prediction of the ductile damage in single point incremental forming process," *Int. J. Mech. Sci.*, vol. 131–132, pp. 546–558, 2017. doi:10.1016/j.ijmecsci.2017.08.026.
27. S. Deokar, P. K. Jain, and P. Tandon, "Formability assessment in single point incremental sheet forming through finite element analysis," *Mater. Today Proc.*, vol. 5, no. 11, pp. 25430–25439, 2018. doi:10.1016/j.matpr.2018.10.348.
28. M. Bambach, M. Cannamela, M. Azaouzi, G. Hirt, and J. L. Batoz, "Computer-aided tool path optimization for single point incremental sheet forming," in *Advanced Methods in Material Forming: With 264 Figures and 37 Tables*, 2007. doi: 10.1007/3-540-69845-0_14.
29. M. Rauch, J. Y. Hascoet, J. C. Hamann, and Y. Plenel, "Tool path programming optimization for incremental sheet forming applications," *CAD Comput. Aided Des.*, vol. 41, no. 12, pp. 877–885, 2009. doi:10.1016/j.cad.2009.06.006.
30. Z. Liu, W. J. T. Daniel, Y. Li, S. Liu, and P. A. Meehan, "Multi-pass deformation design for incremental sheet forming: Analytical modeling, finite element analysis and experimental validation," *J. Mater. Process. Technol.*, vol. 214, no. 3, pp. 620–634, 2014. doi:10.1016/j.jmatprotec.2013.11.010.
31. Z. Liu, Y. Li, and P. A. Meehan, "Tool path strategies and deformation analysis in multi-pass incremental sheet forming process," *Int. J. Adv. Manuf. Technol.*, vol. 75, no. 1–4, pp. 395–409, 2014. doi:10.1007/s00170-014-6143-6.
32. J. Wang, L. Li, P. Zhou, X. Wang, and S. Sun, "Improving formability of sheet metals in incremental forming by equal diameter spiral tool path," *Int. J. Adv. Manuf. Technol.*, vol. 101, no. 1–4, pp. 225–234, 2019. doi:10.1007/s00170-018-2911-z.
33. T. J. Grimm and L. Mears, "Investigation of a radial toolpath in single point incremental forming," *Procedia Manuf.*, vol. 48, pp. 215–222, 2020. doi:10.1016/j.promfg.2020.05.040.
34. J. R. Duflou, B. Callebaut, J. Verbert, and H. De Baerdemaeker, "Laser assisted incremental forming: Formability and accuracy improvement," *CIRP Ann. - Manuf. Technol.*, vol. 56, no. 1, pp. 273–276, 2007. doi:10.1016/j.cirp.2007.05.063.
35. G. Ambrogio, L. Filice, and F. Gagliardi, "Formability of lightweight alloys by hot incremental sheet forming," *Mater. Des.*, vol. 34, pp. 501–508, 2012. doi:10.1016/j.matdes.2011.08.024.
36. D. Xu, B. Lu, T. Cao, J. Chen, H. Long, and J. Cao, "A comparative study on process potentials for frictional stir- and electric hot-assisted incremental sheet forming," *Procedia Eng.*, vol. 81, no. October, pp. 2324–2329, 2014. doi:10.1016/j.proeng.2014.10.328.
37. A. Mohammadi, H. Vanhove, A. Van Bael, and J. R. Duflou, "Towards accuracy improvement in single point incremental forming of shallow parts formed under laser

assisted conditions," *Int. J. Mater. Form.*, vol. 9, no. 3, pp. 339–351, 2016. doi:10.1007/s12289-014-1203-x.
38. D. K. Xu et al., "Enhancement of process capabilities in electrically-assisted double sided incremental forming," *Mat. Des.*, vol. 92, pp. 268–280, 2016. doi:10.1016/j.matdes.2015.12.009.
39. M. Vahdani, M. J. Mirnia, M. Bakhshi-Jooybari, and H. Gorji, "Electric hot incremental sheet forming of Ti-6Al-4V titanium, AA6061 aluminum, and DC01 steel sheets," *Int. J. Adv. Manuf. Technol.*, vol. 103, no. 1–4, pp. 1199–1209, 2019. doi:10.1007/s00170-019-03624-2.
40. X. Xiao, C. Il Kim, X. D. Lv, T. S. Hwang, and Y. S. Kim, "Formability and forming force in incremental sheet forming of AA7075-T6 at different temperatures," *J. Mech. Sci. Technol.*, vol. 33, no. 8, pp. 3795–3802, 2019. doi:10.1007/s12206-019-0722-2.
41. S. Zhang, G. H. Tang, W. Wang, and X. Jiang, "Evaluation and optimization on the formability of an AZ31B Mg alloy during warm incremental sheet forming assisted with oil bath heating," *Meas. J. Int. Meas. Confed.*, vol. 157, p. 107673, 2020. doi:10.1016/j.measurement.2020.107673.
42. H. Iseki, "Flexible and Incremental Bulging of Sheet Metal using High Speed Water Jet," Nihon Kikai Gakkai Ronbunshu, C Hen/Transactions of the Japan Society of Mechanical Engineers, Part C, vol. 65, no. 635. pp. 2912–2918, 1999. doi:10.1299/kikaic.65.2912.
43. B. Jurisevic, K. Kuzman, and M. Junkar, "Water jetting technology: An alternative in incremental sheet metal forming," *Int. J. Adv. Manuf. Technol.*, vol. 31, no. 1–2, pp. 18–23, Oct. 2006. doi:10.1007/s00170-005-0176-9.
44. J. Li, K. He, S. Wei, X. Dang, and R. Du, "Modeling and experimental validation for truncated cone parts forming based on water jet incremental sheet metal forming," *Int. J. Adv. Manuf. Technol.*, vol. 75, no. 9–12, pp. 1691–1699, 2014. doi:10.1007/s00170-014-6222-8.
45. R. Malhotra, J. Cao, F. Ren, V. Kiridena, Z. Cedric Xia, and N. V. Reddy, "Improvement of geometric accuracy in incremental forming by using a squeezing toolpath strategy with two forming tools," *J. Manuf. Sci. Eng.*, vol. 133, no. 6, pp. 1–9, December 2011. doi:10.1115/1.4005179.
46. J. Smith, R. Malhotra, W. K. Liu, and J. Cao, "Deformation mechanics in single-point and accumulative double-sided incremental forming," *Int. J. Adv. Manuf. Technol.*, vol. 69, no. 5–8, pp. 1185–1201, 2013. doi:10.1007/s00170-013-5053-3.
47. S. Ai, R. Dai, and H. Long, "Investigating formability enhancement in double side incremental forming by developing a new test method of tension under cyclic bending and compression," *J. Mater. Process. Technol.*, vol. 275, no. July 2019, p. 116349, 2020. doi:10.1016/j.jmatprotec.2019.116349.
48. D. Möllensiep, T. Gorlas, P. Kulessa, and B. Kuhlenkötter, "Real-time stiffness compensation and force control of cooperating robots in robot-based double sided incremental sheet forming," *Prod. Eng.*, no. 0123456789, 2021. doi:10.1007/s11740-021-01052-4.
49. P. Li et al., "Evaluation of forming forces in ultrasonic incremental sheet metal forming," *Aerosp. Sci. Technol.*, vol. 63, no. December 2016, pp. 132–139, 2017. doi:10.1016/j.ast.2016.12.028.
50. V. C. Do, D. C. Ahn, and Y. S. Kim, "Formability and effect of hole bridge in the single point incremental forming," *Int. J. Precis. Eng. Manuf.*, vol. 18, no. 3, pp. 453–460, 2017. doi:10.1007/s12541-017-0054-z.
51. F. A. Lora, D. Fritzen, R. Alves de Sousa, and L. Schaffer, "Studying formability limits by combining conventional and incremental sheet forming process," *Chin. J. Mech. Eng.* (English Ed.), vol. 34, no. 1, 2021. doi:10.1186/s10033-021-00562-7.

17 Deformation Mechanism of Polymers, Metals, and Their Composites in Dieless Forming Operations

Saurabh Thakur and Parnika Shrivastava
National Institute of Technology, Hamirpur, India

CONTENTS

17.1 INTRODUCTION

The cost associated with conventional sheet forming operations, employing dies and punches, makes such setups suitable only for mass production. The incremental sheet forming (ISF) process, in which a tool that is computer numerical control (CNC)–controlled, has been found ideal for rapid prototyping and producing customized parts in small batches. Forming processes such as spinning, flow forming, shear forming, and arguably rolling can be characterized as incremental forming processes. The primary difference in these processes is in the thickness of the part produced [1]. ISF, although not a direct replacement of the traditional manufacturing process, has been established as a high potential process for rapid manufacturing, low volume production, and certain key applications. The material formability is also enhanced in metal sheets compared to traditional sheet forming processes. A local deformation takes place in ISF; that is, at a given time, only a small location at is the product is actually being formed. The local deformation moves over the entire product and the final shape is formed [2].

DOI: 10.1201/9781003284574-19

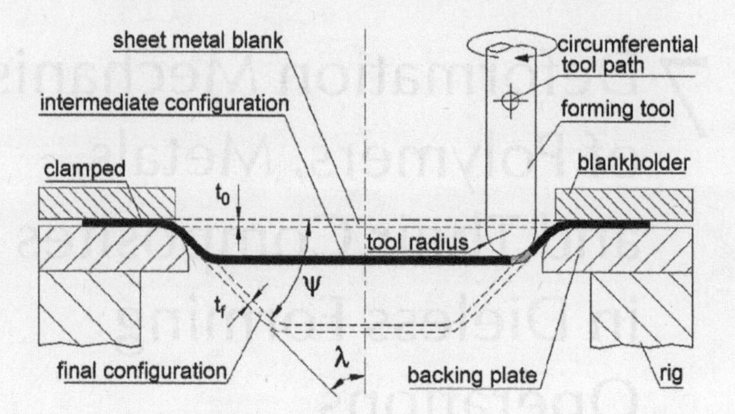

FIGURE 17.1 Schematic representation of single-point incremental sheet forming.

Reprinted from [4] with permission from Elsevier.

Single-point ISF (SPIF) is one of the ISF processes for sheet metal manufacturing that has gained much popularity in recent years. In SPIF, the sheet metal part is formed in a stepwise manner incrementally over a sheet clamped peripherally. The tool used is a CNC-controlled tool and generally is in a hemispherical shape (Figure 17.1). The process removes the need for a die to operate as in traditional forming processes, so the degree of flexibility is higher. The cost of tooling is reduced, making it cheaper, and SPIF is relatively fast for small production quantities.

A major disadvantage of ISF, when compared to processes such as stamping and deep drawing, is the time involved in the manufacturing, which is greater when compared to these traditional forming processes. Also, steep wall angles are difficult to form, and multiple passes are required to produce a part. Some applications of the ISF process include manufacturing prototypes. A great number of advantages have in explored and reported in batch manufacturing systems in which complex components can be produced in small batches in medical and aeronautical fields [3]. SPIF is one of the very few processes in which the advantage of solving even the simplest problem experimentally outweighs theoretical and numerical analysis [4]. Because of its rapid and economic ways and shorter lead and setup times, the process has already found its way in automobile industries.

To take advantage of this high formability and other benefits of the process it is necessary to understand the deformation mechanism. Several parametric approaches have tried to maximize the limits through manipulating parameters. Kim and Yang showed that the increase in formability is deformation occurring mostly by shear deformation, making it possible to predict the final thickness of the product by using a simple trigonometric relationship [5]. Over time, various process parameters have been studied, and their effect on formability was also investigated by performing a series of forming tests or modeling through finite element methods (FEM) After all these years, there are still contradictions regarding the deformation mechanism in ISF. Furthermore, the enhanced material advancements and process capabilities have led to the numerous modifications in the process, and ISF has also been extended to be applied on polymers and composites. The aim of this study is to report each deformation mechanisms suggested by the literature for ISF of sheet metals and then compare it to the deformation mechanisms when ISF is applied to polymers or composites.

17.2 DEFORMATION MECHANISM IN METALS AND METAL ALLOYS

The deformation in ISF has always been a subject of controversy. Some researchers believe that the governing mechanics for the deformation is shearing, and some are of the opinion that it takes place due to stretching. All have sufficient research work backing their respective claims. Two approaches have commonly been used for the analysis of the deformation mechanism, numerical and experimental. The experimental method involves the calculation of strains, and the numerical approach generally involves estimations using finite element methods. Although sometimes contradictory, the available work has provided valuable insights and this review tries to consolidate all that work to have a clear understanding into the governing mechanism of deformation in ISF.

In the earlier development of the mechanics of the ISF process was assumed to be similar to the conventional spinning process, that is, shear-dominant deformation. Kim and Yang developed a formulation of the deformation mechanism and used FEM to find the final thickness strain [5]. It was assumed, as an engineering approximation, that all deformation occurred by only shear deformation. Iseki proposed a plane strain model (Figure 17.2) for deformation and used FEM to calculate the deformation of a shell bulged incrementally, assuming that the sheet metal in contact with the tool stretches uniformly. Both calculations were found to be in reasonable agreement validating the plane strain model [6]. Shim and Park carried an

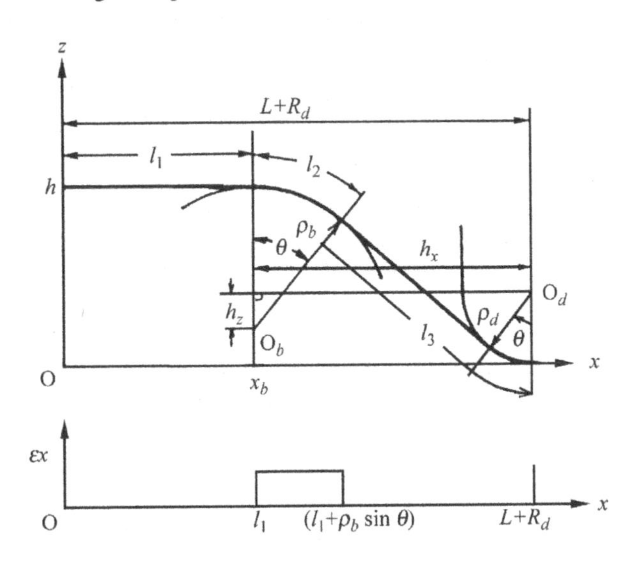

FIGURE 17.2 Plane-strain deformation model.

Reprinted from [6] with permission from Elsevier.

investigation of the forming limit curve by imposing various strain paths and measuring major and minor strains. Near equi-biaxial stretching was reported to occur at the corners and near-plane-strain stretching along the side [7].

Several forming strategies have also been proposed and compared with previously used linear bending methods. This was done with the hope of overcoming the existing limitations to the process, but the deformation mode, again, was assumed to be very close to plane strain [8]. Filice, Fratini, and Micari, while measuring material formability obtained a forming limit curve quite different from a traditional one. A wide range of typical straining conditions were taken into consideration and ISF was characterized by local stretching deformation mechanics [9]. Young and Jeswiet describes ISF as a constant volume process and used sine law to predict the final thickness. The thickness of the sheet for different wall-angled cones was measured, and a conclusion was drawn that, primarily, deformation was because of shear after an initial bending deformation [10].

A detailed study of the process from its genesis to conceptualization to its recent practical utilization was done by Jeswiet et al. The work provided a design and manufacturing guideline by dwelling into every aspect of the process. The effect of the increment step size and the draw angle, accompanied by formability, force, plasticity, and springback analyses, was studied [11]. The formability mechanism, for previous and subsequent collaborated studies [10, 12–14] for analyzing wall thickness variations or constructing forming limit diagrams, was assumed to be primarily shear and a constant volume law; that is, the sine law was favored when determining the final thickness of the material.

Wilko C. Emmens and van den Boogaard discussed the shear-forming mechanism in ISF and explored the possibility of stretching. The shear-deformation mechanism had its roots in conventional spinning processes, and there was no direct experimental evidence. Experimental evidence showed that the constantly changing strain path condition leads to extremely complex material behavior and that the assumption that the forming processes were supposed to act under forming by shear is not true in its entirety [15]. J. M. Allwood, Shouler, and Tekkaya stated that the possibility of sine law and deformation primarily based on vertical shear being wrong and reported a significant through-thickness shear acting parallel to the tool path. Further research explored the forming limits of ISF by considering any proportional loading, including six components of symmetric stress tensors [16]. Previous research was extended by proposing new generalized forming limit diagram (FLD) based on observation of through-thickness shear while the old FLDs were developed assuming that deformation occurs in plane stress [17].

Silva, Skjoedt, et al. also stated the obvious confusion in deciding the governing deformation mechanism in ISF and, to help understand the fundamentals better, proposed a theoretical model for SPIF based on membrane analysis [18]. Further research [4, 19] to investigate the formability limits of SPIF assumed stretching to be the principal mode of deformation. Jackson and Allwood experimentally observed the deformation mechanism in specially prepared copper sheets in the form of truncated cones. The measurements showed that the deformation mechanism was a combination of bending, stretching, and shear. In the direction of the tool, the greatest strain component for the ISF process was shear, and both stretching and shear of similar magnitudes were observed perpendicular to the tool direction [20]. Bambach proposed a new model for the kinematics of the ISF process, assuming that the deformation between intermediate shapes proceeds by displacements along the surface normal of the current shape. The thickness estimates using a new model were better than the estimates derived from the sine law. The new model worked exceptionally well for strains parallel to the tool direction in the nonflat parts of the sheet [21].

Major research interest includes developing a general analytical model for deformation that is capable of predicting strain distribution for arbitrary geometries. For simplicity, analysts have modeled deformation on the basis of plane strain while equi-biaxial state has been modeled only by a few. Cui et al. developed a mathematical model of deformation as an extension of the sine law for any part shape, assuming that the material only undergoes pure stretching [22]. An open-loop multistage deformation pass model to achieve a uniform thickness distribution has been developed only by considering deformation due to shear [23]. Several finite element analysis (FEA) models have also been developed after the advances in computational methods and speeds, even though finite modeling still remains a challenging task. Although the contribution of shearing, bending, and stretching varied quantitatively, the results were sufficiently supported to conclude that the deformation mechanism was a combination of all three [24]. Ai et al. concluded that bending is a major mode of deformation in ISF [25]. A new analytical model was presented by focusing on the deformation stability and its effect on the metal sheet fracture and critical strain of deformation instability were obtained [26].

Some of the recent approaches try to quantify the contribution of every type of deformation mechanism in the SPIF process. This can be done by splitting the plastic energy dissipation into contribution made by membrane stretching, bending, and through-thickness shear deformation [27]. This revealed that dominant deformation mode depends on selected process variables, such as tool diameter, toll stepdown, blank thickness, and friction. Yazar et al. studied the deformation mechanism by observing changes in the microstructure and lattice rotation and found that deformation is majorly plane strain with a contribution of through-thickness shear, whose magnitude strictly depends on vertical step size and tool diameter [28].

The results of the literature suggest that the deformation mechanism in the ISF process is a complex combination of bending, stretching, and shearing. Recent research has focused on the analytical and numerical modeling of the process and established geometrical accuracies as a direct consequence of the dominant deformation mechanism.

17.3 ISF OF POLYMERS AND COMPOSITES: FEASIBILITY AND DEFORMATION MECHANISMS

Conventionally, the process was limited to the forming of sheet metals and alloys. However, for broader industrial use, hybrid ISF processes have been developed, and the scope of the process has increased from metals to composites and polymers. This was only in the first decade of the 20th century that the ISF of polymer sheets was made possible. With the advances in the process, a substantial number of studies have been carried out to successfully form polymers and metal composites.

17.3.1 ISF Studies in Polymers

Polymers are gaining huge popularity in manufacturing owing to their advantages over metal counterparts. They have a high strength-to-mass ratio and are easily formable under elevated temperatures. Their thermal properties, impact resistance, and bearing capabilities are also comparable to metals and alloys. In processing polymers, temperature plays a very important role as shaping is done using by first heating to a rubbery state and then cooling to form predefined shapes. This process is energy-intensive, making it suitable only for mass production. ISF process has been identified to work exceptionally well in small-batch manufacturing scenarios and rapid product development.

Franzen et al. focused on the possibility of cold-forming commercial PVC using SPIF. The wall angles obtained indicated very high formability and possibility of forming a wide range of geometries. An analysis of the failure modes suggested the presence of shear and bending [29]. Martins et al. formed five different thermoplastic materials using ISF. The formability and accuracy of the parts produced were assessed and forming was supposed to be performed under a plane-strain condition [30]. Silva, Alves, and Martins used membrane analysis [31] to model the deformation in polymers and checked the efficacy of the model. The three modes of failure in polymers were analyzed, and the deformations associated with them were found to be stretching and shear. The feasibility of the process was evaluated and found to be achievable even under room temperature. The schematic diagram of membrane analysis, showing all the stresses at the contact area between tool and sheet, is shown in Figure 17.3.

Sy and Nam used a modified viscoelasticity model to simulate deformation for the estimation of geometrical accuracy and thickness distribution. The experimental results were in good agreement with the numerical estimation [33]. Alkas Yonan et al. reported that a through-thickness shear mechanism was not analyzed in thermoplastics and that it is a major reason for increased forming limits in ISF of metals. The work used an extension of the visco-plastic model to simulate the ISF of thermoplastics with consideration to through-thickness shear [34]. Alkas Yonan et al. for analyzing stress fields for high values of strain proposed a methodology for categorizing plastic flow and failure. Equations relating in-plane strain increments and stresses applied were constituted, assuming deformation to be isotropic and under a plane-strain condition [35].

Bagudanch et al. used optical microscopy to carry out fractographic analysis to understand the deformation of polymers better. The formability of biocompatible polymers was also studied [36]. In 2018, Durante, Formisano, and Lambiase researched the effect of tool type and tool path on the forming characteristics of PVC. The three modes of failure as already stated by previous researchers were taken for explaining deformation and failure [37]. Durante, Formisano, and Lambiase later, in 2019, found that wrinkling affected the surface quality of polymers and stated that the analyzing techniques for formability in metals and polymers should be different [38].

Major studies have been performed to understand the failure mode of polymers in ISF. Three modes, along the interface between the corner and the inclination of the

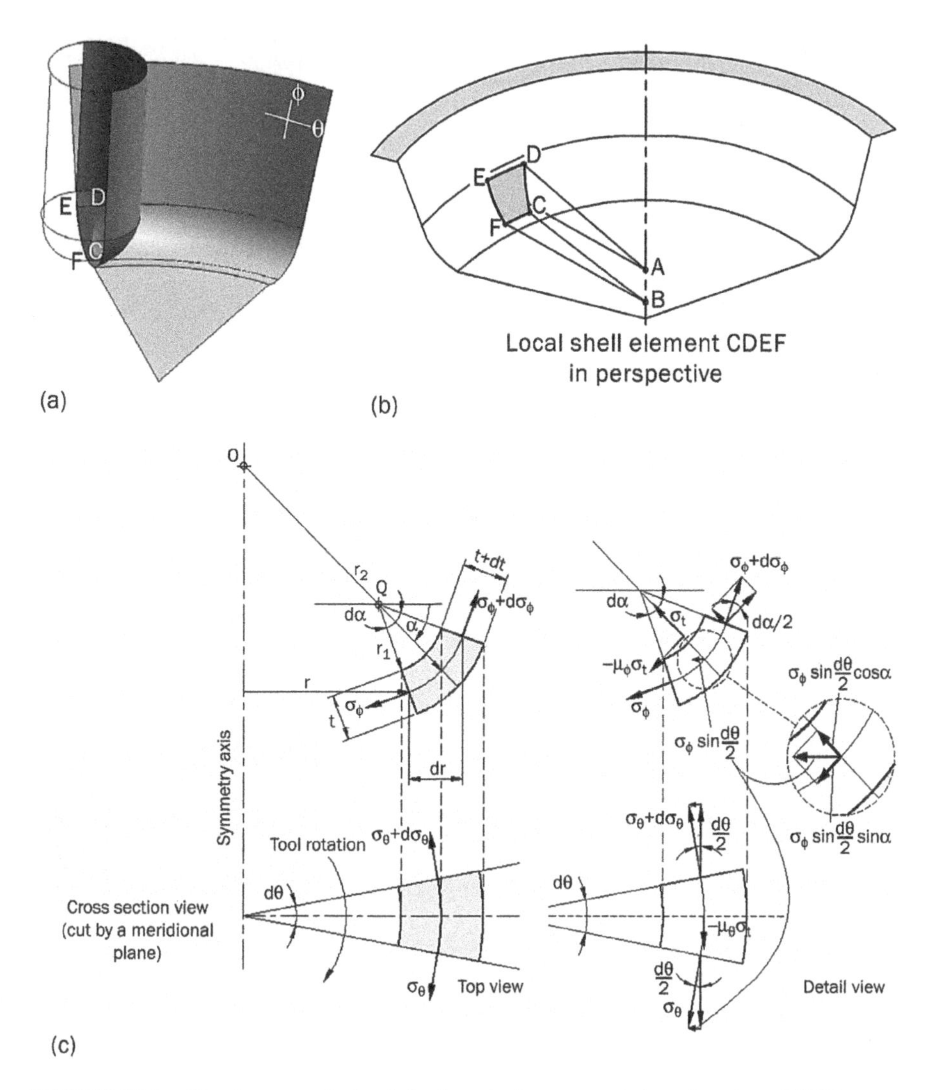

FIGURE 17.3 Membrane analysis of single-point incremental sheet forming of PVC.

Reprinted from [32] with permission from Elsevier.

wall and the direction along the circumference, have been identified, and the deformation mechanism leading to these three failure modes are discussed: Mode 1: crack along the circumferential direction, Mode 2: wrinkles along the inclined wall of a part, and Mode 3: oblique crack on the inclined wall of a part (Figures 17.4 and 17.5). Initial research on the deformation mechanism focused on Mode 1 of failure, assuming deformation under the plane-strain condition. Later, biaxial stretching and shear mechanisms were also accepted as deforming mechanisms in the ISF process of polymers.

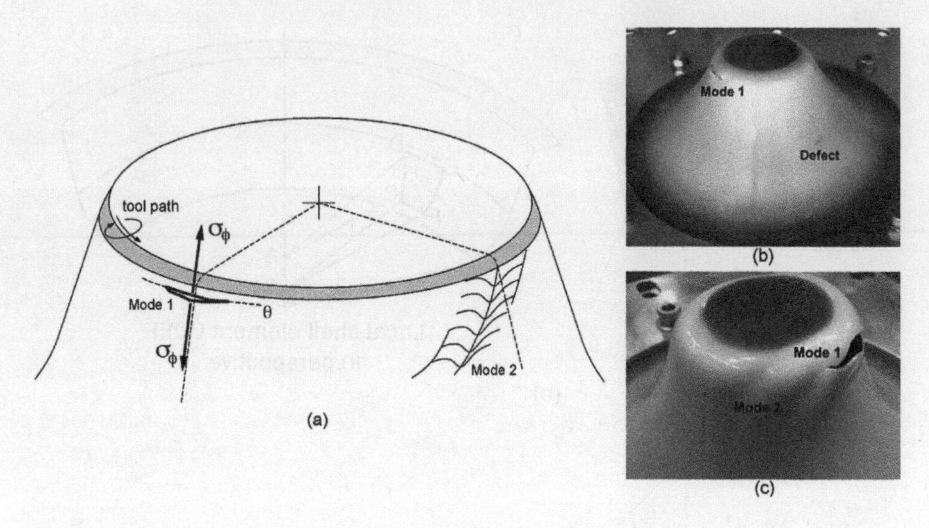

FIGURE 17.4 Modes of failure in incremental sheet forming of polymers (Mode 1 and 2). **Reprinted from [29] with permission from Elsevier.**

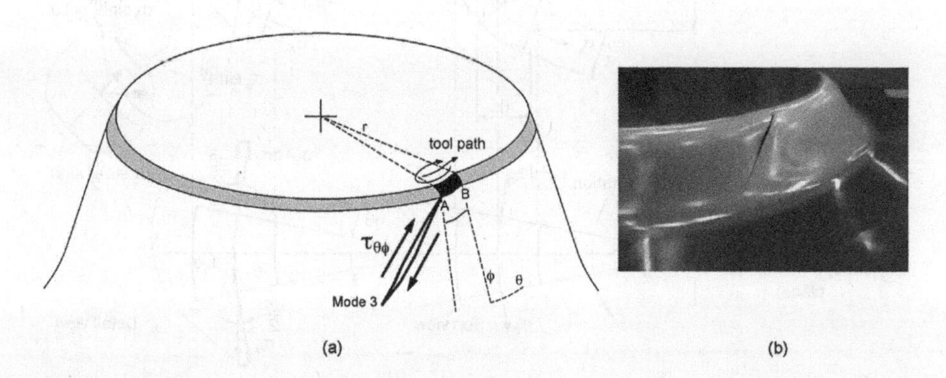

FIGURE 17.5 Modes of failure in incremental sheet forming of polymers (Mode 3). **Reprinted from [29] with permission from Elsevier.**

17.3.2 ISF Applied to Composites

Several studies have investigated the effect of adding reinforcement in a matrix of metals and composites and concluded that there was a significant increase in desirable properties. For polymers, few researchers have explored the possibility of employing the ISF process to composites. Lozano-Sánchez et al. used SPIF to form a composite with polypropylene as the matrix and multiwalled carbon nanotubes as the reinforcement. The research was more oriented on the feasibility of the process when applied to composites rather than the mechanism involved [39]. Al-Obaidi, Kunke, and Kräusel formed glass fiber–reinforced plastic using SPIF to benefit from the same advantages offered by the process when applied to metals or polymers. Conical parts that were formed and had wall angles greater than 50° suffered internal cracks, voids, and wrinkles [40]. The deformation mechanism reported between the woven fibers was reported to be shear. The studies done primarily focus on the applicability of SPIF to composites. The studies regarding SPIF of composites are limited, and a deep investigation related to process parameters or material properties is required.

17.4 CONCLUSION

The study aimed to make a comparison between the ISF of metals and polymers and composites. Polymers are rapidly making their way into the industries owing to their advantages over metals as already discussed in the literature. So, it becomes imperative to study the applicability and the mechanics involved in the process to polymers that have already been successfully applied and studied in metals. The fundamental differences between metals and polymers cause differences in forming characteristics, and these characteristics are influenced by process parameters in both metals and polymers. Polymers have a low elastic modulus and have a higher stress/strain relaxation than metals, due to which springback occurs more in polymers. Furthermore, a decrease in yield strength and load-bearing capacity has also been associated with the ISF of polymers. The deformation of polymers and metals differs at molecular levels in mechanism, molecular chain fragmentation in polymers, and dislocation or twining in metals [41]. The primary modes of deformation in polymers are reported to be plain strain and equi-biaxial stretching. In metals, the membrane analysis clearly shows uniaxial tension to be one major mode of deformation. The yielding of polymers is also different and is based on pressure criteria. Further differences are observed in the failure criteria of both materials, but the study here is limited to only deformation mechanics.

In conclusion, a significant number of differences are observed in the ISF of polymers and metals. For industrial applicability, a database is required to be set up listing all the process parameters, working conditions, material requirements, and any other modification to the equipment or the process best suited to perform forming operations. The ISF process provides great flexibility in manufacturing, particularly in the biomedical field, where the need for rapidly manufactured customized products is paramount. Biocompatible polymers are gaining increasing popularity because of aforesaid advantages. Any progress or recent development in the ISF of metals needs to be tracked and checked for its application to polymers and other materials.

REFERENCES

1. S. Gatea, H. Ou, and G. McCartney, "Review on the influence of process parameters in incremental sheet forming," *Int. J. Adv. Manuf. Technol.*, vol. 87, no. 1–4, pp. 479–499, 2016, doi:10.1007/s00170-016-8426-6.
2. W. C. Emmens, G. Sebastiani, and A. H. van den Boogaard, "The technology of Incremental Sheet Forming-A brief review of the history," *J. Mater. Process. Technol.*, vol. 210, no. 8, pp. 981–997, 2010, doi:10.1016/j.jmatprotec.2010.02.014.

3. M. Fiorotto, M. Sorgente, and G. Lucchetta, "Preliminary studies on single point incremental forming for composite materials," *Int. J. Mater. Form.*, vol. 3, no. Suppl. 1, pp. 951–954, 2010, doi:10.1007/s12289-010-0926-6.
4. P. A. F. Martins, N. Bay, M. Skjoedt, and M. B. Silva, "Theory of single point incremental forming," *CIRP Ann. - Manuf. Technol.*, vol. 57, no. 1, pp. 247–252, 2008, doi:10.1016/j.cirp.2008.03.047.
5. T. J. Kim and D. Y. Yang, "Improvement of formability for the incremental sheet metal forming process," *Int. J. Mech. Sci.*, vol. 42, no. 7, pp. 1271–1286, 2000, doi:10.1016/S0020-7403(99)00047-8.
6. H. Iseki, "An approximate deformation analysis and FEM analysis for the incremental bulging of sheet metal using a spherical roller," *J. Mater. Process. Technol.*, vol. 111, no. 1–3, pp. 150–154, 2001, doi:10.1016/S0924-0136(01)00500-3.
7. M. S. Shim and J. J. Park, "The formability of aluminum sheet in incremental forming," *J. Mater. Process. Technol.*, vol. 113, no. 1–3, pp. 654–658, 2001, doi:10.1016/S0924-0136(01)00679-3.
8. G. Hirt, J. Ames, M. Bambach, R. Kopp, and R. Kopp, "Forming strategies and process modelling for CNC incremental sheet forming," *CIRP Ann. - Manuf. Technol.*, vol. 53, no. 1, pp. 203–206, 2004, doi:10.1016/S0007-8506(07)60679-9.
9. L. Filice, L. Fratini, and F. Micari, "Analysis of material formability in incremental forming," *CIRP Ann. - Manuf. Technol.*, vol. 51, no. 1, pp. 199–202, 2002, doi:10.1016/S0007-8506(07)61499-1.
10. D. Young and J. Jeswiet, "Wall thickness variations in single-point incremental forming," *Proc. Inst. Mech. Eng. Part B J. Eng. Manuf.*, vol. 218, no. 11, pp. 1453–1459, 2004, doi:10.1243/0954405042418400.
11. J. Jeswiet, F. Micari, G. Hirt, A. Bramley, J. Duflou, and J. Allwood, "Asymmetric single point incremental forming of sheet metal," *CIRP Ann. - Manuf. Technol.*, vol. 54, no. 2, pp. 88–114, 2005, doi:10.1016/s0007-8506(07)60021-3.
12. J. Jeswiet and D. Young, "Forming limit diagrams for single-point incremental forming of aluminium sheet," *Proc. Inst. Mech. Eng. Part B J. Eng. Manuf.*, vol. 219, no. 4, pp. 359–364, 2005, doi:10.1243/095440505X32210.
13. M. Ham and J. Jeswiet, "Single point incremental forming and the forming criteria for AA 3003," *CIRP Ann. - Manuf. Technol.*, vol. 55, no. 1, pp. 241–244, 2006, doi:10.1016/S0007-8506(07)60407-7.
14. M. Ham and J. Jeswiet, "Forming limit curves in single point incremental forming," *CIRP Ann. - Manuf. Technol.*, vol. 56, no. 1, pp. 277–280, 2007, doi:10.1016/j.cirp.2007.05.064.
15. W. C. Emmens and A. H. van den Boogaard, "Strain in Shear, and Material Behaviour in Incremental Forming," *Key Eng. Mater.*, vol. 344, pp. 519–526, 2007, doi:10.4028/www.scientific.net/kem.344.519.
16. J. M. Allwood, D. R. Shouler, and A. E. Tekkaya, "The Increased Forming Limits of Incremental Sheet Forming Processes," *Key Eng. Mater.*, vol. 344, pp. 621–628, 2007, doi:10.4028/www.scientific.net/kem.344.621.
17. J. M. Allwood and D. R. Shouler, "Generalised forming limit diagrams showing increased forming limits with non-planar stress states," *Int. J. Plast.*, vol. 25, no. 7, pp. 1207–1230, 2009, doi:10.1016/j.ijplas.2008.11.001.
18. M. B. Silva, M. Skjoedt, P. A. F. Martins, and N. Bay, "Revisiting the fundamentals of single point incremental forming by means of membrane analysis," *Int. J. Mach. Tools Manuf.*, vol. 48, no. 1, pp. 73–83, 2008, doi:10.1016/j.ijmachtools.2007.07.004.

19. M. B. Silva, M. Skjoedr, A. G. Atkins, N. Bay, and P. A. F. Martins, "Single-point incremental forming and formability-failure diagrams," *J. Strain Anal. Eng. Des.*, vol. 43, no. 1, pp. 15–35, 2008, doi:10.1243/03093247JSA340.
20. K. Jackson and J. Allwood, "The mechanics of incremental sheet forming," *J. Mater. Process. Technol.*, vol. 209, no. 3, pp. 1158–1174, 2009, doi:10.1016/j.jmatprotec.2008.03.025.
21. M. Bambach, "A geometrical model of the kinematics of incremental sheet forming for the prediction of membrane strains and sheet thickness," *J. Mater. Process. Technol.*, vol. 210, no. 12, pp. 1562–1573, 2010, doi:10.1016/j.jmatprotec.2010.05.003.
22. Z. Cui, Z. Cedric Xia, F. Ren, V. Kiridena, and L. Gao, "Modeling and validation of deformation process for incremental sheet forming," *J. Manuf. Process.*, vol. 15, no. 2, pp. 236–241, 2013, doi:10.1016/j.jmapro.2013.01.003.
23. Z. Liu, W. J. T. Daniel, Y. Li, S. Liu, and P. A. Meehan, "Multi-pass deformation design for incremental sheet forming: Analytical modeling, finite element analysis and experimental validation," *J. Mater. Process. Technol.*, vol. 214, no. 3, pp. 620–634, 2014, doi:10.1016/j.jmatprotec.2013.11.010.
24. Y. Li, W. J. T. Daniel, and P. A. Meehan, "Deformation analysis in single-point incremental forming through finite element simulation," *Int. J. Adv. Manuf. Technol.*, vol. 88, no. 1–4, pp. 255–267, 2017, doi:10.1007/s00170-016-8727-9.
25. S. Ai, B. Lu, J. Chen, H. Long, and H. Ou, "Evaluation of deformation stability and fracture mechanism in incremental sheet forming," *Int. J. Mech. Sci.*, vol. 124–125, pp. 174–184, 2017, doi:10.1016/j.ijmecsci.2017.03.012.
26. S. Ai, R. Dai, and H. Long, "Investigating formability enhancement in double side incremental forming by developing a new test method of tension under cyclic bending and compression," *J. Mater. Process. Technol.*, vol. 275, no. July 2019, p. 116349, 2020, doi:10.1016/j.jmatprotec.2019.116349.
27. F. Maqbool and M. Bambach, "Revealing the Dominant Forming Mechanism of Single Point Incremental Forming (SPIF) by Splitting Plastic Energy Dissipation," *Procedia Eng.*, vol. 183, pp. 188–193, 2017, doi:10.1016/j.proeng.2017.04.018.
28. K. U. Yazar, S. Mishra, K. Narasimhan, and P. P. Date, "Deciphering the deformation mechanism in single point incremental forming: experimental and numerical investigation," *Int. J. Adv. Manuf. Technol.*, vol. 101, no. 9–12, pp. 2355–2366, 2019, doi:10.1007/s00170-018-3131-2.
29. V. Franzen, L. Kwiatkowski, P. A. F. Martins, and A. E. Tekkaya, "Single point incremental forming of PVC," *J. Mater. Process. Technol.*, vol. 209, no. 1, pp. 462–469, 2009, doi:10.1016/j.jmatprotec.2008.02.013.
30. P. A. F. Martins, L. Kwiatkowski, V. Franzen, A. E. Tekkaya, and M. Kleiner, "Single point incremental forming of polymers," *CIRP Ann. - Manuf. Technol.*, vol. 58, no. 1, pp. 229–232, 2009, doi:10.1016/j.cirp.2009.03.095.
31. M. B. Silva, L. M. Alves, and P. A. F. Martins, "Single point incremental forming of PVC: Experimental findings and theoretical interpretation," *Eur. J. Mech. A/Solids*, vol. 29, no. 4, pp. 557–566, 2010, doi:10.1016/j.euromechsol.2010.03.008.
32. M. B. Silva, T. Marques, and P. A. F. Martins, "Single-point incremental forming of polymers," *Mechatronics Manuf. Eng.*, pp. 293–331, Jan. 2012, doi: 10.1533/9780857095893.293.
33. L. V. Sy and N. T. Nam, "A numerical simulation of incremental forming process for polymer sheets," *Int. J. Model. Simul.*, vol. 32, no. 4, pp. 265–272, 2012, doi:10.2316/Journal.205.2012.4.205-5752.

34. S. Alkas Yonan, C. Soyarslan, P. Haupt, L. Kwiatkowski, and A. E. Tekkaya, "A simple finite strain non-linear visco-plastic model for thermoplastics and its application to the simulation of incremental cold forming of polyvinylchloride (PVC)," *Int. J. Mech. Sci.*, vol. 66, pp. 192–201, 2013, doi:10.1016/j.ijmecsci.2012.11.007.
35. S. Alkas Yonan, M. B. Silva, P. A. F. Martins, and A. E. Tekkaya, "Plastic flow and failure in single point incremental forming of PVC sheets," *Express Polym. Lett.*, vol. 8, no. 5, pp. 301–311, 2014, doi:10.3144/expresspolymlett.2014.34.
36. I. Bagudanch, G. Centeno, C. Vallellano, and M. L. Garcia-Romeu, "Revisiting formability and failure of polymeric sheets deformed by Single Point Incremental Forming," *Polym. Degrad. Stab.*, vol. 144, pp. 366–377, 2017, doi:10.1016/j.polymdegradstab.2017.08.021.
37. M. Durante, A. Formisano, and F. Lambiase, "Incremental forming of polycarbonate sheets," *J. Mater. Process. Technol.*, vol. 253, no. September 2017, pp. 57–63, 2018, doi:10.1016/j.jmatprotec.2017.11.005.
38. M. Durante, A. Formisano, and F. Lambiase, "Formability of polycarbonate sheets in single-point incremental forming," *Int. J. Adv. Manuf. Technol.*, vol. 102, no. 5–8, pp. 2049–2062, 2019, doi:10.1007/s00170-019-03298-w.
39. L. M. Lozano-Sánchez et al., "Mechanical and structural studies on single point incremental forming of polypropylene-MWCNTs composite sheets," *J. Mater. Process. Technol.*, vol. 242, pp. 218–227, 2017, doi:10.1016/j.jmatprotec.2016.11.032.
40. A. AL-Obaidi, A. Kunke, and V. Kräusel, "Hot single-point incremental forming of glass-fiber-reinforced polymer (PA6GF47) supported by hot air," *J. Manuf. Process.*, vol. 43, no. May, pp. 17–25, 2019, doi:10.1016/j.jmapro.2019.04.036.
41. M. Hassan, G. Hussain, H. Wei, A. Qadeer, and M. AlKahtani, "Progress on single-point incremental forming of polymers," *Int. J. Adv. Manuf. Technol.*, vol. 114, no. 1–2, 2021, doi:10.1007/s00170-021-06620-7.

18 Sustainable Polishing of Directed Energy Deposition–Based Cladding Using Micro-Plasma Transferred Arc

Pravin Kumar, Neelesh Kumar Jain and Abhay Tiwari

Department of Mechanical Engineering, Indian Institute of Technology Indore, Simrol, India

CONTENTS

18.1 INTRODUCTION

American Society for Testing and Materials (ASTM) designation F2792–12a defines directed energy deposition (DED) as an additive manufacturing process in which a heat energy source, such as a laser beam, electron beam, plasma arc, or any other arc, is focused to fuse and deposit the deposition materials over the substrate or previously deposited layer [1]. In this process, a deposition head with an arrangement of

DOI: 10.1201/9781003284574-20

heat source and deposition material feed system is attached on a multi-axis arm to deposit material at a specific position and predefined pattern. Deposition material in various forms, such as powder, wire, particulate, and strips, is fed to the molten puddle to form a deposition. DED is found to be suitable for depositing various materials such as metal matrix composites (MMCs), metallic alloys, functional graded materials, and cermets. This process is widely utilized for freeform manufacturing of complex parts, surface texturing, cladding, restoring, rapid tooling, and rapid prototyping for numerous industrial applications [2]. Their specific applications include different sporting gears, postinjury restoration, biomedical, automobile, aerospace, gas turbines, marine, and gas and oil extraction.

Despite substantial advances in the direction of different material exploration, process optimization, online control, and inspection in DED processes, several unsolved challenges are associated with the parts deposited using the process. One such significant concern of the process is poor deposition quality in terms of accuracy and surface roughness. Investigations made by various researchers reveal that it is due to the accumulations of semi-molten powder that adhere to the wall surface, solidification lines developed due to melt pools, molten pool overflow over layer beneath, and formation of periodic menisci [3]. This leads to further postprocessing and conditioning, such as ultrasonic treatment, sandblasting, heat treatment, polishing, grinding, and milling [4]. The part manufactured by the DED process has a surface roughness value in the range of 1.5–3000 μm, depending on the selection of a heat source. A laser beam is a precise energy source that can deposit a layer of 0.2–2-mm thickness and have a lower surface roughness value. In contrast, arc-based processes such as metal inert gas (MIG) can deposit a layer of thickness 1–6 mm, yielding a higher surface roughness value [5]. So accordingly, the process of postprocessing is decided, which involves an additional arrangement of setup or machine to achieve the desired dimensions, thus increasing the production time and cost. With the commercialization of the additive manufacturing process, different researchers have explored various postprocessing techniques to optimize the aspects of the developed surface finish.

18.1.1 Ultrasonic Surface Treatment

This is one of the processes used for improving surface finish and reducing surface waviness. In this process, ultrasonic vibration is utilized to cause an impact on the surface, causing surface modification. The process considers variable parameters such as static load, amplitude of vibration, number of impacts per unit area, and ball-tip diameter [6]. It imparts additional advantages of grain refinement, hardness improvement, wear resistance, and gain in fatigue strength. Similarly, ultrasonic cavitation abrasive finishing is also used for surface modification. In this process, the manufactured DED part is immersed in deionized water mixed with hard abrasive. An ultrasonic horn tip is maintained at a specific height above the workpiece surface, and an ultrasonic wave of high frequency is generated in the solution to cause cavitation (development, growth, collapse of bubbles). A schematic diagram of the ultrasonic cavitation abrasive finishing process is also shown in Figure 18.1. Cavitation produces micro-jets with velocities ranging from 200–700 m/s, and the available

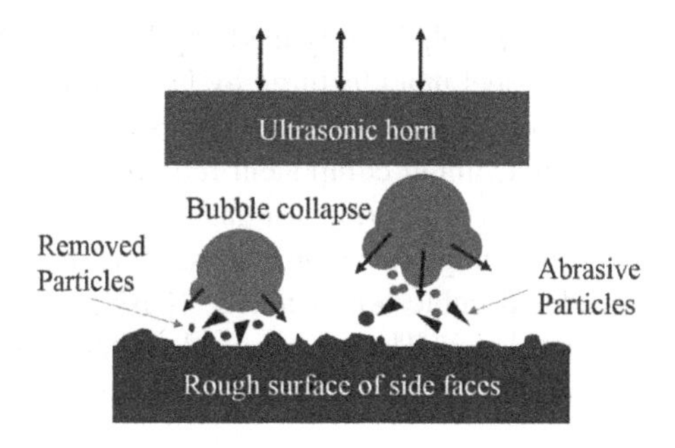

FIGURE 18.1 Schematic diagram of ultrasonic cavitation abrasive finishing.

abrasive particles in the solution finish the surface. It results in improvement of surface roughness in the range of a few microns (1–5 μm) [7]. The previously discussed processes are suitable for finishing up to a few microns and are successfully being used to finish the parts manufactured by a laser-based DED.

18.1.2 Machining

Another approach that is widely been adopted by DED industries is successive milling, grinding, or turning after deposition. DED setup is integrated with a computer numerical controlled (CNC) machine (hybrid layered manufacturing) or the deposited parts are machined separately on CNC machine. The concept of a hybrid layered manufacturing process has been extensively used to achieve high-quality components manufactured by either laser- or arc-based DED processes and found to be suitable for metals possessing lower hardness. A schematic diagram of hybrid layered manufacturing process is also shown in Figure 18.2. Materials such as Co-Cr-, Ni-, Ti-, and Fe-based alloys are widely used in the DED process for manufacturing corrosion- and abrasion-resistant parts and high-strength and -temperature applications. This material poses hardness value ranges from 240 to 450 HV based on different heat sources. Moreover, it has been evident that the DED process involves rapid cooling, thus yielding higher hardness comparative to other conventional manufacturing processes [4]. So, indeed, the process of machining has to compromise with the

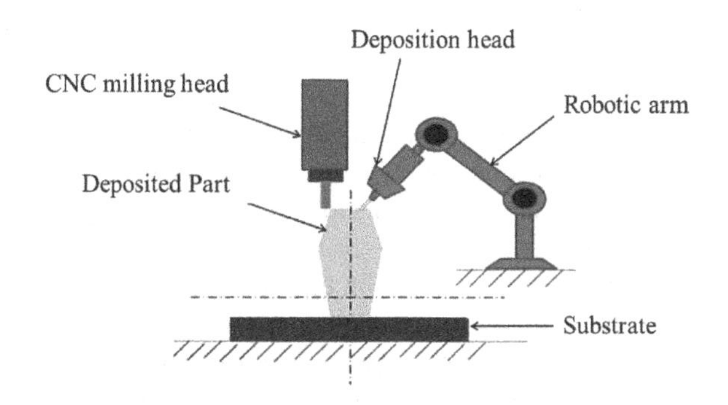

FIGURE 18.2 Schematic diagram of hybrid layered manufacturing process.

higher wear rate of the cutting tool. A study made by Bai et al. [8] reported that the component of ASTM A131 steel manufactured by laser-based DED yielded microhardness as 31.53% higher than that of the hot-rolled sample. Thus, from the result, it can be inferred that the machining component requires higher cutting forces and yields higher tool wear. Oyelola et al. [9] investigated the machinability of functional graded part of Ti6Al4V and tungsten carbide (WC) manufactured by laser-based DED. Their study considered cubic boron nitride (CBN) and polycrystalline diamond inserts for turning operation. The study revealed that the deposited parts have poor machinability that causes a higher wear rate of the tool inserts.

18.1.3 Surface Finishing Using Energy Beam Irradiation

Energy beams, such as laser beams, electron beams, and ion beams, are predominantly used in remelting and polishing the surface of parts manufactured by additive manufacturing. These processes have been widely investigated and used due to their potential to eliminate the limitations of conventional methods, such as finishing complex shapes, higher tool wear, poor machinability, higher time consumption, higher cost, and environmental impacts. The laser beam–based polishing process known as laser polishing (LP) has been used for postprocessing to finish the surface of components manufactured by the DED process. In this process, a laser beam is irradiated over the surface to produce a shallow molten pool, and the laser scanning strategy is maintained such that material from the convex peaks is melted and dispersed to fill the valley region between two adjacent tracks to yield a plain surface. The scanning strategy has a significant role in attaining minimal surface roughness [10]. Different researchers have explored different scanning strategies as shown in Figure 18.3, such as (a) raster, (b) modified raster, (c) bidirectional, (d) offset-in/out and (e) pendulum. LP requires no separate setup arrangement and can be carried out on the same setup used for the deposition process. It is found to be suitable for improving the surface roughness by a range of 5–20 μm [11, 12].

Electron beam polishing (EBP) is a process in which irradiated electrons are used to remelt the surface and generate a plain surface with improved mechanical properties. Proskurovsky et al. [13] explored an electron beam to alter the surface of metallic materials at the end of 20th century. This process can be further classified as continuous electron beam polishing (CEBP) and pulsed electron beam polishing (PEBP). In the case of CEBP, a continuous electron beam is used to scan the surface whereas, in PEBP, an electron beam is used to scan the surface in a pulse mode of several microseconds. Furthermore, electron beam can be focused on a specific area by the magnetic lens or function without a magnetic lens for a comparatively larger unfocused area. It is found to be suitable for improving the surface roughness by 5–10 μm [14]. The EBP process is conducted within a vacuum space, which ensures the elimination of oxidation. As per the available literature, there is no significant investigation for polishing DED parts using the EBP process, but it has been used to polish parts manufactured by other additive manufacturing processes.

Ion beam polishing (IBP) is another energy beam–based process developed in the late 1960s used for highly precise polishing technology compared to laser and electron [15]. It is also called as ion beam milling or ion beam etching. Beam-based

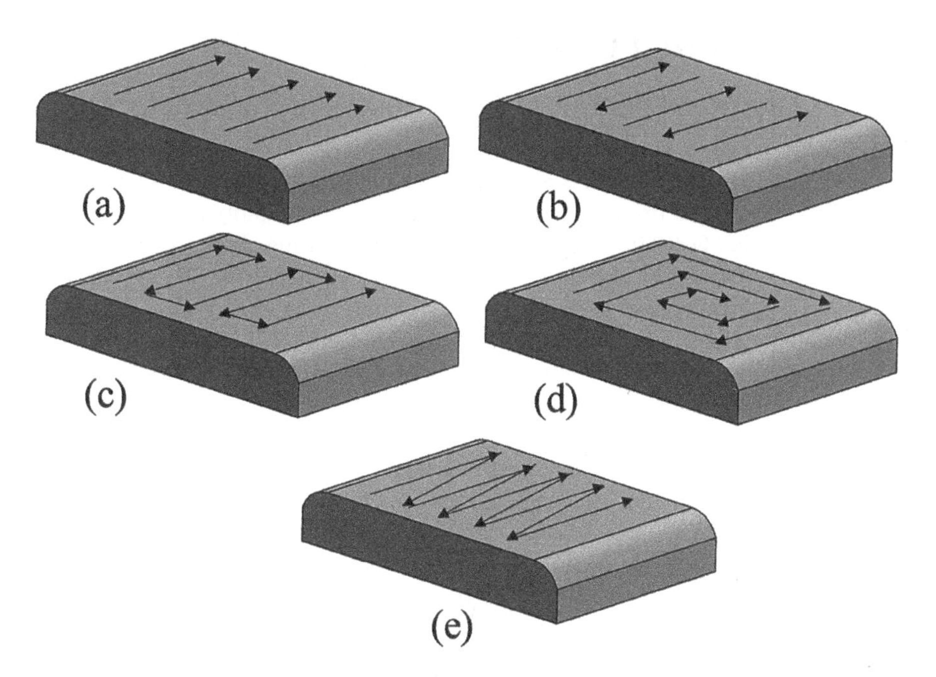

FIGURE 18.3 Different scanning pattern generally used during the laser polishing process.

processing techniques are characterized by melting or evaporating the top surface of the functional part. IBP utilizes the direct transferring of kinetic energy by an elastic collision between the incident ions and the atomic nucleus of the top surface. It is found to be suitable for improving the surface roughness in the range of a few nanometers [16, 17]. Polishing using IBP process is conducted in a vacuum chamber. In general, two types of scanning strategies, raster and bidirectional scans, are used in the IBP process. The major constraint of the IBP and EBP processes is the size of the vacuum workspace, which restricts the surface modification of large parts. Similar to the EBP process, there is no significant investigation for polishing DED parts using the IBP process.

It can be concluded from the review of past works that (1) the machining process is highly preferred for finishing DED parts to achieve the desired dimensional limits followed by a laser polishing process. This is certainly due to the fact that a DED process has higher surface roughness and waviness. (2) Other factors, such as the necessity of a vacuum chamber or the requirement of a controlled environment, also limit the process of surface modification using the EBP and IBP processes. (3) There is a constant effort to finish DED parts in a way that removes the need for additional setup and finishes the intricate parts generated by any DED method, including arc and laser; and is suitable for a wide range of materials. Considering the preceding requirement and research gaps, in the present work, a μ-PTA-based polishing (μ-PTAbP) was explored, and detailed investigations were performed to evaluate polishing effects on microstructure, microhardness, wear resistance, and surface morphology of the clad surface.

18.2 EXPERIMENTAL DETAILS

18.2.1 Experimental Apparatus and Materials Used

A μ-PTAbP process involves a μ-plasma transferred arc, which is initiated from an electrode inside the nozzle and is ejected from the fine nozzle hole, thus narrowing the arc to a small area resulting in higher energy density, stability at a lower value of current, smaller heat-affected zone, and minimal distortion. These characteristics indicate its suitability for a polishing application. To investigate the process capability, an experimental apparatus of μ-PTA-DED process, shown in Figure 18.4, was used for both surface cladding and polishing the clad surface. It comprises five different systems: (1) 440 W (maximum) power source unit, (2) a 5-axis manipulation system, (3) a μ-plasma torch, (4) a deposition head, and (5) a wire/powder feeder. A Co-based alloy, Stellite-6 wire with a diameter of 1.2 mm, was considered as a deposition material. It has wide application in surface cladding to achieve abrasion and corrosion resistance at elevated temperatures. Its chemical composition (by wt.%) consists of 27–32% Cr, 4–6% W, and 0.9–1.4% C, with additions of Ni, Fe, Si, Mn, Mo, and Co as balance. An AISI 4130 plate was used as a substrate of dimensions 200 mm in length, 50 mm in width, and 6 mm in thickness.

18.2.2 Process Parameters for Experimentation

The experimentation preceded in four steps. First, the development of clad samples was followed by a pilot and a main experiment, and finally, a polishing operation was performed using the optimum parameter selected from the main experiments. A μ-PTA-DED process was used to perform Stellite6 cladding over AISI 4130 plate using the process parameters of a μ-plasma power of 407 W, travel speeds of the μ-plasma torch of 1 mm/s, a feed rate of deposition material of 19.79 mm^3/s, a

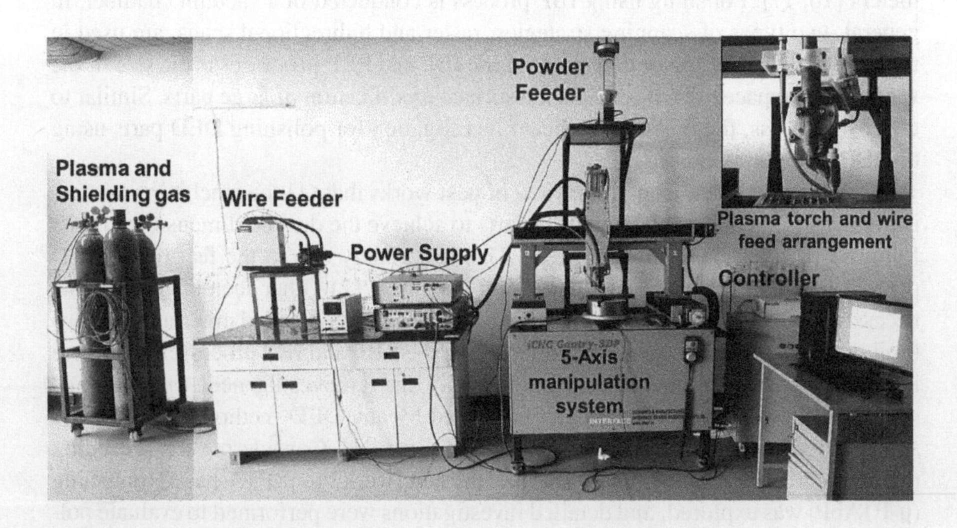

FIGURE 18.4 Experimental setup used for the μ-PTA-DED and μ-PTAbP processes.

stepover distance between two tracks at 1.5 mm, a plasma gas supply rate of 0.3 Nl/min, a shielding gas supply rate of 5 Nl/min, a stand-off distance of 10 mm, and a wire feed angle of 60° [18]. Sixty-four clad samples 20 mm in length and 15 mm in width were prepared to perform the pilot, main, and final experiments of the μ-PTAbP process. Pilot experiments were done to identify the feasible range of process parameters for the μ-PTAbP process, whereas main experiments were done to find the optimum process parameters using criteria from a visual inspection for over-melting and un-melting of the clad surface, and finally, the optimum process parameters from the main experiment were used to perform the μ-PTAbP process and the surface characteristics evaluated for the polished samples. In both pilot and main experiments, raster scanning strategies were adopted and parameters such as micro-plasma power, P; travel speed of μ-plasma torch, v; and stand-off distance, Z, were selected as variable process parameters, whereas the plasma gas supply rate of 0.3 Nl/min, the shielding gas supply rate of 5 Nl/min, and a wire feed angle of 60° were kept constant as per Kumar et al. (2020). Pilot experiments resulted in the identification of (1) μ-plasma power as 242, 264, and 286 W; (2) travel speeds of the μ-plasma torch as 1, 1.08, and 1.16 mm/s; and (3) stand-off distance as 8, 9, and 10 mm. These identified variable parameters were further used to accomplish 27 main experiments using a full-factorial method. Finally, the optimum parameters selected from the main experiment are μ-plasma power of 264 W, travel speed of the μ-plasma torch of 1.16 mm/s, and a stand-off distance of 8 mm to evaluate the μ-PTAbP process performance characteristics, such as microstructure, microhardness, resistance to scratch, wear, and effects of polishing on the surface deviation of the clad surface.

18.2.3 Investigation of Performance Characteristics

Samples for metallographic characterization such as microstructure and microhardness were prepared by polishing using emery papers and diamond paste of different grades to attain mirror-finish followed by etching in a solution containing 50 ml HCl, 5 ml HNO3, and 50 ml H_2O at 50 °C. These samples were investigated for their macro- and microstructural features using an optical microscope having image analysis software. Furthermore, microhardness was evaluated as per ASTM standard E92-82 using Vicker's micro-UHL VMH at a loading condition of 500 g for 15 seconds. For the investigation, a scratch test was performed by considering the standard of ASTM G171-03 for the polished and unpolished region of the clad sample for 7-mm sliding distance under a constant normal load of 10 N and a sliding speed of 20 mm/s. The scratched track width was measured by an optical microscope (Leica Microsystems DM 2500 M), and the scratch hardness number (HS_P) was calculated based to ASTM standard G171-03 and subsequently using the formula as mentioned in Equation 18.1, where P is applied normal load (N) and w is the width of the scratch (mm):

$$HS_P = \frac{8P}{\pi w^2}. \tag{18.1}$$

A wear test of the polished and unpolished regions of the clad sample was performed to investigate the microscale tribology, abrasion, and friction wear properties. In order to analyze the tribological behavior, samples 10 mm in length, 10 mm in width, and 6 mm in thickness were cut from the polished and unpolished regions and were further polished to attain roughness of 1 ± 0.12 μm. Test for evaluation of wear characteristic was done on linear reciprocating fretting wear type tribometer (Ducom CM-9104) with WC balls 6 mm in diameter. The test was performed under a load of 12.5 N, an amplitude of 1.5 mm, and a frequency of 75 Hz for 15 minutes. The corresponding loss of mass from the sample due to wear was measured using a micro balance, having an accuracy of 0.01 mg. These values of mass loss were further evaluated in terms of volume loss and Archard's wear equation, as given in Equation 18.2, was used to calculate the wear rate:

$$V_i = k_i F s, \tag{18.2}$$

where V_i is the volume of wear (mm^3), k_i is the specific wear rate (mm^3/Nm), F is the normal load (N), and s is the distance of sliding (m). Afterward, the wear test samples were observed using an optical microscope to find the wear profiles. The surface polished by the μ-PTAbP process was also investigated for surface deviation using a three-dimensional (3D) laser scanner (FARO 350). In this process, the polished and unpolished surfaces were scanned by a laser to acquire point cloud data, which can collect 600,000 points per second. The acquired data are processed by metrology software (Geomagic Control X) to generate a 3D informational model and determine the surface deviations.

18.3 RESULT AND DISCUSSIONS

18.3.1 Microstructure and Microhardness

Figure 18.5 represents optical micrographs showing the microstructure of unpolished (Figure 18.5a) and polished (Figure 18.5b) clad samples at 100× magnification. In

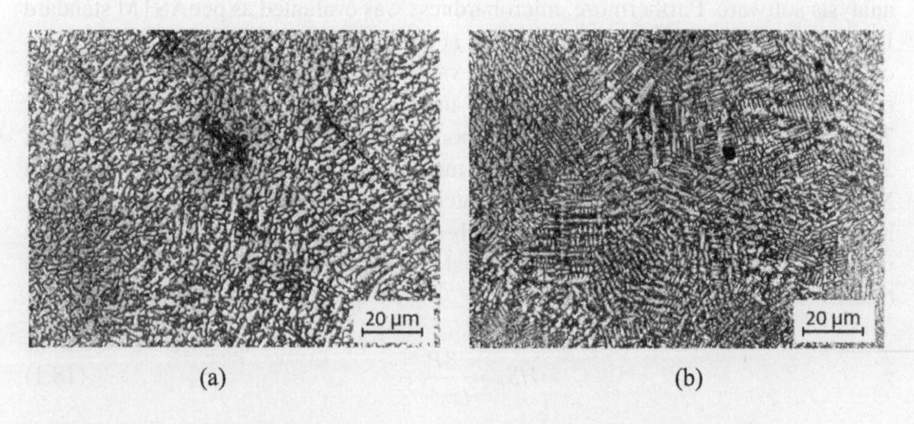

FIGURE 18.5 Microstructure of (a) unpolished and (b) polished clad sample.

both cases, the microstructure has almost uniform grains with columnar dendrites; however, the clad surface of the unpolished sample reveals a coarser microstructure (average grain size being 5 μm) compared to the polished sample (average grain size being 3 μm). This can be described by the fact that the polishing process is carried by the surface remelting, which causes rapid cooling. It has also enhanced the surface hardness, which can be evident by the microhardness test. An average value of microhardness for the unpolished and polished surfaces of clad samples was found to be 557 HV and 596 HV, respectively.

18.3.2 Scratch and Wear Resistance

Figure 18.6a and 18.6b show the microscopic images of scratch track on unpolished and polished clad samples, respectively. The mean scratch track width, w; the coefficient of friction, μ, and the scratch hardness number, HS_P (calculated using Equation 18.1) are listed in Table 18.1. The result shows that the polished sample has lower values of scratch track width and coefficient of friction and a higher scratch hardness number than the unpolished clad sample. Thus, it can be inferred that the polishing operation using the μ-PTAbP process enhances the abrasion resistance. This result can also be allied with the results of microhardness and microstructure, whereas Figures 18.7a and 18.7b present the wear images formed on the unpolished and polished clad samples, respectively, acquired from the optical microscope. The mean values of wear rate coefficient, k_i; the mean values of the coefficient of friction, μ_F; and the computed values of wear volume, V_i, are presented in Table 18.2. It can be observed that the polished sample has lower mean values for the coefficient of friction, μ_F. The graph shown in Figure 18.8 represents a lower coefficient of friction

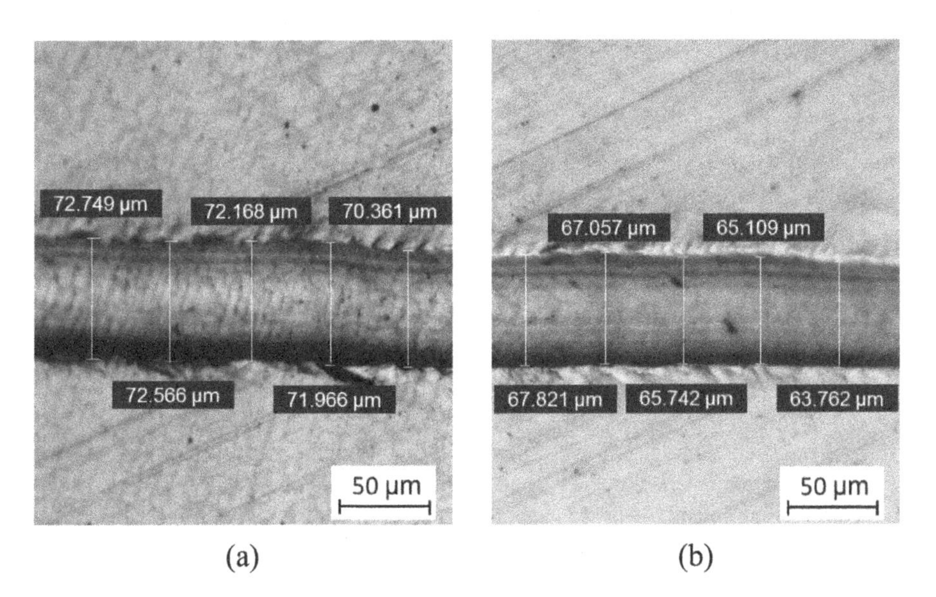

FIGURE 18.6 Optical microscopy images of scratch track formed on the surface of (a) unpolished and (b) polished clad samples.

TABLE 18.1
Results of Scratch Test Presenting Mean Values of Scratch Track Width, Coefficient of Friction, and Scratch Hardness Number

Condition	Mean Value of Scratch Track Width, w (μm)	Mean Value of Coefficient of Friction, μ	Scratch Hardness No., HS_P (GPa)
Before Polishing	71.96	0.141	4.92
After Polishing	65.89	0.126	5.86

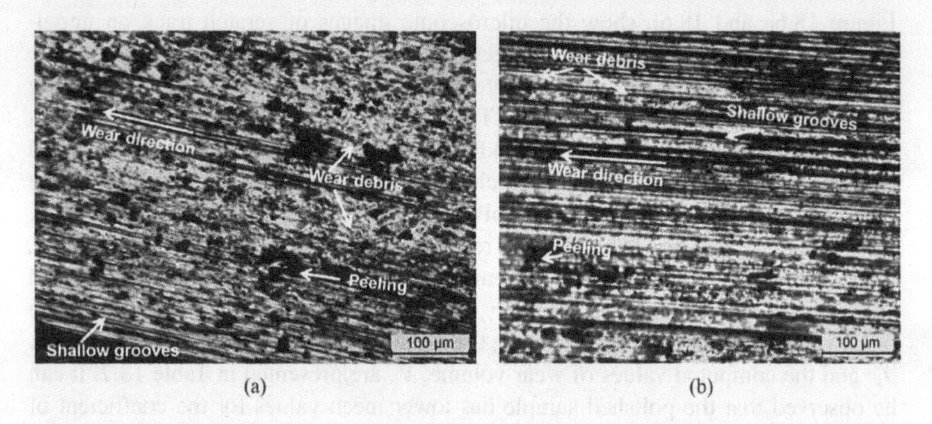

FIGURE 18.7 Optical microscopy images of wear formed on the surface of (a) unpolished and (b) polished clad samples.

TABLE 18.2
Results of Wear Test Presenting Mean Values of Wear Rate Coefficient, Mean Value of Coefficient of Friction and Computed Values of Wear Volume

Condition	Mean Value of Wear Rate Coefficient, k_i	Mean Value of Coefficient of Friction, μ_F	Wear Volume, V_i (mm^3)
Before Polishing	0.483	0.215	0.49
After Polishing	0.303	0.159	0.41

for the polished sample. A lower value of the coefficient of friction yields a lower friction force and surface energy, thus leading to higher abrasion resistance, higher adhesion strength, and lower wear volume, V_i [19]. It can be seen in Figure 18.7a that the polished clad sample has higher wear debris and peeling of the clad material in comparison to the polished sample as shown in Figure 18.7b. The enhancement of wear resistance in the polished zone can be described due to the yielding of a higher hardness of the sample after polishing.

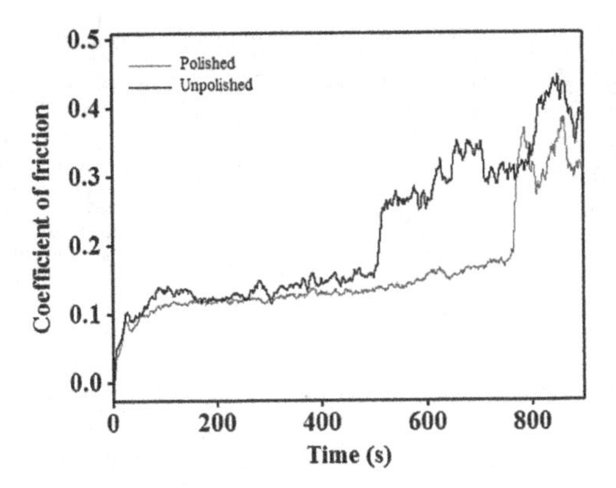

FIGURE 18.8 Graphical representation of variation of the coefficient of friction for unpolished and polished clad samples with respect to time.

18.3.3 Surface Deviation

Figure 18.9 shows the deviation of unpolished and polished surfaces on the same cladded sample using different colored contours. The deviation of the surface was measured from a plane 2 mm above the substrate. Scanned data reveal that the

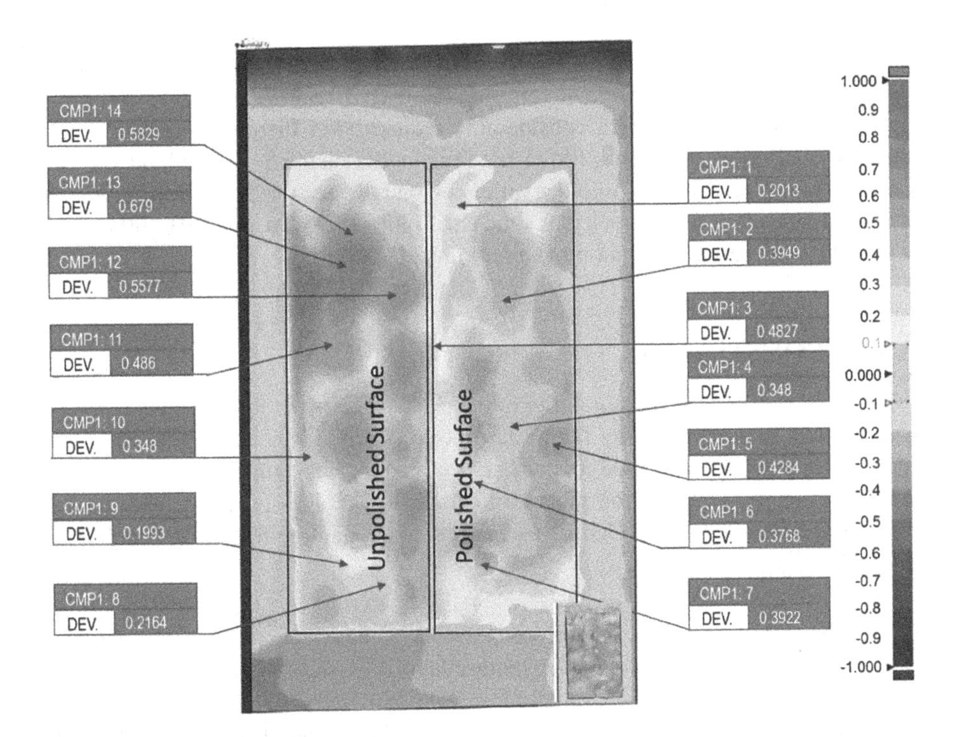

FIGURE 18.9 Surface deviation of the unpolished and polished zones on the same cladded sample using different colored contours.

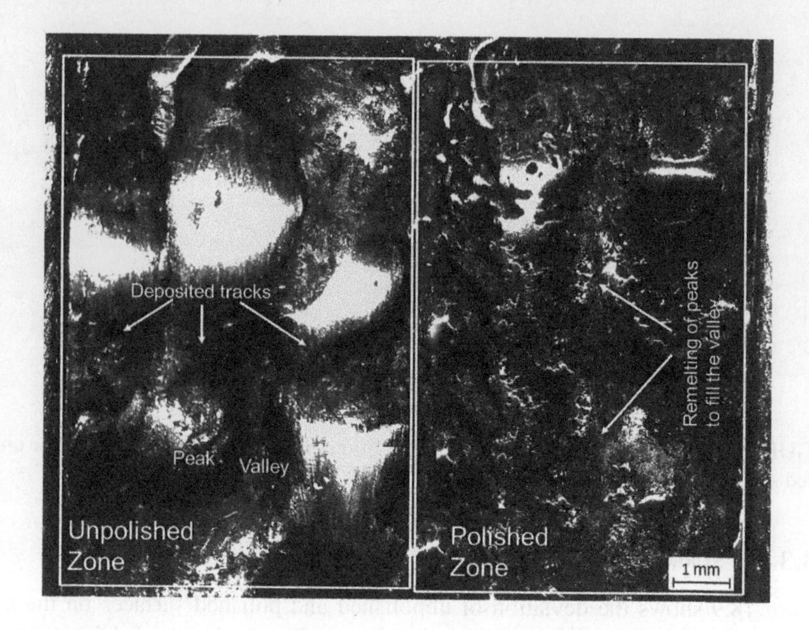

FIGURE 18.10 Microscopic image of unpolished and polished zones on the same clad sample.

unpolished zone has an average surface deviation of 0.564 ± 0.010 mm, and the polished zone has a surface deviation of 0.362 ± 0.010 mm. This is due to the fact that the peak-to-valley height considerably reduces after the polishing process. In the μ-PTAbP process, micro-plasma arc remelts the cladded surface, and the part of the molten material flow into the valley region between the deposition tracks due to the surface tension and gravity, as shown in Figure 18.10. So, the μ-PTAbP process can reduce the surface deviation by approximately 200 μm.

18.4 CONCLUSION

This study investigated the suitability of the μ-PTAbP process for sustainable polishing of parts manufactured by the DED process. An experimental apparatus of the μ-PTA-DED process was used for both surface cladding and polishing. Subsequently, the effect of polishing on the microstructure, microhardness, wear resistance, and surface morphology of the clad surface was studied, and the following conclusions were drawn from the study:

- The μ-PTAbP process is performed on the same apparatus of μ-PTA-DED process, thus eliminating the requirement of a separate arrangement for polishing.
- The microstructure reveals that both unpolished and polished clad surfaces have uniform grains with columnar dendrites; however, the unpolished

sample of the clad surface has a coarser microstructure in comparison to the polished sample. Moreover, an average value of microhardness for the polished clad surface was found to be higher than the unpolished surface of clad samples.

- The study also infers that the polished sample has lower values of scratch track width and coefficient of friction and higher scratch hardness number in comparison to the unpolished clad sample. It can be observed that the polished sample has a lower mean values coefficient of friction. Thus, the polishing operation using the μ-PTAbP process enhances the abrasion and wear resistance of the clad surface.
- Scanned data reveal that the unpolished zone has an average surface deviation of 0.564 ± 0.010 mm and that the polished zone has a surface deviation of 0.362 ± 0.010 mm. So, the μ-PTAbP process can reduce the surface deviation by approximately 200 μm.
- The study reveals the suitability of the μ-PTAbP process for polishing the parts manufactured by the DED process. Furthermore, there is a scope for optimizing the process parameters and further enhancing the capability and repeatability of the process.

REFERENCES

1. ISO/ASTM 52900:2015 (ASTM F2792) Additive manufacturing - General principles – Terminology. https://www.iso.org/standard/69669.html
2. Kumar, P., Jain, N. K., Sawant, M. S. (2020), Modeling of dimensions and investigations on geometrical deviations of metallic components manufactured by μ-plasma transferred arc additive manufacturing process, *The International Journal of Advanced Manufacturing Technology*, 107, 3155–3168. https://doi.org/10.1007/s00170-020-05218-9
3. Gharbi, M., Peyre, P., Gorny, C., Carin, M., Morville, S., Le Masson, P., Fabbro, R. (2013), Influence of various process conditions on surface finishes induced by the direct metal deposition laser technique on a Ti–6Al–4V alloy, *Journal of Materials Processing Technology*, 213, 791–800. https://doi.org/10.1016/j.jmatprotec.2012.11.015
4. DebRoy, T., Wei, H. L., Zuback, J. S., Mukherjee, T., Elmer, J. W., Milewski, J. O., Beese, A. M., Wilson-Heid, A. D., De, A., Zhang, W. (2018), Additive manufacturing of metallic components–process, structure and properties, *Progress in Materials Science*, 92, 112–224. https://doi.org/10.1016/j.pmatsci.2017.10.001
5. Thivillon, L., Bertrand, P., Laget, B., Smurov, I. (2009), Potential of direct metal deposition technology for manufacturing thick functionally graded coatings and parts for reactors components, *Journal of Nuclear Materials*, 385, 236–241. https://doi.org/10.1016/j.jnucmat.2008.11.023
6. Kim, M. S., Park, S. H., Pyun, Y. S., Shim, D. S. (2020), Optimization of ultrasonic nanocrystal surface modification for surface quality improvement of directed energy depositd stainless steel 316L, *Journal of Materials Research and Technology*, 9, 15102–15122. https://doi.org/10.1016/j.jmrt.2020.10.092
7. Tan, K. L., Yeo, S. H. (2017), Surface modification of additive manufactured components by ultrasonic cavitation abrasive finishing, *Wear*, 378, 90–95. http://dx.doi.org/10.1016/j.wear.2017.02.030

8. Bai, Y., Chaudhari, A., Wang, H. (2020), Investigation on the microstructure and machinability of ASTM A131 steel manufactured by directed energy deposition, *Journal of Materials Processing Technology*, 276, 116410. https://doi.org/10.1016/j.jmatprotec.2019.116410
9. Oyelola, O., Crawforth, P., M'Saoubi, R., Clare, A. T. (2018), Machining of functionally graded Ti6Al4V/WC produced by directed energy deposition, *Additive Manufacturing*, 24, 20–29. https://doi.org/10.1016/j.addma.2018.09.007
10. Jaritngam, P., Tangwarodomnukun, V., Qi, H., Dumkum, C. (2020), Surface and subsurface characteristics of laser polished Ti6Al4V titanium alloy, *Optics and Laser Technology*, 126, 106102. https://doi.org/10.1016/j.optlastec.2020.106102
11. Rosa, B., Mognol, P., Hascoët, J. Y. (2015), Laser polishing of additive laser manufacturing surfaces, *Journal of Laser Applications*, 27, S29102. https://doi.org/10.2351/1.4906385
12. Ma, C. P., Guan, Y. C., Zhou, W. (2017), Laser polishing of additive manufactured Ti alloys, *Optics and Lasers in Engineering*, 93, 171–177. https://doi.org/10.1016/j.optlaseng.2017.02.005
13. Proskurovsky, D. I., Rotshtein, V. P., Ozur, G. E., Markov, A. B., Nazarov, D. S., Shulov, V. A., Buchheit, R. G. (1998), Pulsed electron-beam technology for surface modification of metallic materials, *Journal of Vacuum Science & Technology A: Vacuum, Surfaces, and Films*, 16, 2480–2488. https://doi.org/10.1116/1.581369
14. Zhu, X. P., Lei, M. K., Ma, T. C. (2003), Surface morphology of titanium irradiated by high-intensity pulsed ion beam, *Nuclear Instruments and Methods in Physics Research Section B: Beam Interactions with Materials and Atoms*, 211, 69–79. https://doi.org/10.1016/S0168-583X(03)01124-8
15. Townsend, P. D. (1970), Ion beams in optics—an introduction, *Optics Technology*, 2, 65–67. https://doi.org/10.1016/0374-3926(70)90003-7
16. Jie, X. U., Wang, C. J., Bin, G. U. O., Shan, D. B., Sugiyama, Y., Ono, S. (2009), Surface finish of micro punch with ion beam irradiation, *Transactions of Nonferrous Metals Society of China*, 19, 526–530. https://doi.org/10.1016/S1003-6326(10)60102-1
17. Yamada, I. (1999), Low-energy cluster ion beam modification of surfaces, *Nuclear Instruments and Methods in Physics Research Section B: Beam Interactions, Materials and Atoms*, 148, 1–11. https://doi.org/10.1016/S0168-583X(98)00875-1
18. Kumar, P., Jain, N. K. (2020), Effect of material form on deposition characteristics in micro-plasma transferred arc additive manufacturing process, *CIRP Journal of Manufacturing Science and Technology*, 30, 195–205. https://doi.org/10.1016/j.cirpj.2020.05.008
19. Sawant, M. S., Jain, N. K. (2017), Characteristics of single-track and multi-track depositions of Stellite by micro-plasma transferred arc powder deposition process, *Journal of Materials Engineering and Performance*, 26, 4029–4039. https://doi.org/10.1007/s11665-017-2828-y

Index

Note: Pages in **bold** refer to tables.